VOLTAGE STABILITY
OF ELECTRIC POWER SYSTEMS

THE KLUWER INTERNATIONAL SERIES
IN ENGINEERING AND COMPUTER SCIENCE

Power Electronics and Power Systems
Consulting Editors
Thomas A. Lipo and M. A. Pai

Other books in the series:

VOLTAGE STABILITY
OF ELECTRIC POWER SYSTEMS

Thierry VAN CUTSEM
University of Liège
Belgium

Costas VOURNAS
National Technical University
Athens, Greece

KLUWER ACADEMIC PUBLISHERS
Boston/London/Dordrecht

Distributors for North America:
Kluwer Academic Publishers
101 Philip Drive
Assinippi Park
Norwell, Massachusetts 02061 USA

Distributors for all other countries:
Kluwer Academic Publishers Group
Distribution Centre
Post Office Box 322
3300 AH Dordrecht, THE NETHERLANDS

Library of Congress Cataloging-in-Publication Data

A C.I.P. Catalogue record for this book is available
from the Library of Congress.

Printed on acid-free paper.

Printed in the United States of America

CONTENTS

FOREWORD

Angle stability had been the primary concern of the utilities for many decades. In the 80's due to the declining investments in new generation and transmission facilities, the system became stressed, resulting in a new phenomena hitherto largely ignored, namely voltage stability. Thus the role of reactive power in maintaining proper voltage profile in the system began receiving attention. Several instances of voltage collapse around the world only heightened the interest in this topic. Initially, treated as a static concept, the importance of dynamics of the machines, exciters, tap changers as well as dynamics of the load were found to affect voltage stability significantly.

This monograph addresses all these issues in depth from a rigorous analytical perspective as well as practical insight. Besides being a useful resource for engineers in industry, it will serve as a starting point for new researchers in this field. In the evolving scenario of a restructured power industry, the issues of voltage stability will be more complex and challenging to solve.

Profs. T. Van Cutsem and C. Vournas have worked in this research area extensively and have also offered short courses on this topic in the past.

I have great pleasure in welcoming this monograph in our power electronics and power system series.

M. A. Pai
University of Illinois
Urbana, IL

PREFACE

The idea of writing this book has probably its root in the enjoyable discussions we had in the Spring of 1994 during a sabbatical visit in Liège. Two years later, a couple of common papers and the encouragement of Prof. Pai provided the motivation for embarking on this adventure.

We undertook it courageously, in the middle of other obligations and with the help of Internet to bridge the 2500 km that separate our cities. It has been a year-long, difficult but enriching journey and for us it will remain memorable.

And so it will for our families ! Our very first thanks go to them: Marie-Paule, François, Nicolas, and Olivier on the one part, Malvina on the other. Their patience in coping with us and our absences during this period is heartily recognized.

We are grateful to Prof. Pai, the series editor, not only for his kind encouragement, but also for helping us in reviewing the text and doing his best to improve our English.

Sincere thanks are also due to our colleagues and graduate students, in particular to Dr. Patricia Rousseaux, at the University of Liège, George Manos and Basil Nomikos, at NTUA, for carefully reading our draft and suggesting corrections and improvements.

Finally, the valuable technical support of George Efthivoulidis, at NTUA, as well as his help in improving LaTeXstyles, are thankfully acknowledged.

PART I

COMPONENTS AND PHENOMENA

1

INTRODUCTION

*"Je n'ai fait celle-ci plus longue que parce que
je n'ai pas eu le loisir de la faire plus courte"*[1]
Blaise Pascal

1.1 WHY ANOTHER BOOK?

There was a time when power systems, and in particular transmission systems could afford to be overdesigned. However, in the last two decades power systems have been operated under much more stressed conditions than was usual in the past. There is a number of factors responsible for this: environmental pressures on transmission expansion, increased electricity consumption in heavy load areas (where it is not feasible or economical to install new generating plants), new system loading patterns due to the opening up of the electricity market, etc. It seems as though the development brought about by the increased use of electricity is raising new barriers to power system expansion.

Under these stressed conditions a power system can exhibit a new type of unstable behaviour characterized by slow (or sudden) voltage drops, sometimes escalating to the form of a collapse. A number of such voltage instability incidents have been experienced around the world. Many of them are described in [Tay94]. As a consequence, voltage stability has become a major concern in power system planning and operation.

As expected, the power engineering community has responded to the new phenomenon and significant research efforts have been devoted to developing new analysis tools

---

[1] (speaking of a letter) I made this one longer, only because I had not enough time to make it shorter

and controlling this type of instability. Among the early references dealing with the subject are textbooks on power system analysis devoting a section to voltage stability [ZR69, Wee79, Mil82] as well as technical papers [WC68, Nag75, Lac78, BB80, TMI83, BCR84, Cal86, KG86, Cla87, CTF87, Con91]. A series of three seminars on this specific topic [Fin88, Fin91, Fin94] has provided a forum for the presentation of research advances. Several CIGRE Task Forces [CTF93, CTF94a, CTF94b, CWG98] and IEEE Working Group reports [IWG90, IWG93, IWG96] have offered a compilation of techniques for analyzing and counteracting voltage instability. More recently, a monograph [Tay94] as well as one chapter of a textbook [Kun94] have been devoted to this topic.

One important aspect of the voltage stability problem, making its understanding and solution more difficult, is that the phenomena involved are truly nonlinear. As the stress on the system increases, this nonlinearity becomes more and more pronounced. This makes it necessary to look for a new theoretical approach using notions of nonlinear system theory [Hil95].

In this general framework the objective of our book is twofold:

- formulate a unified and coherent approach to the voltage stability problem, consistent with other areas of power system dynamics, and based on analytical concepts from nonlinear systems theory;

- use this approach in describing methods that can be, or have been, applied to solve practical voltage stability problems.

To achieve these two goals, we rely on a variety of power system examples. We start from simple two-bus systems, on which we illustrate the essence of the theory. We proceed with a slightly more complex system that is detailed enough to capture the main voltage phenomena, while still allowing analytical derivations. We end up with simulation examples from a real-life system.

1.2   VOLTAGE STABILITY

Let us now address a fundamental question: *what is voltage stability ?*

Convenient definitions have been given by IEEE and CIGRE Working Groups, for which the reader is referred to the previously mentioned reports. However, at this early point we would like to define voltage *instability* within the perspective adopted throughout this book:

> *Voltage instability stems from the attempt of load dynamics to restore power consumption beyond the capability of the combined transmission and generation system.*

Let us follow this descriptive definition word by word:

- *Voltage*: as already stated, the phenomenon is manifested in the form of large, uncontrollable voltage drops at a number of network buses. Thus the term "voltage" has been universally accepted for its description.

- *Instability*: having crossed the maximum deliverable power limit, the mechanism of load power restoration becomes unstable, reducing instead of increasing the power consumed. This mechanism is the heart of voltage instability.

- *Dynamics*: any stability problem involves dynamics. These can be modelled with either differential equations (continuous dynamics), or with difference equations (discrete dynamics). We will refer later to the misconception of labeling voltage stability a "static" problem.

- *Loads* are the driving force of voltage instability, and for this reason this phenomenon has also been called *load instability*. Note, however, that loads are not the only players in this game.

- *Transmission* systems have a limited capability for power transfer, as is well known from circuit theory. This limit (as affected also by the generation system) marks the onset of voltage instability.

- *Generation*: generators are not ideal voltage sources. Their accurate modelling (including controllers) is important for correctly assessing voltage stability.

One term also used in conjunction with voltage stability problems is *voltage collapse*. In this book we use the term "collapse" to signify a sudden catastrophic transition that is usually due to an instability occurring in a faster time-scale than the one considered. As we will see, voltage collapse may, or may not be the final outcome of voltage instability.

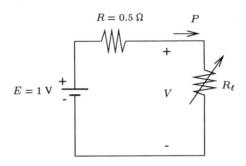

Figure 1.1   DC system

On the rôle of reactive power

The reader may have noticed that we did not include in the above definition of voltage instability the important concept of *reactive power*. It is a well-known fact that in AC systems dominated by reactances (as power systems typically are) there is a close link between voltage control and reactive power. However, by not referring to reactive power in our definition, we intend not to overemphasize its rôle in voltage stability, where *both* active and reactive power share the leading rôle.

The decoupling between active power and phase angles on the one hand, and reactive power and voltage magnitudes on the other hand, applies to normal operating conditions and cannot be extended to the extreme loading conditions typical of voltage instability scenarios.

The following example illustrates that there is no "cause and effect" relationship between reactive power and voltage instability. Consider the system of Fig. 1.1 made up of a DC voltage source E feeding through a line resistance R a variable load resistance R_ℓ.

We assume that R_ℓ is automatically varied by a control device, so as to achieve a power consumption setpoint P_o. For instance it could be governed by the following ordinary differential equation:

$$\dot{R}_\ell = I^2 R_\ell - P_o \qquad (1.1)$$

It is well known that the maximum power that can be transferred to the load corresponds to the condition $R_\ell = R$ and is given by:

$$P_{max} = \frac{E^2}{4R} \qquad (1.2)$$

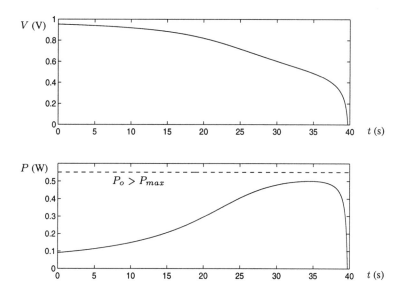

Figure 1.2 Voltage instability in a DC system

If the demand P_o is made larger than P_{max} the load resistance will decrease below R and voltage instability will result after crossing the maximum power point. A typical simulation for this case is shown in Fig. 1.2.

This simple paradigm has the major characteristics of voltage instability, although it does not involve reactive power. In actual AC power systems, reactive power makes the picture much more complicated but it is certainly not the only source of the problem.

1.3 POWER SYSTEM STABILITY CLASSIFICATION

We now place voltage stability within the context of power system stability in general. Table 1.1 shows a classification scheme based upon two criteria: time scale and driving force of instability.

The first power system stability problems encountered were related to generator rotor angle stability, either in the form of undamped electromechanical oscillations, or in the form of monotonic rotor acceleration leading to the loss of synchronism. The former type of instability is due to a lack of *damping* torque, and the latter to a lack of *synchronizing* torque.

Table 1.1 Power System Stability Classification

Time scale	Generator-driven	Load-driven
short-term	rotor angle stability transient · steady-state	short-term voltage stability
long-term	frequency stability	long-term voltage stability

The first type of instability is present even for small disturbances and is thus called *steady-state* or *small-signal* stability. The second one is initiated by large disturbances and is called *transient* or *large-disturbance* stability. For the analysis of steady-state stability it is sufficient to consider the linearized version of the system around an operating point, typically using eigenvalue and eigenvector techniques. For transient stability one has to assess the performance of the system for a set of specified disturbances.

The time frame of rotor angle stability is that of electromechanical dynamics, lasting typically for a few seconds. Automatic voltage regulators, excitation systems, turbine and governor dynamics all act within this time frame. The relevant dynamics have been called *transient dynamics* in accordance with transient stability, generator transient reactances, etc. However, this may create misinterpretations, since "transient" is also used in "transient stability" to distinguish it from "steady-state stability", which also belongs to the same time frame. For this reason we prefer to refer to the above time frame of a few seconds as the *short-term time scale*.

When the above mentioned short-term dynamics are stable they eventually die out some time after a disturbance, and the system enters a slower time frame. Various dynamic components are present in this time frame, such as transformer tap changers, generator limiters, boilers, etc. The relevant transients last typically for several minutes. We will call this the *long-term time scale*.

In the long-term time scale we can distinguish between two types of stability problems:

1. *frequency* problems due to generation–load imbalance irrespective of network aspects within each connected area;

2. *voltage* problems, which are due to the electrical distance between generation and loads and thus depend on the network structure.

In modern power systems, frequency stability problems can be encountered after a major disturbance has resulted in islanding. Since we have assumed that the electromechanical oscillations have died out, frequency is common throughout each island and the problem can be analyzed using a single-bus equivalent, on which all generators and loads are connected. The frequency instability is related to the active power imbalance between generators and loads in each island[2].

Voltage stability, on the other hand, requires a full network representation for its analysis. This is a main aspect separating the two classes of long-term stability problems. Moreover, as suggested by the definition we gave in Section 1.2, voltage instability is load driven.

Now, when referring to voltage stability we can identify dynamic load components with the tendency to restore their consumed power in the time-frame of a second, i.e. in the short-term time scale. Such components are mainly induction motors and electronically controlled loads, including HVDC interconnections. We have thus to introduce a *short-term voltage stability* class alongside generator rotor angle stability. Since these two classes of stability problems belong to the same time scale, they require basically the same complexity of component models and sometimes distinction between the two in meshed systems becomes difficult [VSP96]. In other words, in the short-term time scale, there is not a clear-cut separation between load-driven and generator-driven stability problems, as there is as between frequency and long-term voltage stability.

It should be noted that the identification of the driving force for an instability mechanism in Table 1.1 does not exclude the other components from affecting this mechanism. For instance, load modelling does affect rotor angle stability, and, as we will show in this book, generator modelling is important for a correct voltage stability assessment.

Each of the four major stability classes of Table 1.1 may have its own further subdivisions, like the ones we have already seen in the case of generator rotor angle stability. We can thus identify small-signal and large-disturbance forms of voltage stability. Note, however, that this distinction is not as important as in the case of rotor angle stability, where transient and steady-state stability relate to different problems. Thus, although the small-signal versus large-disturbance terminology exists and is in accordance with the above stability classification we will not use it extensively in this book. We see voltage stability as a single problem on which one can apply a combination of both linearized and nonlinear tools.

---

[2]Note that the counterpart of frequency stability in the short-term time scale is rotor angle stability, since in this time scale there is no common frequency

Another point to be made here deals with the distinction between dynamic and "static" aspects. In fact, long-term voltage stability has been many times misunderstood as a "static" problem. The misconception stems from the fact that static tools (such as modified power flow programs) are acceptable for simpler and faster analysis. Voltage stability, however, is dynamic by nature, and in some cases one has to resort to dynamic analysis tools (such as time-domain methods). One should thus avoid to confuse means with ends in stability classification.

1.4   STRUCTURE OF THIS BOOK

The book consists of two parts.

Part I deals with phenomena and components. It includes Chapters 2, 3, and 4, each dealing with one of the three major aspects of the voltage stability problem according to our definition of Section 1.2.

We start with transmission aspects in *Chapter 2*, because it is the limits on power transfer that set up the voltage stability problem. In this chapter we review the problem of maximum deliverable power in AC systems and concentrate on a number of transmission components that are linked to voltage stability, such as compensation, off-nominal tap transformers, etc.

Chapter 3 reviews the basics of generator modelling, including significant details, such as the effect of saturation on capability limits. Frequency and voltage controls are also reviewed, as well as the various limiting devices that protect generators from overloading. We finally consider how generator limits affect the maximum deliverable power of the system.

In *Chapter 4* we focus on the driving force of voltage instability, i.e. load dynamics. We first give a general framework of load restoration and then we proceed with the analysis of three major components of load restoration, namely induction motors, load tap-changers and thermostatic load. Finally we discuss aggregate generic load models.

Part II of the book deals with the description of voltage instability mechanisms and analysis methods.

We first provide in *Chapter 5* a summary of the mathematical background from non-linear system theory necessary for the analysis of later chapters. This includes the notions of bifurcation, singularity, and time-scale decomposition.

In *Chapter 6* we discuss general modelling requirements for voltage stability analysis, and illustrate them using a simple but fully detailed example.

Chapter 7 gives the basic voltage stability theory in terms of three closely linked concepts: loadability limits, bifurcations, and sensitivities. For the most part, this chapter deals with smooth parameter changes. The effect of discontinuities, especially those caused by the overexcitation limiters of synchronous generators is explicitly taken into account.

In *Chapter 8* we concentrate on large, abrupt disturbances and describe one by one the possible mechanisms of losing stability, whether in the long-term, or the short-term time scale. We also concentrate on countermeasures applicable to each type of instability. The detailed example introduced in Chapter 6 is used to illustrate some of the key instability mechanisms.

Finally, in *Chapter 9* we give a representative sample of criteria and computer methods for voltage stability analysis. After a brief review of security concepts, we consider methods for contingency evaluation, loadability limit computation and determination of secure operation limits. We end up with examples from a real-life system.

At the end of some chapters we provide problems. Some of them are straightforward applications of the presented methods. Other problems refer to the examples and test cases given in the text. Finally, some are at the level of research topics. The authors would be pleased to receive suggestions and exchange views on all these.

1.5   NOTATION

We give below a short list of notation conventions used in this book.

- Phasors are shown as capital letters with an overline, e.g. $\bar{I}, \bar{V}$.

- Phasor magnitudes are shown by the same capital letter without the overline, e.g. I, V.

- Lowercase bold letters, e.g. $\mathbf{x}, \mathbf{y}$, correspond to column vectors. Superscript T denotes transpose. Therefore row vectors are written as $\mathbf{x}^T, \mathbf{y}^T$.

- A collection of phasors in a column vector is represented as a capital bold letter with an overline, e.g. $\bar{\mathbf{I}}$.

- Matrices are normally shown as bold capital letters, e.g. $\mathbf{A}, \mathbf{J}$.

- Jacobian matrices are shown as a bold letter (indicating the vector function) with a bold subscript (indicating the vector with respect to which we differentiate). Thus:
$$\mathbf{f_x} = \left[\frac{\partial f_i}{\partial x_j}\right]$$

- Time derivatives appear with a dot, e.g. $\dot{x}$.

2

TRANSMISSION SYSTEM ASPECTS

"Maybe I can't define stability, but I know it when I see it !"[1]

Carson W. Taylor

In this chapter we analyze the rôle played by the transmission system in voltage stability.

We first deal with two basic notions: the maximum power that can be delivered to loads and the relationship between load power and network voltage. Then we briefly and qualitatively explain how these two basic properties may result in voltage instability. Next, we discuss the effect of components that affect the transmission capability, series and shunt compensation on one hand, transformers with adjustable tap ratio on the other hand. We also introduce the notion of VQ curves that express the relationship between voltage and reactive power at a given bus.

Most of the material of this chapter is based on the analysis of a simple single-load infinite-bus system, which allows easy analytical derivations and provides insight into the problem. Basic concepts introduced in this chapter will be generalized in later chapters to large system of arbitrary complexity.

2.1 SINGLE-LOAD INFINITE-BUS SYSTEM

We consider the simple system of Fig. 2.1, which consists of one load fed by an infinite bus through a transmission line. By definition, the voltage magnitude and frequency

---

[1] Panel Session presentation at the 1997 IEEE/PES Winter Power Meeting

Figure 2.1   Single-load infinite-bus system

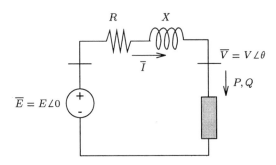

Figure 2.2   Circuit representation

are constant at the infinite bus. We assume balanced 3-phase operating conditions, so that the per phase representation is sufficient. We also consider steady-state sinusoidal operating conditions, characterized by phasors and complex numbers. The phase reference is arbitrary and need not be specified at this stage.

This leads to the circuit representation of Fig. 2.2. The infinite bus is represented by an ideal voltage source E. The transmission line is represented by its series resistance R and reactance X, as given by the classical pi-equivalent. The line shunt capacitance is neglected for simplicity (the effects of shunt capacitors are considered later in Section 2.6.2). The transmission impedance is:

$$Z = R + jX$$

Alternatively, we may think of E and Z as the Thévenin equivalent of a power system as seen from one bus. Note that, because power generators are not pure voltage sources, the Thévenin emf somewhat varies as more and more power is drawn from the system; we will however neglect this variation in a first approximation and consider a constant emf E as mentioned previously.

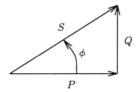

Figure 2.3 Definition of angle ϕ

Finally, let us recall that the *load power factor* is given by:

$$\text{PF} = \frac{P}{S} = \frac{P}{\sqrt{P^2 + Q^2}} = \cos\phi$$

where P, Q and S are the active, reactive and apparent powers and ϕ is the angle defined in Fig. 2.3.

2.2   MAXIMUM DELIVERABLE POWER

As pointed out in the Introduction, voltage instability results from the attempt of loads to draw more power than can be delivered by the transmission and generation system. In this section we focus on determining the maximum power that can be obtained at the receiving end of the simple system of Fig. 2.2, under various constraints.

2.2.1   Unconstrained maximum power

For the sake of simplicity we start by assuming that the load behaves as an impedance. In fact we will show later on that this choice does not affect the results. We denote the load impedance by:

$$Z_\ell = R_\ell + jX_\ell$$

where R_ℓ and X_ℓ are the load resistance and reactance, respectively.

We first revisit a classical derivation of circuit theory known as the load adaptation problem [CDK87] or maximum power transfer theorem: assuming that both R_ℓ and X_ℓ are free to vary, find the values which maximize the *active* power consumed by the load.

The current $\bar{I}$ in Fig. 2.2 is given by:

$$\bar{I} = \frac{\bar{E}}{(R + R_\ell) + j(X + X_\ell)}$$

and the active power consumed by the load:

$$P = R_\ell I^2 = \frac{R_\ell E^2}{(R + R_\ell)^2 + (X + X_\ell)^2} \tag{2.1}$$

Maximizing P over the two variables R_ℓ and X_ℓ, the necessary extremum conditions are:

$$\frac{\partial P}{\partial R_\ell} = 0$$

$$\frac{\partial P}{\partial X_\ell} = 0$$

which after some calculations yields:

$$(R + R_\ell)^2 + (X + X_\ell)^2 - 2R_\ell(R + R_\ell) = 0$$
$$-R_\ell(X + X_\ell) = 0$$

The solution to these equations, under the constraint $R_\ell > 0$, is unique:

$$R_\ell = R \tag{2.2a}$$
$$X_\ell = -X \tag{2.2b}$$

or in complex form:

$$Z_\ell = Z^*$$

One easily checks that this solution corresponds to a maximum of P. In other words:

> load power is maximized when the load impedance is the complex conjugate of the transmission impedance.

Under the maximum power conditions, the impedance seen by the voltage source is $R + R_\ell + jX + jX_\ell = 2R$, i.e. it is purely resistive and the source does not produce any reactive power. The corresponding load power is:

$$P_{max} = \frac{E^2}{4R} \tag{2.3}$$

and the receiving-end voltage:

$$V_{maxP} = \frac{E}{2}$$

where the subscript $max P$ denotes a value under maximum active power condition.

The unconstrained case is not well suited for power system applications. The first problem is that in a transmission system the resistance R can be negligible compared to the reactance X. Now, making R tend to zero, the optimal load resistance (2.2a) also goes to zero, while the maximum power (2.3) goes to infinity. The two results might seem in contradiction: however, as R and R_ℓ go to zero, the current I goes to infinity (since $X + X_\ell = 0$) and so does the power $R_\ell I^2$! This is obviously unrealistic.

Even when taking into account the nonzero transmission resistance R, the above result is not directly applicable to power systems. Indeed, a highly capacitive load would be required to match the dominantly inductive nature of the system impedance. A modified derivation, closer to power system applications is made by assuming that the power factor of the load is specified. This case is dealt with in the next subsection.

2.2.2 Maximum power under a given load power factor

Specifying the load power factor $\cos \phi$ is equivalent to having a load impedance of the form:

$$Z_\ell = R_\ell + jX_\ell = R_\ell + jR_\ell \tan \phi$$

which now leaves R_ℓ as the single degree of freedom for maximizing the load power.

The current $\bar{I}$ is now given by:

$$\bar{I} = \frac{\bar{E}}{(R + R_\ell) + j(X + R_\ell \tan \phi)}$$

and the load active power by:

$$P = R_\ell I^2 = \frac{R_\ell E^2}{(R + R_\ell)^2 + (X + R_\ell \tan \phi)^2} \tag{2.4}$$

The extremum condition is:

$$\frac{\partial P}{\partial R_\ell} = 0$$

or, after some calculations:

$$(R^2 + X^2) - R_\ell^2(1 + \tan^2 \phi) = 0 \tag{2.5}$$

which is equivalent to:

$$|Z_\ell| = |Z|$$

The second derivative is given by:

$$\frac{\partial^2 P}{\partial R_\ell^2} = -2R_\ell(1 + \tan^2 \phi)$$

which is always negative, thereby indicating that the solution is a maximum. In other words:

> under constant power factor, load power is maximized when the load impedance becomes equal in magnitude to the transmission impedance.

The optimal load resistance and reactance are thus given by:

$$R_{\ell max P} = |Z| \cos \phi$$
$$X_{\ell max P} = |Z| \sin \phi = R_{\ell max P} \tan \phi$$

As an illustration, Fig. 2.4 shows the load power P, the voltage V and the current magnitude I as a function of R_ℓ. An infinite R_ℓ corresponds to open-circuit conditions. As R_ℓ decreases, V drops while I increases. As long as R_ℓ remains larger than $R_{\ell max P}$, the increase in I^2 gains over the decrease in R_ℓ and hence P increases. When R_ℓ becomes smaller than $R_{\ell max P}$ the reverse holds true. Finally, $R_\ell = 0$ corresponds to short-circuit conditions.

Lossless transmission

Let us come back to the case where $R = 0$. The optimal load resistance under constant power factor is, according to (2.5):

$$R_{\ell max P} = X \cos \phi$$

Substituting in (2.4) yields the maximum active power:

$$P_{max} = \frac{\cos \phi}{1 + \sin \phi} \frac{E^2}{2X} \tag{2.6}$$

with the corresponding reactive power:

$$Q_{max P} = \frac{\sin \phi}{1 + \sin \phi} \frac{E^2}{2X} \tag{2.7}$$

and receiving-end voltage:

$$V_{max P} = \frac{E}{\sqrt{2}\sqrt{1 + \sin \phi}} \tag{2.8}$$

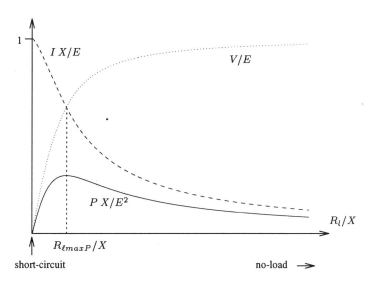

Figure 2.4 P, V and I as a function of R_ℓ, for a lossless system ($R = 0$) and under constant power factor ($\tan \phi = 0.2$)

Lossless transmission and unity power factor

If we assume furthermore that the load is perfectly compensated, so that $\cos \phi = 1$, the optimal resistance, maximum power and receiving-end voltage become respectively:

$$R_{\ell max P} = X$$
$$P_{max} = \frac{E^2}{2X}$$
$$V_{max P} = \frac{E}{\sqrt{2}} \simeq 0.707 E$$

Extensions to multiport systems

Some generalizations of the above results to multiport systems are given in [Cal83]. Let a multiport circuit be characterized by

$$\bar{V} = \bar{E} + Z\bar{I}$$

where $\bar{V}$ is the vector of terminal voltages, $\bar{E}$ the vector of open-circuit voltages, $\bar{I}$ the vector of injected currents and Z the (short-circuit) impedance matrix.

If the circuit is purely reactive, and characterized by $\mathbf{Z} = j\mathbf{X}$, it can be shown that the total active power delivered is maximized when a purely *resistive* network with impedance matrix $\mathbf{Z}_\ell = \mathbf{X}$ is connected to the multiport. The corresponding maximum power is easily obtained.

Furthermore if all the elements of the multiport matrix $\mathbf{Z}$ have the same argument ζ and all the elements of the loading matrix $\mathbf{Z}_\ell$ the same argument ϕ, i.e.

$$\mathbf{Z} = \mathbf{N}e^{j\zeta} \quad \text{and} \quad \mathbf{Z}_\ell = \mathbf{L}e^{j\phi}$$

the total active power is maximum when $\mathbf{N} = \mathbf{L}$.

Note that the individual load powers are not constrained with respect to each other in this derivation. If a pattern of load increase is specified, the maximum power delivered will be smaller. These aspects will be further discussed in Chapters 7 and 9.

Remark on load characteristics

Note that the maximum deliverable power given by either (2.3) or (2.6) depends only on the network parameters (R, X) and is independent of the load characteristic which was assumed to be that of an impedance for simplicity. This will be verified in the sequel, where no assumption will be made as to the nature of the load. For this purpose we now adopt a formulation in terms of powers.

2.2.3   Maximum power derived from load flow equations

For the sake of simplicity, we neglect the transmission resistance R (see Fig. 2.2). We also take the ideal voltage source as the phase reference by setting $\bar{E} = E \angle 0$. We denote the load voltage magnitude and phase angle by V and θ respectively.

One easily obtains from Fig. 2.2:

$$\bar{V} = \bar{E} - jX\bar{I}$$

The complex power *absorbed* by the load is:

$$
\begin{aligned}
S &= P + jQ = \bar{V}\,\bar{I}^\star = \bar{V}\,\frac{\bar{E}^\star - \bar{V}^\star}{-jX} \\
&= \frac{j}{X}(EV\cos\theta + jEV\sin\theta - V^2)
\end{aligned}
\tag{2.9}
$$

which decomposes into:

$$P = -\frac{EV}{X} \sin\theta \qquad (2.10a)$$

$$Q = -\frac{V^2}{X} + \frac{EV}{X} \cos\theta \qquad (2.10b)$$

Equations (2.10a,b) are the *power flow* or *load flow* equations of the lossless system. For a given load (P, Q), they have to be solved with respect to V and θ, from which all other variables can be computed. Let us determine for which values of (P, Q) there is one solution.

Eliminating θ from (2.10a,b) gives:

$$(V^2)^2 + (2QX - E^2)V^2 + X^2(P^2 + Q^2) = 0 \qquad (2.11)$$

This is a second-order equation with respect to V^2. The condition to have at least one solution is:

$$(2QX - E^2)^2 - 4X^2(P^2 + Q^2) \geq 0$$

which can be simplified into:

$$-P^2 - \frac{E^2}{X}Q + (\frac{E^2}{2X})^2 \geq 0 \qquad (2.12)$$

The equality in (2.12) corresponds to a parabola in the (P, Q) plane, as shown in Fig. 2.5. All points "inside" this parabola satisfy (2.12) and thus lead to two load flow solutions. Outside there is no solution while on the parabola there is a single solution.

This parabola is the locus of all maximum power points. Points with negative P correspond to a maximum generation while each point with positive P corresponds to the maximum load under a given power factor, as derived in the previous section.

The locus is symmetric with respect to the Q-axis (i.e. with respect to changing P into $-P$). In other words, the maximum power that can be injected at the load end is exactly equal to the maximum power that can be absorbed. However, this symmetry disappears if one takes into account the line resistance.

Setting $P = 0$ in (2.12) one obtains:

$$Q \leq \frac{E^2}{4X}$$

Noting that E^2/X is the short-circuit power at the load bus, i.e. the product of the no-load voltage E by the short-circuit current E/X, the maximum of purely reactive load is one fourth of the short-circuit power.

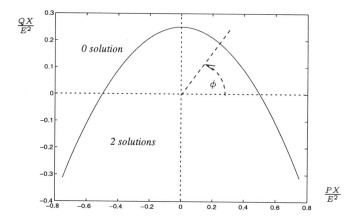

Figure 2.5   Domain of existence of a load flow solution

Similarly, by setting $Q = 0$ in (2.12) one gets:

$$P \leq \frac{E^2}{2X}$$

which is the same power limit we derived for a lossless line with unity power factor, and corresponds to half the short-circuit power.

As can be seen, there is a fundamental difference between the active and reactive powers: any active power can be consumed provided that enough reactive power is injected at the load bus ($Q < 0$), while the reactive load power can never exceed $E^2/4X$. This difference comes from the inductive nature of the transmission system and further illustrates the difficulty of transporting large amounts of reactive power. Note that in practice the large reactive support that is required for large active power will finally result in unacceptably high load bus voltage.

2.3   POWER-VOLTAGE RELATIONSHIPS

Assuming that condition (2.12) holds, the two solutions of (2.11) are given by:

$$V = \sqrt{\frac{E^2}{2} - QX \pm \sqrt{\frac{E^4}{4} - X^2 P^2 - XE^2 Q}} \qquad (2.13)$$

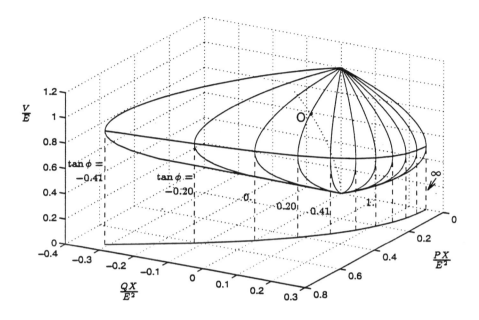

Figure 2.6 Voltage as a function of load active and reactive powers

In the (P, Q, V) space, equation (2.11) defines a two dimensional surface shown in Fig. 2.6. The upper part of this surface corresponds to the solution with the plus sign in (2.13), or the higher voltage solution, while the lower part corresponds to the solution with the minus sign, which is the low voltage one. The "equator" of this surface, along which the two solutions are equal corresponds to the maximum power points as given by (2.6, 2.7, 2.8). The projection of this limit curve onto the (P, Q) plane coincides with the parabola of Fig. 2.5.

The "meridians" drawn with solid lines in Fig. 2.6 correspond to intersections with vertical planes $Q = P \tan \phi$, for ϕ varying from $-\pi/8$ to $\pi/2$ by steps of $\pi/16$. Projecting these meridians onto the (P, V) plane provides the curves of load voltage as a function of active power, for the various $\tan \phi$. These famous curves, shown in Fig. 2.7, are generally referred to as the *PV curves* or *nose curves*. They play a major role in understanding and explaining voltage instability.

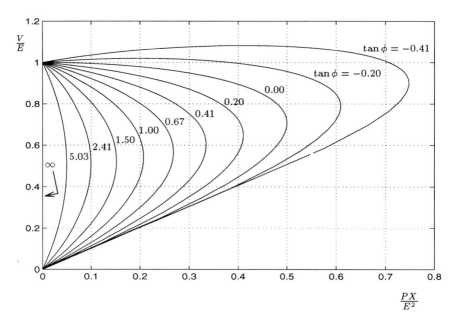

Figure 2.7   The famous PV curves

Although they are probably the most popular, the PV curves are not the only possible projection of the surface of Fig. 2.6 onto a plane. We could similarly:

- project the meridians onto the (Q, V) plane, thereby producing QV curves

- take the apparent power $S = \sqrt{P^2 + Q^2}$ as the abscissa, and consider SV curves

- consider QV curves corresponding to constant active power P

- or PV curves under constant reactive power Q.

All these curves have basically the shape shown in Fig. 2.7, the only difference being that curves drawn under constant P or constant Q do not go through zero voltage (except, of course, when the power held constant is equal to zero).

The following observations can be made regarding the curves of Fig. 2.7:

1. For a given load power below the maximum, there are two solutions: one with higher voltage and lower current, the other with lower voltage and higher current. The former corresponds to "normal" operating conditions, with voltage V closer to the generator voltage E. Permanent operation at the lower solutions is unacceptable, as will be discussed in the next section.

2. As the load is more and more compensated (which corresponds to smaller $\tan \phi$), the maximum power increases. However, the voltage at which this maximum occurs also increases. This situation is dangerous in the sense that maximum transfer capability may be reached at voltages close to normal operation values. Also, for a high degree of compensation and a load power close to the maximum, the two voltage solutions are close to each other and without further analysis it may be difficult to decide if a given solution is the "normal" one.

3. For over-compensated loads ($\tan \phi < 0$), there is a portion of the upper PV curve along which the voltage increases with the load power. The explanation is that under negative $\tan \phi$, when more active power is consumed, more reactive power is produced by the load. At low load, the voltage drop due to the former is offset by the voltage increase due to the latter. The more negative $\tan \phi$ is, the larger is the portion of the PV curve where this takes place.

2.4 GENERATOR REACTIVE POWER REQUIREMENT

In this chapter, generators are treated as voltage sources of constant magnitude. As will be discussed in the next chapter the main defect of this assumption lies in the limited reactive power capability of generators. It is therefore of interest to determine how the reactive generation increases with load.

Pursuing the example of Fig. 2.2, in the lossless case $R = 0$, we express the generator reactive production as the sum of the load and the network losses:

$$Q_g = Q + XI^2 \tag{2.14}$$

where the line current I relates to the generator apparent power S_g through:

$$I = \frac{S_g}{E} = \frac{\sqrt{P_g^2 + Q_g^2}}{E}$$

Substituting I in (2.14) and noting that $P_g = P$ in the absence of real power losses, we get:

$$Q_g = Q + \frac{X}{E^2}(P^2 + Q_g^2)$$

which can be reordered into:

$$Q_g^2 - \frac{E^2}{X}Q_g + \frac{E^2}{X}Q + P^2 = 0 \tag{2.15}$$

Solving this equation with respect to Q_g yields:

$$Q_g = \frac{E^2}{2X} \pm \sqrt{\left(\frac{E^2}{2X}\right)^2 - \frac{QE^2}{X} - P^2} \tag{2.16}$$

Note that (2.15) has a solution only when the condition (2.12) is satisfied. Equation (2.15) defines a surface in the (P, Q, Q_g) space. Cutting this surface with constant power factor planes - as we did in Fig. 2.6 - one obtains the $P\,Q_g$ curves shown in Fig. 2.8. These curves are similar to the PV curves, except that normal operating points now lie on the *lower* part of the curves. Starting from open-circuit conditions $(P = 0, Q_g = 0)$ and increasing the load, the reactive generation increases nonlinearly with P up to the maximum power. Beyond this point, P decreases while reactive losses continue to increase, up to the point $(P = 0, Q_g = \frac{E^2}{X})$ which corresponds to a short-circuit at the load bus. Note finally that all the maximum power points are characterized by:

$$Q_{g\,maxP} = \frac{E^2}{2X}$$

whatever the load power factor be.

2.5   A FIRST GLANCE AT INSTABILITY MECHANISMS

The purpose of this section is to emphasize why the existence of a maximum deliverable power may result in system instability and voltage collapse. We propose here some intuitive views, keeping a more rigorous analysis for later chapters.

2.5.1   Network vs. load PV characteristics

The power consumed by loads varies with voltage and frequency. In this book we will concentrate mainly on variations with voltage. We call *load characteristic* the expression of the load active and reactive power as a function of voltage V and an independent variable z, which corresponds to the amount of connected equipment. We

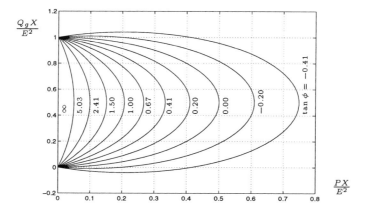

Figure 2.8   Generator reactive production as a function of load power

call z the *load demand*. Thus the load characteristic takes on the general form:

$$P = P(V, z) \tag{2.17a}$$
$$Q = Q(V, z) \tag{2.17b}$$

For a specified demand z, equations (2.17a,b) define a curve in the (P, Q, V) space. This curve intersects the $V(P, Q)$ surface at one or more points. These are possible operating points for the specified demand. When the latter changes, the intersection points move on the surface. If we project the set of intersection points for all values of the demand onto the (P, V) plane, we obtain what we call the *network PV characteristic* as opposed to the *load PV characteristic* given by (2.17a). Alternatively, we may project the set on the (Q, V) plane and consider the *load QV characteristic*. Note that the network characteristic cannot be defined without considering how the load power varies with voltage.

Consider for instance the widely used load characteristic known as the *exponential load model*:

$$P = zP_o \left(\frac{V}{V_o}\right)^\alpha \tag{2.18a}$$

$$Q = zQ_o \left(\frac{V}{V_o}\right)^\beta \tag{2.18b}$$

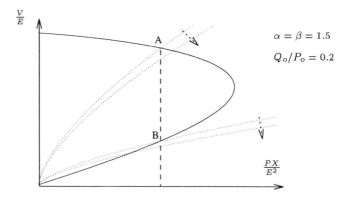

Figure 2.9 Network and load PV curves

In this model P_o (resp. Q_o) is the active (resp. reactive) power consumed for $z = 1$ and a voltage V equal to the reference voltage V_o. As an example, the dotted curve shown in Fig. 2.6 corresponds to (2.18a,b) with $\alpha = \beta = 1.5$ and $Q_o/P_o = 0.2$. It intersects the $V(P, Q)$ surface at point O and at the origin. As the demand z changes, so does the intersection point O. The set of points O for all possible demands, projected on the (P, V) plane is the solid line in Fig. 2.9. This is the network characteristic *corresponding to the assumed change in the load active and reactive power components.* In the above specific example:

$$\frac{Q}{P} = \frac{Q_o}{P_o} \left(\frac{V}{V_o} \right)^{\beta - \alpha}$$

and since $\alpha = \beta$ in this example, the load power factor is constant whatever the voltage. Hence, the network PV curve is merely the curve of Fig. 2.7 corresponding to $\tan \phi = 0.2$. This shortcut is no longer possible when $\alpha \neq \beta$.

2.5.2 Instability scenarios

Each dotted line in Fig. 2.9 is the load PV curve for some value of P_o. A and B are two operating points characterized by the same power P but different demands z.

Consider the effect of a small increase in demand z, as depicted in Fig. 2.9. At point A, the higher demand causes some voltage drop but results in a higher load power. This is the expected mode of operation of a power system. At point B however, the larger demand is accompanied by a decrease in *both* the voltage *and* the load power. If the load is purely static, operation at point B is possible, although perhaps non-viable due to low voltage and high current; this is however a matter of viability, not of stability. On the other hand, by assuming a load controller, or some inherent mechanism built in the load, that tends to increase the demand in order to achieve a specified power consumption, the operating point B becomes *unstable*. It will be shown in Chapter 4 that induction motors, load tap changers and heating thermostats are typical components which exhibit, directly or indirectly, the above behaviour.

Consider now a load which, following some disturbance, behaves instantaneously according to the dotted PV characteristic of Fig. 2.9 but tends dynamically to a constant power characteristic as given by the dashed line in the same figure. Anticipating a little about the dynamic notions of Chapter 5, we will say that this dashed vertical line is the *load equilibrium characteristic*, or *load steady-state characteristic*. Similarly, the network PV curve, if properly determined, corresponds to the equilibrium condition of the generation and transmission systems.

An obvious prerequisite to stable system operation is the existence of an equilibrium, given by the intersection of both characteristics. It happens precisely that an important class of voltage instability scenarios corresponds to changes in system parameters that lead to the disappearance of an equilibrium.

A first mechanism is illustrated in Fig. 2.10.a: an increase in demand causes the load equilibrium characteristic to change until finally it does not intersect the network characteristic.

A second, practically even more important scenario corresponds to a large disturbance. Disturbances of concern are the loss of transmission and/or generation equipments. In our two-bus example this corresponds to an increase in X and/or a decrease in E. The instability mechanism is depicted in Fig. 2.10.b: the large disturbance causes the network characteristic to shrink drastically so that the post-disturbance network PV curve does no longer intersect the (unchanged) load characteristic. Voltage collapse results from the loss of an equilibrium in the post-disturbance network.

Figure 2.11 illustrates the same two scenarios for a load characterized by $\alpha = \beta = 0.7$ (instead of $\alpha = \beta = 0$) at equilibrium.

Assuming a smooth load increase as in Fig. 2.10.a and 2.11.a, the point where the load characteristic becomes tangent to the network characteristic defines the *loadability limit* of the system. As mentioned above, a load increase beyond the loadability limit results in loss of equilibrium, and the system can no longer operate. In Fig. 2.10.a the point where the load and network PV curves are tangent coincides with the maximum

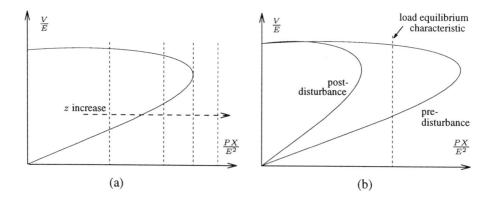

Figure 2.10 Instability mechanisms illustrated with PV curves; load equilibrium characteristic with $\alpha = \beta = 0$

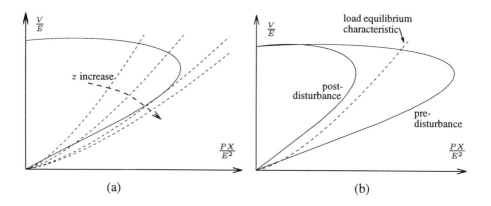

Figure 2.11 Instability mechanisms illustrated with PV curves; load equilibrium characteristic with $\alpha = \beta = 0.7$

deliverable power, because the load is assumed to restore to constant power, an important case in practice. However, a loadability limit does not necessarily coincide with the maximum deliverable power, since it depends on the load characteristic. This can be seen from Fig. 2.11.a. Note also that for certain load characteristics (e.g. the one in Fig. 2.9) there is no loadability limit, i.e. there is an operating point for all demands. Of course, some of these operating points may be infeasible for other reasons, such as unacceptably low voltage.

The load characteristics will be analyzed further in Chapter 4, while a more thorough discussion on loadability limits is left for Chapter 7.

The above scenarios do not tell us the course of events that occur as a result of the loss of equilibrium. They only tell us that, as far as network and load PV curves are the equilibrium characteristics of the system dynamics, system operation will experience a disruption. An in-depth investigation of the instability mechanism requires that we consider the dynamic behaviour of each component. Moreover, there are instability mechanisms that cannot be foreseen from purely static characteristics.

2.6 EFFECT OF COMPENSATION

Generally speaking, compensation consists of injecting reactive power to improve power system operation, more specifically keep voltages close to nominal values, reduce line currents and hence network losses, and contribute to stability enhancement [Mil82].

Most often compensation is provided by capacitors, counterbalancing the predominantly inductive nature of either the transmission system, or the loads. It may also consist of reactors where reactive power absorption is of concern.

Regarding voltage stability, the effects of *load compensation* have been discussed in Section 2.3. In this section we focus on *network compensation*, which may consist of either capacitors installed in series with transmission lines or shunt elements connected to system buses.

2.6.1 Line series compensation

Series compensation is used basically to decrease the impedance of transmission lines carrying power over long distances, as shown by the simple equivalent of Fig. 2.12 (the

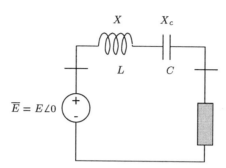

Figure 2.12   Series compensation

latter does not take into account the series capacitors location, e.g. at the mid-point or 1/3 or 1/4 points of the line).

The line net reactance is given by:

$$X_{net} = X - X_c = \omega L - \frac{1}{\omega C}$$

with the degree of compensation

$$\frac{X - X_{net}}{X} = \frac{X_c}{X}$$

being usually in the range $0.3 - 0.8$.

Replacing X by X_{net} in (2.6, 2.8) it is clearly seen that the maximum deliverable power is increased, while the voltage under maximum power is left unchanged.

Series compensation addresses a fundamental aspect of voltage instability, namely the electrical distance between generation and load centers. In this respect it is a very efficient countermeasure to instability .

2.6.2   Shunt compensation

The connection of shunt capacitors (or reactors) is probably the simplest and most widely used form of compensation. To investigate its effect in some detail, we consider the simple system of Fig. 2.13, which combines the effect of line charging (susceptance B_l) with that of an adjustable shunt compensation (susceptance B_c).

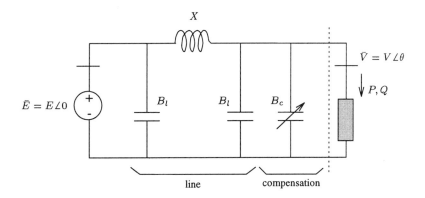

Figure 2.13 Network capacitances and shunt compensation

The Thévenin equivalent as seen by the load (i.e. to the left of the dotted line in Fig. 2.13) has the following emf and reactance:

$$E_{th} = \frac{1}{1 - (B_c + B_l)X} E$$

$$X_{th} = \frac{1}{1 - (B_c + B_l)X} X$$

Replacing E by E_{th} and X by X_{th} in (2.6, 2.8) gives the maximum deliverable power (under power factor $\cos \phi$):

$$P_{max} = \frac{\cos \phi}{1 + \sin \phi} \frac{E_{th}^2}{2X_{th}} = \frac{1}{1 - (B_c + B_l)X} \frac{\cos \phi}{1 + \sin \phi} \frac{E^2}{2X}$$

and the corresponding load voltage:

$$V_{maxP} = \frac{E_{th}}{\sqrt{2}\sqrt{1 + \sin \phi}} = \frac{1}{1 - (B_c + B_l)X} \frac{E}{\sqrt{2}\sqrt{1 + \sin \phi}}$$

A quick comparison with (2.6, 2.8) shows that both P_{max} and V_{maxP} increase by the same percentage when network capacitances are taken into account and/or capacitive compensation is added.

Figure 2.14 shows a situation where as load power increases, more shunt compensation has to be added in order to keep the voltage within the limits shown by the dotted lines (typically 0.95 and 1.05 pu respectively). The resulting PV curve is shown in heavy line in Fig. 2.14. Note that the addition of shunt compensation may come from an

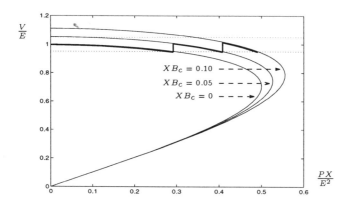

Figure 2.14   PV curves for various compensation levels

operator action or an automatic device. In the latter case, capacitors may be either mechanically switched or thyristor controlled.

The figure also illustrates a factor of critical importance in voltage instability. As the load grows in areas lacking generation, more and more shunt compensation is used to keep voltages in the normal operating range. By so doing, normal operating points progressively approach maximum deliverable power and in stressed conditions, the scenarios depicted by Figs. 2.10 and 2.11 could become a real threat.

Similarly, in systems with large capacitive effects, shunt reactors ($B_c < 0$) must be connected under light load conditions to avoid overvoltages. This is often the case in Extra High Voltage (EHV) systems where power transfers over long distances, limited by stability considerations, are below surge impedance loading. This requires shunt reactors to absorb the excess reactive power generated.

2.6.3   Static Var Compensators

Simply stated, a *Static Var Compensator* (SVC) is a voltage controlled shunt compensation device. In transmission system applications, the shunt susceptance connected to a Medium Voltage (MV) bus is quickly varied so as to maintain the voltage at a High Voltage (HV) or EHV bus (nearly) constant. SVCs are fast devices, acting typically over several cycles. The significantly higher cost of an SVC is justified when speed of

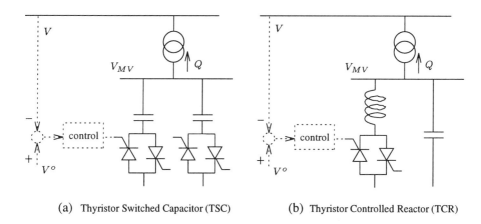

(a) Thyristor Switched Capacitor (TSC)   (b) Thyristor Controlled Reactor (TCR)

Figure 2.15 Schematic representation of SVCs

action is required for stability improvement. This is the case in angle instability and short-term voltage instability problems. Beside voltage control, SVCs can be also used to damp rotor angle oscillations through additional susceptance modulation [Mil82].

The following are the main two techniques used to obtain a variable susceptance:

■ in the *Thyristor Switched Capacitor* (TSC) (see Fig. 2.15.a) a variable number of shunt capacitor units are connected to the system by thyristors used as switches;

■ in the *Thyristor Controlled Reactor* (TCR) (see Fig. 2.15.b), the firing angle of thyristors connected in series with a reactor is adjusted to vary the fundamental frequency component of the current flowing into this reactor, while the harmonics are filtered out by different techniques. This is equivalent to having a variable shunt reactor in parallel with a fixed capacitor.

In steady-state conditions, the reactive power produced by the SVC is given by:

$$Q = B \, V_{MV}^2 \tag{2.19}$$

where V_{MV} is the MV-bus voltage and B the variable susceptance. The latter obeys:

$$B = K(V_o - V) \tag{2.20}$$

subject to:

$$B^{min} \leq B \leq B^{max} \tag{2.21}$$

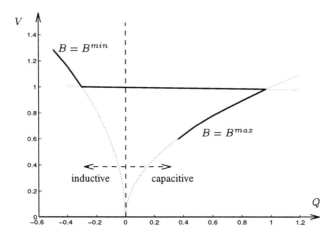

Figure 2.16   Steady-state characteristic of an SVC (B^{min} =- 0.3, B^{max} = 1., K = 50, V_o= 1 , in pu on the compensator rating)

where K is the SVC gain, V_o the voltage reference and B^{min}, B^{max} correspond to extreme thyristor conduction conditions.

The corresponding QV characteristic is shown with solid line in Fig. 2.16. The step-up transformer impedance has been neglected for simplicity (hence making $V = V_{MV}$ in per unit) but is taken into account in detailed simulation. The almost flat portion of the characteristic corresponds to (2.19) and (2.20). It is very close to a straight line with a small droop, due to the high value of K (of the order of 25–100 on the SVC rating). The parabolic parts correspond to (2.19) with B at one of the limits (2.21).

The term *Static Var System* (SVS) is used to designate the combination of an SVC with a mechanically switched capacitor [Mil82, Kun94]. Most often the rôle of the latter is to reset the SVC operating point so that the compensator is left with an adequate reactive reserve to face sudden disturbances.

Coming back to voltage stability considerations, consider the system of Fig. 2.13 with the adjustable capacitor replaced by an automatic SVC. With a TSC the network PV characteristic is close to that shown in Fig. 2.14, with the small steps corresponding to capacitor units successively switched in. With the continuously acting TCR, the characteristic becomes the heavy line in Fig. 2.17.

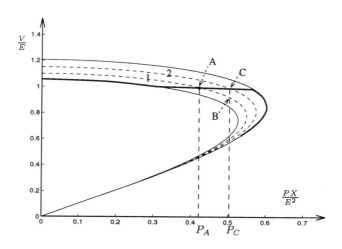

Figure 2.17 PV curves in the presence of an SVC

Assume for instance that the system is operating initially at point A on the dashed PV curve numbered 1 and that the load power is increased from P_A to P_C. In the absence of SVC reaction, the new operating point would be B. However this causes a voltage drop that the SVC will counteract by increasing its susceptance. In accordance with Fig. 2.14, the resulting network characteristic is the dashed PV curve numbered 2 and the new operating point is C. All points like A, C, etc. fall on the slightly sloping line, which corresponds to voltage control by the SVC. The slope of this line is dictated by the gain K. The two PV curves shown with solid line correspond to the susceptance limits (2.21).

As can be seen, the SVC significantly affects the shape of the network characteristic. Similar discontinuities, caused by generator reactive power limits, will be discussed in Section 3.6.

When limited, the SVC behaves as a mere shunt capacitor (or reactor), with the reactive power proportional to the square of the voltage. Comparatively, a better reactive support is offered by a synchronous generator or condenser under limit. Also more favorable, the recently proposed GTO-thyristor based STATic synchronous COMpensator (STATCOM) exhibits a constant current characteristic under limit [Gyu94].

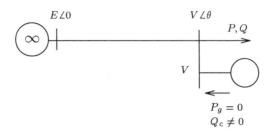

Figure 2.18   Use of a fictitious generator to produce VQ curves

2.7 VQ CURVES

A VQ curve expresses the relationship between the reactive support Q_c at a given bus
and the voltage at that bus. It can be determined by connecting a fictitious generator
with zero active power and recording the reactive power Q_c produced as the terminal
voltage V is being varied [CTF87, MJP88]. Because it does not produce active power,
this fictitious generator is often referred to as a synchronous condenser. Since voltage
is taken as the independent variable, it is a common practice to use V as the abscissa
and produce VQ instead of QV curves, as was done for loads earlier in this chapter.
We will conform to this practice.

We illustrate the technique on the 2-bus example sketched in Fig. 2.18. The load flow
equations (2.10a,b) become :

$$P \;=\; -\frac{EV}{X}\sin\theta \tag{2.22a}$$

$$Q - Q_c \;=\; -\frac{V^2}{X} + \frac{EV}{X}\cos\theta \tag{2.22b}$$

It must be noted at this point that the VQ curve is a characteristic of both the network and
the load. As the curve aims at characterizing the steady-state operation of the system,
the load must be accordingly represented through its steady-state characteristic. In this
simple example we assume a constant power load.

For each value of the voltage V, θ is first obtained from (2.22a), then the reactive power
Q_c is computed from (2.22b). Three such VQ curves are shown in Fig. 2.19. Curve 1
refers to system operation far below the maximum power. The two intersection points
with the V-axis correspond to no compensation. Referring to a previous discussion, the
higher voltage solution (marked O in Fig. 2.19) is the normal operating point. As can

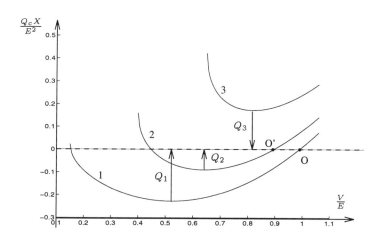

Figure 2.19 VQ curves

be seen, the VQ curve does not depart very much from a straight line around this point. Curve 2 refers to a more loaded situation. The operating point without compensation is O', where the curvature of the VQ curve is more pronounced. The Q_1 and Q_2 values shown in the figure are reactive power margins with respect to the loss of an operating point. These correspond to the minimum amount of reactive load increase (or equivalently generation decrease) for which there is no operating point any more. Finally curve 3 corresponds to a situation where the system cannot operate without reactive power injection. It might result from a severe disturbance that increases X. The shown margin Q_3 is negative and provides a measure of the Mvar distance to system operability.

VQ curves can help determining the amount of shunt compensation needed to either restore an operating point or obtain a desired voltage. We start for instance from Curve 3 of Fig. 2.19 and consider how an operating point can be restored using either a shunt capacitor or an SVC.

The case of introducing a shunt capacitor is shown in Fig. 2.20. The parabola $Q_c = BV^2$ corresponds to the minimal compensation needed to restore an operating point (denoted O) while the parabola $Q_c = B'V^2$ corresponds to the compensation needed to get the desired voltage V_d (point O'). In the latter case, the figure shows the reactive power reserve made available by the larger compensation used.

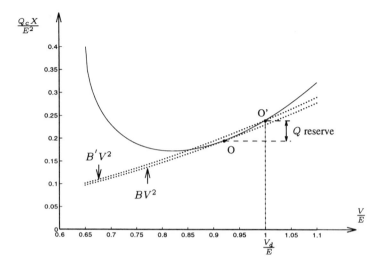

Figure 2.20   Shunt capacitor sizing based on VQ curves

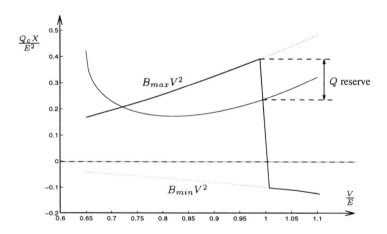

Figure 2.21   SVC sizing based on VQ curves

Note that when the reactive power source available is not producing a constant amount of Mvars the relation between Q_c and V must be taken into account to establish the reactive reserve. Thus, in the shunt capacitor case, point O does not correspond to the

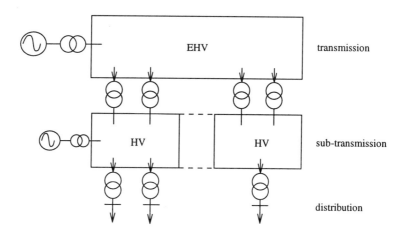

Figure 2.22   Two-level LTC structure

minimum of the VQ curve and the reactive reserve is not measured with respect to this minimum, but rather with respect to point O.

The use of an SVC is considered in Fig. 2.21. The steady-state characteristic of a device with production ($B = B_{max} > 0$) and absorption ($B = B_{min} < 0$) capabilities is shown with heavy lines. The chosen limit B_{max} leaves some reactive power reserve, as indicated in the figure.

2.8   EFFECT OF ADJUSTABLE TRANSFORMER RATIOS

Most contemporary power transmission systems are separated into different voltage levels. For instance a system may have a main transmission grid at EHV level, ranging typically from 220 to 735 kV, and a secondary transmission or sub-transmission network at HV level, with a nominal voltage from 60 to 150 kV. This two-level structure is sketched in Fig. 2.22.

It is common to have the transformers connecting the various levels equipped with Load Tap Changers (LTCs), i.e. devices which allow the turns ratio of the transformer to be adjusted without interrupting the power flow in the apparatus. Depending on the system, these can be found on :

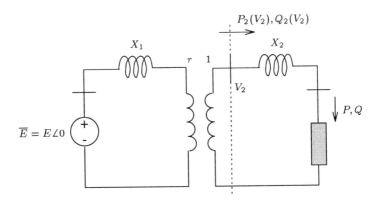

Figure 2.23   equivalent circuit showing the effect of transformer ratios

1. transformers feeding the distribution systems

2. transformers connecting sub-transmission to transmission

3. transformers connecting two transmission voltage levels

4. generator step-up transformers.

The first type of LTC is an important component of load dynamics and as such will be dealt with in Chapter 4. In this section we focus on the last three types and investigate their effects on network characteristics.

We first consider manual adjustments of the ratio, such as those performed remotely by control center operators.

Consider for this purpose the simple circuit of Fig. 2.23. The reactance X_1 on the "primary side" may represent either a transmission system equivalent reactance (Cases 2 and 3 above) or it may account for the generator voltage droop effect (Case 4). Similarly, X_2 may represent transmission and/or sub-transmission reactances. The transformer is assumed ideal, by incorporating its leakage reactance to X_2. In normal operating conditions, the ratio r is decreased (resp. increased) when an increase (resp. decrease) in voltage V_2 is sought.

The Thévenin equivalent as seen by the load has the following emf and reactance :

$$E_{th} = \frac{E}{r}$$

$$X_{th} = \frac{X_1}{r^2} + X_2$$

Replacing E by E_{th} and X by X_{th} into (2.6, 2.8) gives the maximum deliverable power (under power factor $\cos\phi$) :

$$P_{max} = \frac{1}{2} \frac{\cos\phi}{1 + \sin\phi} \frac{E^2}{r^2 X_2 + X_1} \tag{2.23}$$

and the corresponding voltage :

$$V_{maxP} = \frac{E}{r\sqrt{2}\sqrt{1 + \sin\phi}} \tag{2.24}$$

Comparing to the case without transformer, which corresponds to $r = 1$, we conclude that by decreasing r, so as to increase the secondary voltage V_2, more power can be delivered to the load. The larger the X_2/X_1 ratio, the more pronounced this effect. Formula (2.23) also shows that decreasing r is equivalent to decreasing the net impedance between the source and the load.

We consider now the case of an automatic LTC adjusting r in order to keep the secondary voltage V_2 equal to a setpoint value V_2^0. We neglect here the deadband and discrete step effects that characterize the real device and we ignore the limits on r. Conditions for steady-state operation of this system can be derived as follows.

The reactance X_2 together with the load make up a voltage sensitive load with power $P_2(V_2) + jQ(V_2)$ as shown in Fig. 2.23. By restoring the voltage V_2 to its setpoint V_2^0, the LTC restores the above power to the constant value :

$$P_2(V_2^0) = P_2^0$$
$$Q_2(V_2^0) = Q_2^0$$

The same power enters the primary side of the (ideal) transformer, leading to the situation depicted by the left circuit in Fig. 2.24. This is possible only if the point (P_2^0, Q_2^0) lies within the corresponding feasible domain, which is limited by the parabola shown at the bottom left of Fig. 2.24 and derived as explained in Section 2.2.3.

Assuming that this condition is met, the voltage V_2 is equal to V_2^0 in steady-state. This is equivalent to replacing the transformer and its primary side by a voltage source V_2^0, as shown by the right circuit in Fig. 2.24. Again, the condition for this subsystem to operate is to have the point (P, Q) within the feasibility domain, which is limited by the parabola shown at the bottom right of Fig. 2.24.

As can be seen, the effect of the voltage controlling LTC is to "break" the electrical distance between the source and the load. Some systems have more than one level

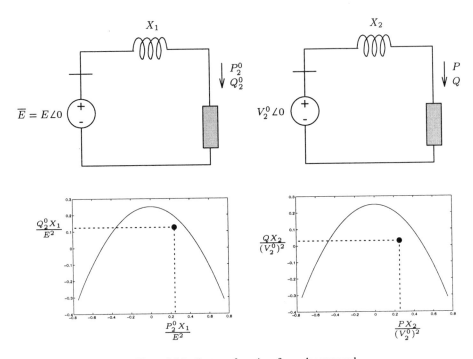

Figure 2.24 System of previous figure decomposed

of LTCs controlling voltages at intermediate points. The above reasoning applies to each level. The LTCs make it possible to operate the system with electrical distances between generators and loads that would otherwise not allow the power to be delivered to loads.

The dynamic interaction between various levels of LTCs in cascade will be discussed in Section 4.4.

2.9 PROBLEMS

2.1. Show that the extremum given by (2.2a,b) is indeed a maximum of P with respect to R_ℓ and X_ℓ.

2.2. Referring to Fig. 2.4, show that the value of the load resistance R_ℓ which maximizes the active power P also corresponds to the intersection of the IX/E and

Table 2.1 Typical data for transmission lines (2 conductors per phase)

nominal frequency	50 Hz	60 Hz
nominal voltage	380 kV	500 kV
X (see Fig. 2.13)	0.3 Ω/km	0.37 Ω/km
B_l (see Fig. 2.13)	1.5 μS/km	2.05 μS/km

V/E curves. Assume a lossless system and a constant load power factor, as in the figure.

2.3. In the (P, Q) plane of Fig. 2.5, determine the locus of operating points characterized by a given voltage magnitude V at the load bus. Show that, for $V > 0.5E$, this locus is tangent to the parabola of the figure at two points; determine these points. *Hint*: the second part does not require any calculation !

2.4. Derive the load flow equations of the 2-bus system when the transmission resistance R is not neglected. Determine the condition of existence of a solution. Draw the corresponding boundary in the (P, Q) plane.

2.5. With the load flow equations (2.10a,b) written in matrix form:

$$\mathbf{f(x)} = \mathbf{0} \qquad \text{with} \ \ \mathbf{x} = \begin{bmatrix} V \\ \theta \end{bmatrix}$$

the load flow Jacobian $\mathbf{J}$ is defined as the matrix of partial derivatives of $\mathbf{f}$ with respect to $\mathbf{x}$.
Show that this matrix is singular under maximum power conditions, for any load power factor (or equivalently, the Jacobian is singular at any point of the parabola of Fig. 2.5).

2.6. Consider the typical line characteristics given in Table 2.1. Assuming a 1 pu sending-end voltage and a unity power factor load, compute the maximum deliverable power as a function of line length. The correction for long lines will be neglected in a first approximation.
Compare with the line surge impedance loading.
Repeat for a 1.05 pu sending-end voltage.

2.7. Using the same data as in the previous problem, with a 1.05 pu sending-end voltage, determine the load power which results in a receiving-end voltage $V=0.95$ pu as a function of line length.
Repeat for various decreasing values of V.

2.8. Consider a lossless transmission line with a 1 pu sending-end voltage and shunt compensation at its receiving end so that the load voltage is always equal to 1 pu. Furthermore assume that the load dynamics is such that operation on the lower part of PV curve is unstable. What is the maximum admissible load power ?

2.9. Consider the 4-bus system shown in Fig. 3.20 (see Section 3.6). Its load flow data (on 100-MVA base) are as follows:

$$
\begin{array}{rl}
\text{line A-B:} & X = 0.056250 \text{ pu} \\
& R \text{ and } B_l \text{ neglected} \\
\text{line B-L:} & X = 0.005625 \text{ pu} \\
& R \text{ and } B_l \text{ neglected} \\
\text{step-up transformer:} & X = 0.032 \text{ pu} \\
& \text{open-circuit ratio } V_B/V_G = 1.04 \text{ pu} \\
& \text{series resistance and shunt susceptance neglected} \\
\text{shunt at bus L:} & B_c = 0.25 \text{ pu} \\
\text{generator G:} & \text{nominal apparent power} = 250 \text{ MVA}
\end{array}
$$

Run a base case load flow corresponding to the following operating point:

$$
\begin{array}{rl}
\text{load L:} & P = 2 \text{ pu}, Q = 0.5 \text{ pu} \\
\text{generator G:} & P = 2 \text{ pu}, V = 1 \text{ pu} \\
\text{generator } G_\infty: & \text{slack-bus with } V = 1.04 \text{ pu}
\end{array}
$$

Assuming that generator G (resp. G_∞) can be represented by a PV (resp. a slack-) bus and ignoring (presently: see Problem 3.6) any reactive power limit, obtain the PV curve at bus L using repeated load flow calculations. The load power factor will be kept constant and the slack-bus will compensate for the active power.

Hints: Use small enough steps to approach the maximum power point with reasonable accuracy. If your load flow allows for constant impedance load modelling, use this option to produce the lower part of the PV curve. The solution is the solid curve of Fig. 3.21, when using the generator equilibrium equations instead of the PV-bus approximation.

2.10. For the same system, draw the VQ curve at bus L for a load of resp. 200, 700 and 1400 MW (under constant power factor). Use the fictitious generator technique of Section 2.7, with Q_c corresponding to the reactive power injected by the shunt capacitor. Determine the amount of compensation needed to restore (i) a 1 pu voltage under the 700 MW load; (ii) an operating point at the 1400 MW level.

3

GENERATION ASPECTS

"Everything should be made as simple as possible. . . but not simpler !"
Albert Einstein

Synchronous generators are a primary source of reactive power and are to a great extent responsible for maintaining a good voltage profile across a power system. Consequently their characteristics and their limitations are of major importance for the analysis of voltage stability. It is worth noting that in almost all voltage instability incidents, one or several crucial generators were operating with limited reactive capability.

In this chapter we first review the basics of synchronous machine theory to arrive at dynamic and steady-state models appropriate for voltage stability analysis. Following this, we describe frequency and voltage control (including secondary voltage control, used or planned in some countries) and pay some attention to limiting devices that affect voltage instability. Next, we analyze the generator modes of operation through voltage-reactive power characteristics, as well as capability curves. Finally, we come back to notions discussed in Chapter 2 and investigate the impact of generator reactive power limitations on power-voltage relationships and maximum deliverable power.

3.1   A REVIEW OF SYNCHRONOUS MACHINE THEORY

3.1.1   Basic modelling assumptions

The synchronous machine is modelled as represented symbolically in Fig. 3.1.

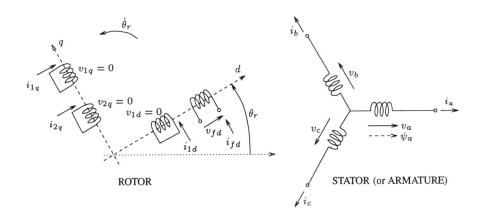

Figure 3.1 Circuit representation of a synchronous machine

The stator circuits consist of three-phase armature windings with voltages v_a, v_b, v_c and currents i_a, i_b, i_c respectively. We adopt the generator convention shown in Fig. 3.1 for the relative orientation of currents and voltages. The terms *stator* and *armature* will be used interchangeably throughout the whole book.

One degree of modelling refinement relates directly to the number of rotor circuits used to account for various phenomena. It is common in modern computer programs to use detailed models with 4 rotor windings for round-rotor machines (usually driven by steam of gas turbines) and 3 windings for salient-point machines (usually driven by hydro turbines). The rotor circuits are located along the *direct* and *quadrature* axes respectively. The direct axis coincides with the axis of the field (or excitation) winding, denoted by fd. The quadrature axis lies 90 degrees ahead along the direction of rotation shown in Fig. 3.1. The windings labelled $1d$ and $1q$ represent amortisseur (or damper) bar effects, while the $2q$ winding accounts for eddy currents in round-rotor machines. In the field winding the voltage and current are denoted by v_{fd} and i_{fd} respectively and are oriented as shown in Fig. 3.1. All other rotor windings are permanently short-circuited.

It is usually assumed that the magnetic field produced by a winding has a sinusoidal distribution along the machine air gap. Under this assumption, and neglecting cross-magnetization effects in saturated iron, two coils with perpendicular axes are not magnetically coupled.

The rotor motion is characterized by the electrical angle θ_r between the rotor direct axis and the axis of the armature phase a, as shown in Fig. 3.1. When the machine

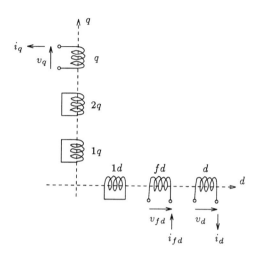

Figure 3.2 Machine windings after Park transformation

rotates at the nominal speed one has:

$$\theta_r = \omega_o t + \theta_r^o$$

where θ_r^o is an arbitrary constant and ω_o is the nominal angular frequency (in rad/s), related to nominal frequency f through:

$$\omega_o = 2\pi f$$

3.1.2 Park equations

The essential mathematical tool to study the synchronous machine is the *Park transformation* originally proposed in [Par29, Par33]. This transformation consists in replacing the a, b, c windings by three fictitious windings labelled d, q, o. The d and q windings rotate together with the machine rotor, with the d (resp. q) winding lying along the direct (resp. quadrature) axis, as shown in Fig. 3.2. Expectedly, in steady-state operation, direct currents in d and q windings correspond to balanced three-phase currents in a, b, c circuits both producing a magnetic field rotating with synchronous speed. The o winding is not magnetically coupled with the other two circuits and plays a rôle under unbalanced conditions only. It will not be considered in the sequel where we will restrict ourselves to balanced conditions.

The main advantage of the Park transformation is that all windings in Fig. 3.2 are fixed with respect to each other, thereby leading to constant self and mutual inductances. This results in considerably simpler equations in terms of d, q, o variables than in terms of phase quantities. The *Park equations*, relative to stator voltages, take on the form:

$$v_d \;=\; -R_a i_d - \dot{\theta}_r \psi_q + \dot{\psi}_d \tag{3.1a}$$

$$v_q \;=\; -R_a i_q + \dot{\theta}_r \psi_d + \dot{\psi}_q \tag{3.1b}$$

$$v_o \;=\; -R_a i_o + \dot{\psi}_o \tag{3.1c}$$

where v_d (resp. v_q)   is the d (resp. q) winding voltage
i_d (resp. i_q)   is the d (resp. q) winding current
ψ_d (resp. ψ_q, ψ_o)   is the flux linkage in the d (resp. q, o) winding
$\dot{\theta}_r$ is the electrical angular velocity
R_a is the armature resistance.

The $\dot{\theta}_r \psi_d$ and $\dot{\theta}_r \psi_q$ terms in (3.1a,b) result from the rotating field and therefore are called *speed voltages*. The $\dot{\psi}_d$ and $\dot{\psi}_q$ are referred to as *transformer voltages*.

The rotor circuits, left unchanged by the Park transformation, are described by the basic relationships:

$$v_{fd} \;=\; R_{fd} i_{fd} + \dot{\psi}_{fd} \tag{3.2a}$$

$$0 \;=\; R_{1d} i_{1d} + \dot{\psi}_{1d} \tag{3.2b}$$

$$0 \;=\; R_{1q} i_{1q} + \dot{\psi}_{1q} \tag{3.2c}$$

$$0 \;=\; R_{2q} i_{2q} + \dot{\psi}_{2q} \tag{3.2d}$$

where R_{fd} is the field circuit resistance, ψ_{fd} its flux linkage, and similarly for the other circuits.

Synchronous machines are subject to the following transients:

- stator transients: associated with the transformer voltages. Soon after a change is imposed on the system, the transformer voltages vanish and the speed voltages dominate in the system response. For instance, following a short-circuit, the transformer voltages are responsible for DC components of the stator phase currents, which die out in a fraction of a second, a relatively short period compared to the typical time interval of interest in stability studies. Therefore a usual simplification consists in neglecting the transformer voltages in the stator equations (3.1a,b).

- rotor electrical transients: associated with the $\dot{\psi}_{fd}, \dot{\psi}_{1q}$, etc. terms of the rotor winding equations. Two types of dynamics can be distinguished:

- the subtransient dynamics: associated with damper windings and eddy currents
- the transient dynamics: associated with field winding

■ mechanical transients associated with the shaft motion.

Reference [SP97] provides a detailed analysis for the reduction of the generator model. With the transformer voltages neglected, i_d and i_q take on the form of DC currents varying with time according to subtransient and transient time constants. It follows that armature currents are sinusoidal with frequency $\dot{\theta}_r$ and a time varying magnitude. In most cases of interest for stability studies, $\dot{\theta}_r$ does not depart very much from the nominal angular frequency ω_o. This justifies the *quasi-sinusoidal*, or *nominal frequency* assumption, in which the system dynamics is seen as a sinusoidal regime at frequency ω_o, with time varying magnitudes and phase angles for voltages and currents. Clearly this simplification is valid as long as these time variations are slow compared to the nominal frequency of the (50 or 60 Hz) "carrier".

Similarly, the fast network transients, which take place in the same time scale as the machine stator transients, can be neglected in most stability studies. This, together with the quasi-sinusoidal assumption, allows to represent the network with constant impedances, at nominal frequency, instead of resorting to differential equations.

3.1.3   Motion dynamics

For stability analysis purposes it is more convenient to refer the rotor position to a synchronously rotating reference. We define the *rotor angle* (in rad) as the electrical angle between the machine quadrature axis and a synchronous reference:

$$\delta = \theta_r - \omega_o t - C \tag{3.3}$$

where C is an arbitrary constant. When expressed in terms of δ and in per unit on the machine base, the equation of motion of the generator-turbine rotating masses takes on the form:

$$\frac{2H}{\omega_o}\ddot{\delta} = T_m - T_e \tag{3.4}$$

where H is the inertia constant (in s)
   T_m is the mechanical torque produced by the turbine (in per unit)
   T_e is the electromagnetic torque of the generator (in per unit)
The above equation is often referred to as the *swing equation* of the machine.

3.1.4 Simplified modelling with excitation only

The modelling approaches used in stability studies usually rely on the following two assumptions:

- the transformer voltages are neglected, as previously discussed

- the usual speed deviations are small compared to ω_o. Hence the rotor speed $\dot{\theta}_r$ is taken equal to the nominal angular frequency ω_o. (Note however that these variations are sufficient to produce significant rotor angle deviations.)

We recall hereafter a further simplified model that will be used conveniently for later discussions in this book. This model basically neglects all rotor winding except the field circuit. Expectedly the main drawback of this simplification is an underestimation of the real rotor oscillation damping. However, this damping effect is not primarily related to voltage stability and up to some extent it can be compensated for by introducing a damping term $D\delta$ in the motion equation (3.4). The following additional assumptions are made:

- the armature resistance, which is very small, is neglected for simplicity

- magnetic saturation is neglected; this will be considered in more detail in Section 3.1.8.

Assuming a constant speed, equal to nominal, and neglecting the armature resistance, the per unit electromagnetic torque is equal to the active power P produced by the machine while the per unit mechanical torque is equal to the mechanical power P_m. Equation (3.4) can thus be rewritten as:

$$\frac{2H}{\omega_o}\ddot{\delta} + \frac{D}{\omega_o}\dot{\delta} = P_m - P \tag{3.5}$$

where D is the damping coefficient in per unit.

On the other hand, under the above assumptions, the stator equations (3.1a,b) amount to:

$$v_d = -\omega_o\psi_q \tag{3.6a}$$
$$v_q = \omega_o\psi_d \tag{3.6b}$$

while for the three remaining windings, flux linkages are related to currents through:

$$\psi_d = -L_d i_d + L_{ad} i_{fd} \tag{3.7a}$$

$$\psi_q = -L_q i_q \tag{3.7b}$$

$$\psi_{fd} = -L_{ad} i_d + L_{fd} i_{fd} \tag{3.7c}$$

where L_d (resp. L_q) is the d (resp. q) winding self inductance
       L_{fd} is the field winding self inductance
       L_{ad} is the mutual inductance between the field and d windings.

Introducing (3.7a,b) into (3.6a,b) yields:

$$v_d = X_q i_q \tag{3.8a}$$

$$v_q = -X_d i_d + E_q \tag{3.8b}$$

where $X_d = \omega_o L_d$ and $X_q = \omega_o L_q$ are the *direct and quadrature axis synchronous reactances*, respectively, and

$$E_q = \omega_o L_{ad} i_{fd} \tag{3.9}$$

Under no-load condition, $i_d = i_q = 0$ and $v_q = E_q$. Hence E_q is the *no-load emf* or *open-circuit voltage*.

When sudden changes take place in the system, i_d, i_q, ψ_d and ψ_q will change instantaneously due to the neglected transformer voltages. On the other hand, due to (3.2a) the field flux linkage ψ_{fd} cannot change instantaneously. Equation (3.7c) with ψ_{fd} constant and i_d changing abruptly results in a sudden change of the field current i_{fd}. In the real machine a field current component is induced which guarantees the continuity of flux. Because it is proportional to i_{fd}, the emf E_q also changes rapidly in the exact model and abruptly when transformer voltages are neglected. It is therefore more appropriate to derive machine equations in terms of an emf proportional to field flux linkage.

To this purpose we introduce the emf:

$$E'_q = \omega_o \frac{L_{ad}}{L_{fd}} \psi_{fd} \tag{3.10}$$

as well as the *direct-axis transient reactance*:

$$X'_d = \omega_o L'_d = \omega_o (L_d - \frac{L_{ad}^2}{L_{fd}})$$

Using (3.7c) to express i_{fd} in terms of ψ_{fd} and i_d and substituting in (3.7a) and (3.6b) successively, we get:

$$v_q = -X'_d i_d + E'_q \tag{3.11}$$

E'_q is called the *emf behind transient reactance*. Note that (3.8a) is unchanged since there is no rotor winding along the quadrature-axis in this simplified model. The relationship between E_q and E'_q is easily obtained from (3.8b) and (3.11):

$$E_q = (X_d - X'_d)i_d + E'_q \qquad (3.12)$$

The dynamics of E'_q is obtained as follows:

$$\dot{E}'_q = \omega_o \frac{L_{ad}}{L_{fd}} \dot{\psi}_{fd} = \omega_o \frac{L_{ad}}{L_{fd}}(v_{fd} - R_{fd}i_{fd}) = \frac{E_f - E_q}{T'_{do}} \qquad (3.13)$$

where:

$$E_f = \omega_o \frac{L_{ad}}{R_{fd}} v_{fd} \qquad (3.14)$$

is an emf proportional to field voltage and

$$T'_{do} = \frac{L_{fd}}{R_{fd}}$$

is the *open-circuit transient time constant*, i.e. the time constant of the rotor winding when the stator windings are open. Introducing (3.12) into (3.13) yields the equivalent differential equation:

$$\dot{E}'_q = \frac{-E'_q + E_f - (X_d - X'_d)i_d}{T'_{do}} \qquad (3.15)$$

Equations (3.13) or (3.15) are often referred to as the *field flux decay* equations as they express how the flux linkage in the field winding (represented by E'_q) varies under the influence of both the exciter (E_f) and the armature reaction (i_d).

3.1.5 Phasor representation

Equations (3.8a,b) can be interpreted as the projection on the machine d and q axes of the complex equation:

$$\bar{E}_q = \bar{V} + jX_d\bar{I}_d + jX_q\bar{I}_q \qquad (3.16)$$

corresponding to the phasor diagram shown in Fig. 3.3. $\bar{V}$ (resp. $\bar{I}$) is the phasor corresponding to the armature voltage (resp. current). $\bar{I}_d$ and $\bar{I}_q$ are the projections of $\bar{I}$ on the d and q axes respectively, with (see Fig. 3.3):

$$\bar{I} = \bar{I}_d + \bar{I}_q = (i_d + j\,i_q)e^{j(\delta - \pi/2)} \qquad (3.17)$$

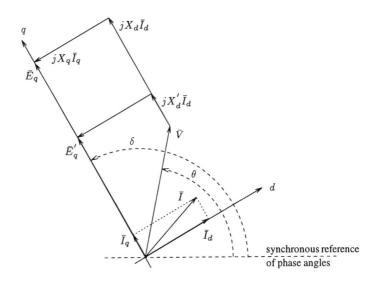

Figure 3.3 Phasor diagram of the synchronous machine

where δ is the previously defined rotor angle. The phasor $\bar{E}_q$ is directed along the q axis and δ is also the phase angle of this emf.

Similarly, (3.8a) and (3.11) can be written in complex form as:

$$\bar{E}'_q = \bar{V} + jX'_d\bar{I}_d + jX_q\bar{I}_q$$

as shown in the same figure.

Following a perturbation, both $\bar{E}_q$ and $\bar{E}'_q$ remain along the q-axis (in the absence of rotor winding along the quadrature axis). However, for reasons already mentioned, E_q changes abruptly while E'_q evolves continuously and smoothly.

3.1.6 Synchronous frame of reference

The Park equations are made simple by the use, for stator voltages and currents, of a reference linked to the rotor of the machine: the (d, q) axes. However, for multimachine system modelling it is necessary to refer all stator voltages and currents to a single, common reference.

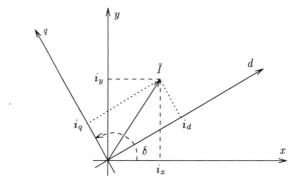

Figure 3.4   Machine and system references

For the latter we consider two orthogonal axes, denoted x and y, and rotating at the synchronous speed, as sketched in Fig. 3.4. The x axis coincides with the dashed line of Fig. 3.3 while the y axis is 90 degrees ahead.

In the (x, y) frame any phasor can be written as a complex number with real and imaginary parts corresponding to its x and y components respectively. For instance the stator current $\bar{I}$ is written as:

$$\bar{I} = i_x + j\, i_y$$

Equation (3.17) shows that the above expression for $\bar{I}$ differs from the one on the (d, q) frame by an angle $\delta - \pi/2$. This can be expressed in matrix notation as:

$$\begin{bmatrix} i_x \\ i_y \end{bmatrix} = \underbrace{\begin{bmatrix} \sin\delta & \cos\delta \\ -\cos\delta & \sin\delta \end{bmatrix}}_{\mathbf{T}(\delta)} \begin{bmatrix} i_d \\ i_q \end{bmatrix} \tag{3.18}$$

where $\mathbf{T}$ is a rotation matrix. A similar relationship applies to the projections v_x, v_y, v_d and v_q of voltage $\bar{V}$.

The stator equations are expressed in the (x, y) frame of reference as follows. Putting equations (3.8a, 3.11) in matrix form as:

$$\begin{bmatrix} v_d \\ v_q \end{bmatrix} = \begin{bmatrix} 0 & X_q \\ -X'_d & 0 \end{bmatrix} \begin{bmatrix} i_d \\ i_q \end{bmatrix} + \begin{bmatrix} 0 \\ E'_q \end{bmatrix}$$

and using the above defined matrix $\mathbf{T}$, one obtains:

$$\begin{bmatrix} v_x \\ v_y \end{bmatrix} = \mathbf{T} \begin{bmatrix} v_d \\ v_q \end{bmatrix} = \mathbf{T} \begin{bmatrix} 0 & X_q \\ -X_d' & 0 \end{bmatrix} \mathbf{T}^{-1} \begin{bmatrix} i_x \\ i_y \end{bmatrix} + \mathbf{T} \begin{bmatrix} 0 \\ E_q' \end{bmatrix} \tag{3.19}$$

Alternatively, one may express currents as functions of voltages and emf as follows:

$$\begin{bmatrix} i_x \\ i_y \end{bmatrix} = \mathbf{T} \begin{bmatrix} 0 & -1/X_d' \\ 1/X_q & 0 \end{bmatrix} \mathbf{T}^{-1} \left\{ \begin{bmatrix} v_x \\ v_y \end{bmatrix} - \mathbf{T} \begin{bmatrix} 0 \\ E_q' \end{bmatrix} \right\} \tag{3.20}$$

Getting the individual expressions of v_x and v_y from (3.19) or i_x and i_y from (3.20) is just a matter of calculation.

3.1.7 Power relationships

The per unit complex power produced by the machine is given by:

$$S = P + jQ = \bar{V}\bar{I}^*$$

and using (3.17):

$$S = (v_d + jv_q)e^{j(\delta - \pi/2)}(i_d - ji_q)e^{-j(\delta - \pi/2)} = (v_d i_d + v_q i_q) + j(v_q i_d - v_d i_q)$$

Using (3.8a,b) and Fig. 3.3, the active and reactive power take on the following form:

$$P = \frac{E_q V}{X_d} \sin(\delta - \theta) + \frac{V^2}{2}\left(\frac{1}{X_q} - \frac{1}{X_d}\right) \sin 2(\delta - \theta) \tag{3.21a}$$

$$Q = \frac{E_q V}{X_d} \cos(\delta - \theta) - V^2 \left(\frac{\sin^2(\delta - \theta)}{X_q} + \frac{\cos^2(\delta - \theta)}{X_d}\right) \tag{3.21b}$$

where $\delta - \theta$ is often called the *internal* or *load angle* and is incidentally the rotor angle with the terminal voltage as reference (see Fig. 3.3). The above expressions are more appropriate to study steady-state operation as they involve E_q. Note that for a round-rotor machine $X_d \simeq X_q$ and the last term in (3.21a) vanishes while the last term in (3.21b) amounts to $-V^2/X_d$.

3.1.8 Modelling of saturation

In the presence of saturation, the various inductances vary with the machine operating point. Using superscript ^{s} to denote saturated values, the previously derived equations become:

$$v_d = X_q^s i_q \tag{3.22a}$$
$$v_q = -X_d^s i_d + E_q^s \tag{3.22b}$$
$$E_q^s = \omega_o L_{ad}^s i_{fd} \tag{3.22c}$$
$$X_d^s = \omega_o L_d^s \tag{3.22d}$$
$$X_q^s = \omega_o L_q^s \tag{3.22e}$$

The saturated inductances L_d^s and L_q^s can be decomposed into:

$$L_d^s = L_\ell + L_{ad}^s \tag{3.23a}$$
$$L_q^s = L_\ell + L_{aq}^s \tag{3.23b}$$

where L_ℓ is the leakage inductance, which we assume identical in both axes and independent of saturation, for the path of the leakage flux is mainly in the air. We now outline how the various saturated values can be determined as a function of the operating point.

Consider first a machine operating at no load. Figure 3.5 shows the *open-circuit saturation characteristic* relating the open-circuit emf E_q^s to the field current i_{fd}. If the machine was unsaturated, the characteristic would be the dotted line corresponding to (3.9) and called the *air-gap line*. Taking saturation into account the emf E_q^s is given by (3.22c) and follows the solid curve shown. We define the *saturation factor*

$$K = \frac{E_q^s}{E_q} = \frac{L_{ad}^s}{L_{ad}} < 1 \tag{3.24}$$

also shown geometrically in Fig. 3.5. Note that in no-load conditions $L_{ad}^s i_{fd}$ is the air-gap flux linkage, denoted ψ_{ag} and hence

$$E_q^s = \omega_o \psi_{ag} \qquad (i_d = i_q = 0)$$

Using the open-circuit characteristic, K can be expressed in terms of ψ_{ag}. Several analytical expressions have been proposed to this purpose. Where explicitly required, we will follow [SH79] and use:

$$K(\omega_o \psi_{ag}) = \frac{1}{1 + m(\omega_o \psi_{ag})^n} \tag{3.25}$$

where m and n are positive real numbers, with $m = 0$ corresponding to no saturation.

Consider now a loaded machine. In the general case we have to consider different saturation effects in the two axes:

$$X_d^s = X_\ell + X_{ad}^s = X_\ell + K_d X_{ad} = X_\ell + K_d(X_d - X_\ell) \tag{3.26a}$$
$$X_q^s = X_\ell + X_{aq}^s = X_\ell + K_q X_{aq} = X_\ell + K_q(X_q - X_\ell) \tag{3.26b}$$

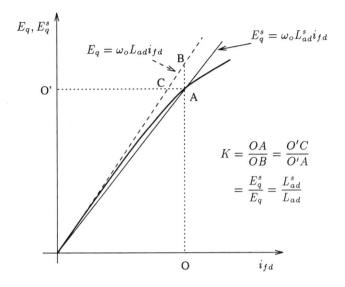

Figure 3.5 Open-circuit saturation characteristic

where X_ℓ is the leakage reactance. For a salient-pole machine the saturation in the q-axis is usually negligible because of the larger air gap, and K_q can be taken equal to 1. For a round-rotor machine, K_d and K_q are also different in principle, although closer to each other. However, in most practical cases a q-axis saturation characteristic is not available and K_d is taken equal to K_q. We adopt this simplification in the sequel. Furthermore the following usual assumptions are made (refer to e.g. [Kun94] and the quoted references for a complete discussion):

■ leakage fluxes do not contribute to iron saturation and hence saturation is determined by the air-gap flux ψ_{ag}

■ the open-circuit saturation characteristic (Fig. 3.5) can be used under loaded conditions to relate ψ_{ag} to i_{fd}

■ any coupling between d and q windings due to saturation is ignored.

The air-gap flux linkage is determined as:

$$\psi_{ag} = \sqrt{\psi_{ad}^2 + \psi_{aq}^2} \tag{3.27}$$

where ψ_{ad} is the d-axis air gap flux given by:

$$\psi_{ad} = -(L_d^s - L_\ell)i_d + L_{ad}^s i_{fd} \tag{3.28}$$

and ψ_{aq} is the q-axis air gap flux given by:

$$\psi_{aq} = -(L_q^s - L_\ell)i_q \tag{3.29}$$

From (3.22b), (3.22c) and (3.28) one easily derives:

$$\omega_o \psi_{ad} = -X_d^s i_d + X_\ell i_d + E_q^s = v_q + X_\ell i_d$$

and similarly from (3.22a) and (3.29):

$$\omega_o \psi_{aq} = -X_q^s i_q + X_\ell i_q = -v_d + X_\ell i_q$$

Introducing these last two results into (3.27):

$$\omega_o \psi_{ag} = \sqrt{\omega_o^2 \psi_{ad}^2 + \omega_o^2 \psi_{aq}^2} = \sqrt{(v_q + X_\ell i_d)^2 + (v_d - X_\ell i_q)^2} = V_\ell$$

where V_ℓ is the magnitude of the voltage:

$$\bar{V}_\ell = \bar{V} + jX_\ell \bar{I} \tag{3.30}$$

often referred to as the *voltage behind leakage reactance*. The corresponding phasor diagram is shown in Fig. 3.6.

To summarize, given a voltage $\bar{V}$ and current $\bar{I}$ at the machine terminal, the saturation is accounted for by computing V_ℓ from (3.30), the saturation factor from (3.25):

$$K = \frac{1}{1 + mV_\ell^n} \tag{3.31}$$

and the saturated reactances from (3.26a,b) with $K_d = K_q = K$.

3.1.9   Steady-state relationships including saturation

In Section 3.1.7 we have derived the expressions of the active and reactive power produced by the synchronous machine as a function of the terminal voltage and two internal variables, namely E_q and $\delta - \theta$. We extend hereafter this result to a saturated machine, using E_q^s as a third internal variable.

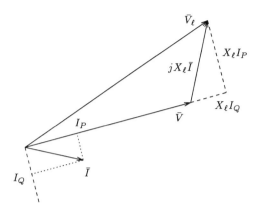

Figure 3.6 Phasor diagram relating $\bar{V}$ and $\bar{V}_\ell$

Adapting (3.21a,b) to account for saturation, we get:

$$P = \frac{E_q^s V}{X_d^s} \sin(\delta - \theta) + \frac{V^2}{2}\left(\frac{1}{X_q^s} - \frac{1}{X_d^s}\right)\sin 2(\delta - \theta)$$

$$Q = \frac{E_q^s V}{X_d^s} \cos(\delta - \theta) - V^2\left(\frac{\sin^2(\delta - \theta)}{X_q^s} + \frac{\cos^2(\delta - \theta)}{X_d^s}\right)$$

Replacing the saturated reactances by their expressions (3.26a,b) and considering the case $K_d = K_q = K$ yields:

$$P = \frac{E_q^s V}{X_\ell + K(X_d - X_\ell)} \sin(\delta - \theta)$$
$$+ \frac{V^2}{2}\left(\frac{1}{X_\ell + K(X_q - X_\ell)} - \frac{1}{X_\ell + K(X_d - X_\ell)}\right)\sin 2(\delta - \theta)$$

$$Q = \frac{E_q^s V}{X_\ell + K(X_d - X_\ell)} \cos(\delta - \theta)$$
$$- V^2\left(\frac{\sin^2(\delta - \theta)}{X_\ell + K(X_q - X_\ell)} + \frac{\cos^2(\delta - \theta)}{X_\ell + K(X_d - X_\ell)}\right)$$

Replacing in turn K by its definition (3.24), we get:

$$P = \frac{E_q E_q^s V}{X_\ell E_q + (X_d - X_\ell)E_q^s} \sin(\delta - \theta) \qquad (3.32a)$$

$$Q = \begin{aligned} &+\frac{E_q V^2}{2}\left(\frac{1}{X_\ell E_q + (X_q - X_\ell)E_q^s} - \frac{1}{X_\ell E_q + (X_d - X_\ell)E_q^s}\right)\sin 2(\delta - \theta) \\ &\frac{E_q E_q^s V}{X_\ell E_q + (X_d - X_\ell)E_q^s}\cos(\delta - \theta) \\ &-E_q V^2\left(\frac{\sin^2(\delta - \theta)}{X_\ell E_q + (X_q - X_\ell)E_q^s} + \frac{\cos^2(\delta - \theta)}{X_\ell E_q + (X_d - X_\ell)E_q^s}\right) \end{aligned} \qquad (3.32b)$$

At this point we introduce the active and reactive currents:

$$I_P = \frac{P}{V} \qquad I_Q = \frac{Q}{V}$$

Referring to Fig. 3.6, the voltage behind leakage reactance can be expressed as:

$$V_\ell = \sqrt{(V + X_\ell I_Q)^2 + (X_\ell I_P)^2}$$

Introducing this result in (3.24) and (3.31) yields the closed-form relationship between E_q and E_q^s :

$$E_q = (1 + m[(V + X_\ell I_Q)^2 + (X_\ell I_P)^2]^{n/2})E_q^s \qquad (3.33)$$

3.1.10   On per unit systems

So far we have assumed all machine equations to be in per unit. In this section we briefly comment on the choice of base quantities.

The per unit system used for the stator is based on the (three-phase) power and voltage rating of the machine. As is well-known, a conversion to a common base power is required for network calculations involving several machines.

In the detailed Park modelling, the *reciprocal per unit system* is used to deal with rotor quantities. This is the only system which allows us to write in per unit, the total inductance as the sum of leakage and mutual inductances, as in (3.7a,c). The reciprocal system however is impractical when dealing with the excitation system. Indeed, since the same base power (in MVA) must be used for both the stator and the field circuit, the base field voltage is very large and usual field voltages become very small in per unit.

When modelling excitation systems (see next section), it is common to choose as the base voltage V_{fB} (resp. base current I_{fB}) the voltage (resp. current) that produces a 1 per unit stator voltage in no-load conditions.

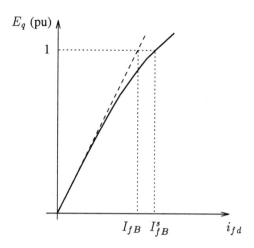

Figure 3.7 Definition of base field current (exciter per unit system)

In order to make the excitation and (detailed) machine models compatible, it is required to perform a base conversion between the two per unit systems. However, as long as we use a machine modelling involving stator quantities only, as in (3.8a,b), (3.11) and (3.13) for instance, per unit rotor quantities do not appear in the equations and the above per unit conversion need not be performed explicitly.

Two possible choices for the base current are shown in Fig. 3.7. I_{fB} is the field current that produces a 1 per unit stator voltage on the air-gap line, while I_{fB}^s produces the same voltage in the real machine, i.e. taking saturation into account. The two bases are linked through:

$$I_{fB} = K(1)\, I_{fB}^s$$

where $K(1)$ is the saturation factor for $V_\ell = 1$. Using the saturation law (3.31):

$$K(1) = \frac{1}{1 + m}$$

Note that another per unit system in use refers to the field current required to operate the machine under rated voltage and powers. All this is just a matter of scaling but the underlying base should be ascertained before using the data in per unit.

In this book, where required, we will use I_{fB} (see Fig. 3.7) as the base current of the field winding. This choice simplifies some previously obtained results, as shown hereafter.

Note first that since V_{fB} is the field voltage that produces the base field current I_{fB} in steady-state conditions, we have $v_{fd} = 1$ pu when $i_{fd} = 1$ pu, which implies that

$$R_{fd} = 1 \quad \text{pu (exciter base)}$$

Consider now the equation (3.9):

$$E_q = \omega_o L_{ad} i_{fd}$$

Since $E_q = 1$ pu when $i_{fd} = 1$ pu, we have:

$$\omega_o L_{ad} = 1 \quad \text{pu (exciter base)}$$

and (3.9) becomes simply:

$$E_q = i_{fd} \quad \text{in pu (exciter base)} \tag{3.34}$$

Consider finally the equation (3.14):

$$E_f = \frac{\omega_o L_{ad}}{R_{fd}} v_{fd}$$

With the previously shown simplifications, this expression becomes merely:

$$E_f = v_{fd} \quad \text{in pu (exciter base)} \tag{3.35}$$

3.2 FREQUENCY AND VOLTAGE CONTROLLERS

3.2.1 Frequency control overview

In large power systems, frequency control is exerted at two levels:

- at the local level, by means of *governors* in power plants. This is also referred to as *primary frequency control*;

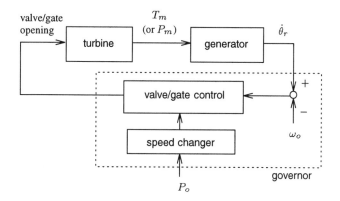

Figure 3.8   Bloc-diagram representation of the rôle of the governor

- at a central level, by means of a *load-frequency control*. Nowadays this takes on the form of a software running in a control center. This control level is also named *secondary frequency control*.

The rôle of a governor is: (i) to keep the generator speed close to its nominal value; (ii) to ensure a quick and automatic participation of the generator to any change in generation required to maintain the active power balance of the system, and (iii) to provide a means to modify the active power production of the unit, through speed-changer adjustments.

The governor adjusts the steam or water input to the turbine. To this purpose it senses the difference between the actual ($\dot{\theta}_r$) and nominal (ω_o) angular frequency and adjusts accordingly the turbine valve or gate, as sketched in Fig. 3.8. In this figure, P_o is the power setpoint of the generator.

Governors are proportional controllers, with the ideal steady-state characteristic shown in Fig. 3.9. The slope of the latter is characterized by the *speed droop R*, i.e. the ratio of per unit speed deviation to per unit power deviation (on the turbine MW rating P_B):

$$R = \frac{\Delta\dot{\theta}_r/\omega_o}{\Delta P/P_B}$$

Typical values of R for units under governor control are 0.04 (e.g. in Europe) or 0.05 (e.g. in North America). Figure 3.9 also illustrates the principle of load sharing by two units operating in parallel. When, following a disturbance (such as a change in

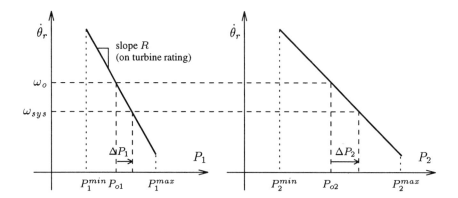

Figure 3.9   Generator load sharing through governor control

load or the loss of a generator), all governors have acted and the system has reached a steady state, all rotor speeds $\dot{\theta}_r$ are equal to ω_{sys}, the system angular frequency. As illustrated by the figure, frequency deviation is the common signal used by the various units to adjust their participations.

Thus the drooping characteristic of the governor takes on the form:

$$\frac{P - P_o}{P_B} = -\frac{1}{R}\frac{\omega_{sys} - \omega_o}{\omega_o}$$

where P_o is the power setpoint, corresponding to the active power produced under nominal frequency. In this expression ω_{sys} and ω_o are in rad/s while P and P_o are in MW. Relating all powers to a common system base S_B (in MVA), the above relationship becomes:

$$P = P_o - \frac{P_B}{S_B R}\frac{\omega_{sys} - \omega_o}{\omega_o} = P_o - \gamma\frac{\omega_{sys} - \omega_o}{\omega_o} \tag{3.36}$$

where γ is a participation factor. In practice, turbine limits must be also taken into account:

$$P^{min} < P < P^{max} \tag{3.37}$$

At the secondary level, the objectives of load-frequency control are: (i) to correct the frequency deviation left after primary control, and (ii) to keep at the scheduled value the net interchange power with neighbouring companies or areas. To this purpose, the frequency and tie-line power errors are combined into a single signal, named *area control error*, which is used by an integral controller adjusting the power setpoints of a certain number of generating units.

A detailed presentation of the above controllers, in particular their dynamics, may be found in many textbooks (e.g. [Ber86, Kun94]). Let us mention here some aspects of frequency control that may interact with voltage phenomena:

- following an incident such as a line or generator outage, grid voltages usually fall, causing the voltage sensitive load powers to decrease correspondingly. The generators react due to governor effects;

- in the case of a generator outage, the location of the generators which compensate for the lost power may play an important rôle. A typical example is when a large power transfer takes place between a sending and a receiving area and the units participating to (primary or secondary) frequency control are mostly located in the sending area. In such a case, a loss of generation in the receiving area will cause an additional power flow over the lines connecting the two areas, which may bring the system closer to its loadability limit. The reaction of the sending-area generators has the same effect as a load increase in the receiving area;

- conversely, following a loss of line between the two areas, generation shedding in the sending area may prove useful provided that there are enough units participating to frequency control (and enough spinning reserve) in the receiving area. The missing power due to generation shedding is then taken up by the receiving-area generators so that the power transfer over the remaining lines is permanently reduced.

3.2.2 Automatic voltage regulators

A schematic description of a typical Automatic Voltage Regulator (AVR), or excitation control system, is shown in Fig. 3.10, where dotted lines are used for blocks that are not necessarily present, and numbered dashed lines show alternate configurations.

The generator terminal voltage V is measured through a Potential Transformer (PT), then rectified and filtered so as to produce a DC signal proportional to the RMS value of this AC voltage.

Optionally the DC signal built can be proportional to:

$$V_c = |\bar{V} \pm (R_c + jX_c)\bar{I}| \tag{3.38}$$

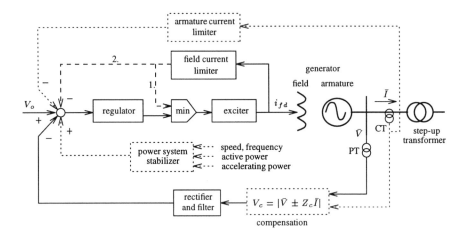

Figure 3.10 Bloc-diagram representation of an automatic voltage regulator

where R_c (resp. X_c) is a compensation resistance (resp. reactance) and $\bar{I}$ is the current flowing out of the generator and measured through a Current Transformer (CT).

■ Using the minus sign in the above expression, a signal proportional to the voltage at some point beyond the generator terminal is obtained and the excitation system regulates the voltage at this point closer to the transmission system. Typically R_c and X_c are set up to 90 % of the step-up transformer impedance (for reasons explained in the next paragraph, a 100 % compensation is not possible when several generators are connected through their own transformers to the same HV bus). This technique is usually called *load*, or *line drop* or *step-up transformer compensation*. By regulating the voltage closer to the loads this compensation improves voltage stability, within the limits of the machine reactive power capability. Because transformer copper losses are comparatively small, R_c can be set to zero. With this simplification and using the active and reactive current components shown in Fig. 3.6, (3.38) becomes:

$$V_c = |\bar{V} - jX_c\bar{I}| = \sqrt{(V - X_cI_Q)^2 + (X_cI_P)^2} \qquad (3.39)$$

■ The plus sign in (3.38) is used to regulate the voltage at a (fictitious) point "within" the generator. Such a compensation ensures proper sharing of the reactive power produced by several generators connected to the same bus, each being equipped with a voltage regulator. If all of them were allowed to control the same voltage, small inevitable differences in generator or regulator parameters could result in large imbalances between the individual reactive power productions. This configuration is typical of hydro power plants where several small units share the same bus and step-up transformer. Clearly, the equal sharing of reactive power is obtained at the expense of deteriorated grid voltage regulation.

We will not consider compensation in the remaining of this book for the sake of simplicity, but it can be easily accounted for by substituting an expression like (3.39) for V in the regulator equations where necessary.

As shown in Fig. 3.10, the V_c signal is compared to the reference V_o and the difference is processed by the regulator whose role is basically to increase the generator excitation voltage in response to a decrease in V_c or an increase in V_o, and conversely. The regulator amplifies the error signal $V_o - V_c$ and brings it to the form suitable for the control of the exciter (e.g. firing of thyristors, etc.). The regulator is usually provided with compensation circuits aimed at meeting dynamic performance and accuracy specifications, in particular counteracting too large an exciter time constant. This compensation uses the generator field current i_{fd} or the exciter field current.

The exciter is an auxiliary device producing the power required by the generator excitation in the form of DC voltage and current that can be quickly varied. Exciters may be classified into two broad categories:

- rotating machines drawing the excitation power from mechanical power, most often through the turbine-generator shaft.

  - DC machines were first used to this purpose and are still in operation in some power plants;
  - AC machines with rectifiers have been preferred since the 60's.

  These machines may be self excited or they may use an auxiliary rotating machine for their own excitation. The latter is called *pilot exciter*. It may take the form of a permanent magnet AC generator followed by a controlled rectifier.

- static excitation systems in which the excitation power is provided by a transformer connected to the machine bus or to an auxiliary bus. The DC power is obtained through a thyristor-controlled rectifier.

Also acting at the summation point of the regulator, the *Power System Stabilizer* (PSS) is a compensation circuit aimed at providing additional damping torque through excitation control. In steady-state operating conditions the PSS has a zero output, leaving the machine terminal voltage unaffected. Figure 3.10 shows various signals which may be used in a PSS.

Finally the excitation control system is provided with several limiting circuits:

■ the *underexcitation limiter* prevents an excessive reduction in the machine excitation (corresponding to reactive power absorption) which would lead to loss of small-disturbance stability (with the machine pulling out of synchronism) or unacceptable heating of stator end region [ITF96c];

■ the *Volts-per-Hertz limiter* protects the generator and its step-up transformer from excessive magnetic flux that would result from either high voltage or low frequency conditions following severe events [ITF96b];

■ the *overexcitation limiter* protects the field winding from an overheating due to excessive current [ITF96b];

■ the *armature current limiter* similarly prevents excessive current in the armature winding.

Among these, the overexcitation limiter and to some extent the armature current limiter will receive particular attention in this book, as being primarily related to voltage instability phenomena.

Depending on the manufacturer and date of construction, many types of excitation control systems are encountered throughout the world. It is not the purpose of this book to detail their modelling in others than the above quoted aspects. The interested reader is referred to e.g. [IWG81, ITF96a, ITF96c, ITF96b] and the quoted references for a sample of representative models suitable for time simulation.

3.2.3   Secondary voltage control

The primary rôle of AVRs is to quickly respond to voltage disturbances occurring in the system. This control is local by nature, since it involves generator buses only [1]. Moreover, the required reactive power will be produced by the generators electrically close to the disturbance[2]. The consequences of this are that : (i) the voltage at non-generator buses in the system may become unacceptable, and (ii) the reactive reserves

---

[1] or points close to them if line drop compensation is used

[2] let us recall that for reactive power there is no counterpart to the frequency deviation that allows active power imbalances to be shared by many units, according to their capabilities

may be unevenly distributed over generators after the disturbance. This situation must be corrected by adjusting the AVR voltage setpoints V_o of the generators.

In many countries, these adjustments are performed manually from a control center. To make them automatic, one could think of sending to generators voltage setpoints computed by an optimal power flow incorporating bus voltage and reactive reserve constraints. However, the performance of such an open-loop control scheme would be affected by errors in power system model, unavailability or inaccuracy of state estimator results, not to mention the computational burden aspect. Instead, the solution implemented in France [PLT87, PCJ90] and Italy [ACN90] since the early 80's (and scheduled for implementation in some other countries) is a dedicated closed-loop control, referred to as *secondary voltage control*.

The first, presently used, generation of secondary voltage control, relies on a division of the network into zones. A zone is a set of buses whose voltages vary in a relatively coherent way and are relatively little affected by controls in other zones. In each zone, a *pilot point* is selected as a representative bus and each participating generator is assigned to the control of a particular pilot point. The objectives of secondary voltage control are : (i) to keep the pilot point voltage at a specified setpoint value, and (ii) to make the reactive power production of each generator proportional to its reactive power capability.

To this purpose, two control levels are added to the primary control by AVRs:

- at the zone level: the difference between the measured and the setpoint values of the pilot point voltage is entered into a Proportional-Integral (PI) controller to obtain a signal N, which is sent to each generator of the zone;

- at the generator level: a reactive power control loop adjusts the AVR voltage reference (by small steps) so that the generator reactive power production Q follows the setpoint $N\,Q^{max}$, where Q^{max} relates to the machine reactive capability[3].

The response time of the various controls are typically in the order of 3 minutes for the zone PI controller, 20 s for a reactive power control loop, and 1 s for AVRs, thereby avoiding interactions between control levels.

With the development of the transmission system, there is a growing interaction between control zones. This, together with some drawbacks of the above scheme, has led to the development of the *coordinated secondary voltage control*, presently under

---

[3] machine reactive power capabilities are thoroughly discussed in the next three sections

test [PCJ90, VPL96]. The latter relies on a network partition into regions, which are much larger than zones and include several pilot points. The new controller still works in closed-loop but directly acts on the AVR voltage setpoints, taking into account interactions between generators within a region. The AVR voltage setpoints, issued each 10 s, are obtained as the solution of an optimization problem, miminizing the sum of squared pilot point voltage deviations and (normalized) generator reactive productions, subject to various inequality constraints. To this purpose, sensitivity matrices of pilot point voltages and generator reactive powers to generator voltages are computed in real-time.

Pilot point voltage setpoints can be adjusted by operators. In the future it is planned to have them determined by a *tertiary level*, which consists of an economic optimization of the whole system, performed each 15 minutes.

Note that secondary voltage control is also in charge of shunt compensation switching, with the objective of maintaining reactive reserves on generators to face incidents.

Secondary voltage control interacts with the following aspects of voltage stability [VPL96, VJM94, BBM96]:

- generally speaking, the increase in generator voltages yields a larger maximum deliverable power. Referring to the simple two-bus system of Chapter 2, the effect of secondary voltage is somewhat similar to that shown in Fig. 2.14, where the various PV curves would now correspond to successive values of the generator voltage. Equivalently, one may also consider that after secondary voltage control action the voltage is constant at the pilot point instead of generator buses, i.e. electrically closer to load buses;

- this control acts in basically the same time scale as long-term load restoration (e.g. by load tap changers: see Section 4.4). As a result, in case of a large disturbance, the two compete against each other and the resulting dynamics have to be taken into account (in particular with respect to oscillations);

- in response to a load increase exceeding system capability, secondary voltage control keeps the voltage profile flat over a longer time interval, but results in a sharper final decrease, because all generators tend to have their reactive reserves exhausted at the same time;

- being more constant, network voltage tends to be a poorer indicator of an insecure situation.

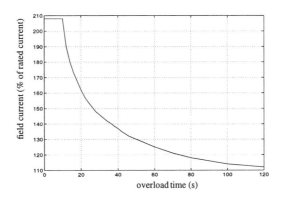

Figure 3.11 Field current overload capability : normalized curve

3.3 LIMITING DEVICES AFFECTING VOLTAGE STABILITY

3.3.1 Overexcitation limiters : description

Overexcitation limiter, maximum excitation limiter and field current limiter are all names for the same limiting device that protects the field winding of a synchronous machine from overheating. In this book we will adopt the term OvereXcitation Limiter (OXL).

For first-swing angle stability enhancement it is necessary to allow maximum field forcing following large disturbances like short-circuits. In such circumstances the field voltage may quickly increase up to its ceiling value and the field current is allowed to reach for a short time a maximum value which is typically twice the permanent admissible current. Such a high value cannot be tolerated for more than a few seconds. However, because rotor heating (and hence the risk of machine damage) is proportional to the integral of the field current squared, the smaller the overload, the longer the time it can be tolerated. This property is described by the ANSI standard C50.13-1977 reproduced in Fig. 3.11.

It is clear that the field current limit is influenced by the machine cooling conditions. In hydrogen-cooled machines for instance an increase in hydrogen pressure allows to reach significantly larger rotor currents. The same remark applies to the armature current.

OXLs are designed to obey the curve of Fig. 3.11 with some security margin. Simplest devices have a fixed current pickup point and a fixed time delay, both corresponding to a point below the characteristic. Most modern OXLs, on the other hand, have an inverse time characteristic which allows smaller overexcitations to last longer.

By taking advantage of the generator overload capability, inverse time characteristics are beneficial in the sense that they leave some time to take other actions in case of voltage emergencies. However by supporting generator voltages longer, they also tend to "hide" a dangerous situation in the time interval following the initiating disturbance.

In older systems the field limitation is performed by switching the machine to manual control, i.e. to a constant excitation voltage. It is important to properly set the value of this voltage, so that the machine is not "overprotected" but rather operates close to its thermal capability. Another technique, used with the older DC excitation systems consists of inserting a resistor in series with the field winding of the exciter so that the maximum voltage produced by the latter is less than or equal to the admissible value.

In modern OXL systems, basically two techniques are used in order to transfer control of the excitation system to the OXL (the numbers refer to the alternate paths of Fig. 3.10):

1. The first one consists in by-passing the normal voltage regulation loop. To this purpose, the exciter is driven by the minimum between the normal AVR and the OXL signals, as indicated by the "minimum" block in Fig. 3.10. Taking into account that the AVR is by-passed in this case, the OXL loop has to be designed so as to ensure stability of the excitation system under limit [ITF96b].

2. In the second technique, the OXL produces a signal that is added to the main summing junction of the AVR with a minus sign (see Fig. 3.10). This signal is equal to zero in normal operating conditions while it forces dynamically the field current to its limit when the OXL is active. This can be seen as a change in the voltage reference V_o.

When the second technique is used, the protection of the field winding relies on the AVR. Hence, a back-up OXL device is needed to protect the generator in case of AVR misoperation [CTF93, IWG96].

Depending on the technique used and the selected settings, the OXL may, or may not, allow some small degree of voltage regulation and auxiliary signals acting on the AVR. The term "auctioneering" or "takeover" [ITF96b] refers to those OXLs that do not allow any signal through the AVR loop. As a consequence, under takeover field

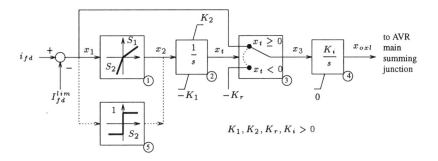

Figure 3.12 Bloc-diagram model of an OXL with integral control of field current

limitation, the PSS becomes inactive and the damping of rotor oscillations may be substantially reduced.

3.3.2 Overexcitation limiters : modelling

Many types of OXLs are encountered in practice. In this section we give two representative OXL models in block diagram form suitable for time simulation programs. These models describe devices that enforce the field limit by injecting a signal into the AVR main summing junction (path 2 in Fig. 3.10). The second scheme, however, can be adapted to model OXLs that act through a minimum block (path 1 in the same figure) as well.

In the sequel we denote by I_{fd}^{lim} the value of the field current that the OXL will enforce. According to our previous description, I_{fd}^{lim} is usually slightly larger than the permanent admissible field current (typically 5 to 10 %).

Figure 3.12 corresponds to a device that tolerates an overload for a variable time and then forces the field current to I_{fd}^{lim} through integral control. Block 1 is a two-slope gain obeying:

$$x_2 = S_1 x_1 \quad \text{if } x_1 \geq 0$$
$$= S_2 x_1 \quad \text{otherwise}$$

with $S_1, S_2 > 0$. The non-windup limited integrator [ITF96a], block 2, behaves as follows:

$$\dot{x}_t = 0 \quad \text{if } (x_t = K_2 \text{ and } \dot{x}_2 \geq 0) \text{ or } (x_t = -K_1 \text{ and } \dot{x}_2 < 0)$$
$$= x_2 \quad \text{otherwise}$$

This integrator is initially at its negative, lower bound $-K_1$. Assume that i_{fd} becomes larger than I_{fd}^{lim} at time t_o. From there on, x_t starts increasing. As soon as it becomes positive, block 3 switches as indicated in the figure. Assuming, for illustration purposes, a constant overload[4]

$$\Delta I = i_{fd} - I_{fd}^{lim}$$

the switching takes place at a time t_{sw} such that:

$$t_{sw} - t_o = \frac{K_1}{S_1 \, \Delta I}$$

which shows the inverse time feature. If block 5 is substituted for block 1, as suggested by the dotted lines, the switching occurs after a fixed delay

$$t_{sw} - t_o = K_1$$

regardless of the field current overload.

The limited integrator of block 4 is initially at its zero lower bound. Right after t_{sw}, its output x_{oxl} increases. This signal is subtracted from the AVR inputs (see Fig. 3.10), which causes i_{fd} to decrease. The system settles down to equilibrium when the input of the integrator vanishes, i.e. when $i_{fd} = I_{fd}^{lim}$.

In this model provision is made for automatic reset of the OXL if system conditions improve. Large values of S_2 and K_r cause x_{oxl} to come back to zero quickly after $i_{fd} < I_{fd}^{lim}$. Note however that some OXLs do not reset automatically but require manual intervention. Block 3 must then be accordingly modified into a non-reversible switch.

Note that in the case of an inverse time characteristic, a single integrator can be used for both timing and limit enforcement; the corresponding, simpler scheme is shown in [CTF93, Tay94].

An example of simulation using the model of Fig. 3.12 is given in Chapter 8, for a simple system detailed in Chapter 6.

Another OXL model is shown in Fig. 3.13, corresponding to a device that ramps down a limiter reference I_{ref} from the instantaneous values I_{ins} to the I_{fd}^{lim} limit and forces the field current to follow this reference by imposing to the AVR a signal proportional to $I_{ref} - i_{fd}$.

---

[4] in practice the latter varies with time

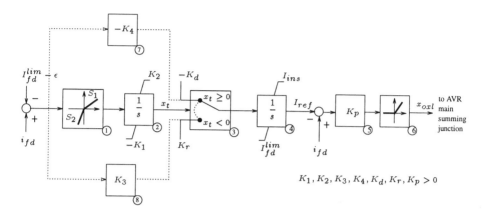

Figure 3.13 Bloc-diagram model of an OXL with proportional control of field current

Blocks 1, 2 and 3 perform essentially the same task as in Fig. 3.12. The limited integrator of block 4 is initially at its upper bound I_{ins}. As soon as switching occurs (due to a positive x_t), the integrator output is decreased at a downward rate K_d. Once I_{ref} becomes smaller than i_{fd}, the x_{oxl} signal starts acting on the AVR, owing to block 6. The small value ϵ may be used to prevent limit cycling if i_{fd} happens to become a little smaller than I_{fd}^{lim}. K_p (block 5) is the gain used to force i_{fd} close to its limit.

In the alternative scheme shown with dotted lines (blocks 7 and 8), ramping down is performed at a rate proportional to the overload $i_{fd} - I_{fd}^{lim}$. This is used in some devices to produce the OXL inverse time characteristics, in which case there is no initial delay (K_1 set to a small value). The limiter then starts ramping down as soon as overexcitation is detected. A system of this type is described in [Kun94] (p. 982).

We have already quoted the wide variety of OXL devices. We finally mention two examples of OXL systems that do not fit exactly the above general models.

1st utility. Older generators are equipped with fixed-delay OXLs that force the field current to I_{fd}^{lim} through integral control as in Fig. 3.12. However, the timer is started once $i_{fd} > I_{fd}^{max}$ where $I_{fd}^{max} > I_{fd}^{lim}$ and I_{fd}^{lim} is chosen slightly above or below the permanent admissible current, depending on the machine. Field currents in between I_{fd}^{lim} and I_{fd}^{max} may be reduced by plant operator action on the AVR voltage reference.

2nd utility. Most OXLs do not take advantage of the thermal overload capability but rather limit the field current almost instantaneously to its permanent admissible value. In case a short-circuit is detected (through a sharp decrease in bus voltage), the field limit is momentarily released to allow field forcing.

3.3.3   Armature current limiters

Automatic armature current limiters are not as common as field current limiters. The main reason is the larger thermal inertia of the armature windings which allows an overload to be taken care by the plant operator. In such circumstances, the plant operator reacts to an alarm indicating excessive armature current by lowering the reactive power output of the machine, through a decrease in voltage reference of the AVR. In some cases the generator active power production can also be reduced.

In some countries, however, generators are provided with automatic armature current limiters, that act on the excitation system as described in the previous section. Some description can be found in [CTF93, JSK94, CTF95] and the quoted references.

3.4   VOLTAGE-REACTIVE POWER CHARACTERISTICS OF SYNCHRONOUS GENERATORS

In Section 2.7 we have introduced the network VQ curve and combined it with the corresponding characteristic of either shunt elements or static var compensators. In the present section we proceed with the synchronous generator, deriving its voltage-reactive power characteristics in various operating conditions. Throughout the whole section, we assume steady-state operation and a constant active power P.

3.4.1   Machine under AVR control

In normal operating conditions, an AVR is characterized by a steady-state relationship:

$$v_{fd} = g_{avr}(V_o, V) \tag{3.40}$$

where the function g_{avr} is monotonically increasing with respect to the voltage reference V_o and decreasing with respect to the stator voltage V. Real AVRs are subject to nonlinearities due, for instance, to the control law of thyristors used to vary the excitation voltage, or the magnetic saturation of the exciter. As an illustration, Fig. 3.14 shows how v_{fd} depends on V for two real excitation systems. The curves have been obtained by simulation, using the detailed AVR models at steady state. In the left diagram, g_{avr} is very close to a linear function of the type:

$$v_{fd} = G(V_o - V) \tag{3.41}$$

where G is the steady-state open-loop gain of the AVR system. A typical range of values for this gain is $20 - 400$ (pu/pu). In right diagram of Fig. 3.14 the curvature is

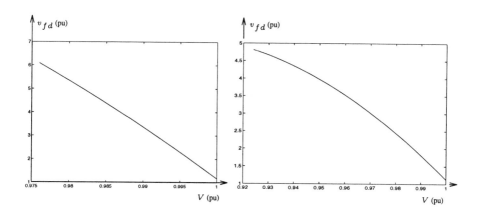

Figure 3.14   Examples of steady-state relationships between v_{fd} and V (brushless excitation systems)

more pronounced and g_{avr} could be more adequately represented by a second-order polynomial.

Consider now the generator behaviour. Under steady-state conditions, $v_{fd} = R_{fd}i_{fd}$ and it is easily seen from (3.9) and (3.14) that:

$$E_f = E_q \tag{3.42}$$

as confirmed by setting the time derivative to zero in the field flux decay equation (3.13). Using the per unit system defined in Section 3.1.10 and putting all together equations (3.35), (3.40) and (3.42) yields:

$$E_q = E_f = v_{fd} = g_{avr}(V_o, V) \tag{3.43}$$

and in the linear case:

$$E_q = E_f = v_{fd} = G(V_o - V) \tag{3.44}$$

Simplified case : unsaturated round-rotor machine

Consider first the simple case of an unsaturated round-rotor machine characterized by $X_d = X_q = X$.

The stator equation in complex form (3.16) becomes:

$$\bar{E}_q = \bar{V} + jX\bar{I}$$

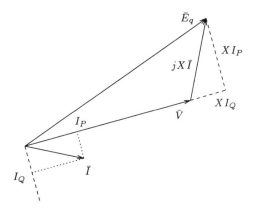

Figure 3.15 Simplified phasor diagram corresponding to $X_d = X_q = X$

and the phasor diagram of Fig. 3.3 becomes the simple one shown in Fig. 3.15. The
latter readily gives:

$$E_q^2 = (V + XI_Q)^2 + (XI_P)^2 \qquad (3.45)$$

Introducing the linear AVR characteristic (3.41) yields:

$$G^2(V_o - V)^2 = (V + X\frac{Q}{V})^2 + (X\frac{P}{V})^2 \qquad (3.46)$$

from which one easily obtains:

$$Q = -\frac{V^2}{X} \pm \frac{1}{X}\sqrt{[GV(V_o - V)]^2 - (XP)^2} \qquad (3.47)$$

The solution with the $-$ sign corresponds to a reactive power absorption beyond the
allowed stability limit (for the given P); hence the sought VQ relationship corresponds
to the $+$ sign in (3.47).

General case : detailed modelling

Consider now the general case with saliency and saturation taken into account.

In Section 3.1.9 we have derived the expressions (3.32a,3.32b) of the machine active
and reactive powers in terms of the voltage V and three internal variables, namely $\delta - \theta$,
E_q and E_q^s. A third equation is provided by the saturation law (3.33) and a fourth
one by the AVR relationship (3.43). We are thus left with four equations involving six

variables (namely $P, Q, V, \delta - \theta, E_q$ and E_q^s). For a specified P, these equations can be solved to obtain Q as a function of V. However they cannot be solved analytically as in the previous simple case. They have to be solved numerically as done in the following example.

Example. A large 50-Hz turbo-generator with a rated apparent power of 1200 MVA and a rated turbine power of 1020 MW has the parameters shown in Table 3.1.

Table 3.1   Data of the example used in §§3.4 and 3.5

$X_d = 2.051$ pu	$m = 0.093$
$X_q = 1.966$ pu	$n = 8.946$
$X_\ell = 0.266$ pu	
$I_{fB} = 2671$ A	$I_{fd}^{lim} = 8300$ A

The corresponding VQ characteristics are shown in Fig. 3.16 for two values of the gain G and two active power outputs. All curves correspond to a zero reactive power at nominal voltage. The curves illustrate the *voltage droop* effect of classical, proportional AVRs. This effect is obviously more pronounced at lower gains. In large power plants where angle stability is a concern, a high gain is used to provide enough synchronizing torque [Kun94] and the voltage droop effect may be neglected. However this is not true for all power plants. The figure also shows that the slope of the characteristic is slightly influenced by the level of active power generation, for a given gain G.

Reference [CG84] derives the expression of the internal reactance X_{mQ} equivalent to the voltage droop effect (see also Problem 3.2), as well as a machine equivalent that can be included in standard load flows and used for contingency evaluation. Applications within the context of voltage stability are shown in [BCR84, VC91a].

Regarding contingency evaluation, many programs assume constant generator voltages (within the limit of the machine capability), which is an acceptable approximation for large gains only. Note however that some AVRs [ITF96a] are equipped with integral control (using PI instead of P controllers), whose effect is to cancel steady-state voltage errors and make the generator appear as truly constant voltage.

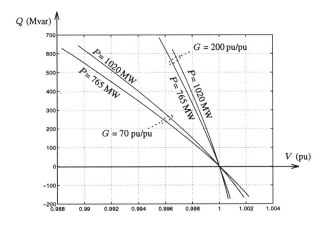

Figure 3.16 Generator VQ characteristics under AVR control

3.4.2 Machine under field current limit

When the machine is under control of the OXL, the AVR steady-state characteristic (3.43) has to be replaced by a relationship of the type:

$$E_q = g_{oxl}(V_o, V) \tag{3.48}$$

For the two types of OXL's described in Section 3.3.2, the above relationship is obtained as follows.

- **OXL using integral control** (see Fig. 3.12). This device is such that in steady state:

$$i_{fd} = I_{fd}^{lim}$$

Therefore the unsaturated emf E_q is held at a constant value:

$$E_q = \omega_o L_{ad} i_{fd} = \omega_o L_{ad} I_{fd}^{lim}$$

In per unit, this relationship amounts to:

$$E_q = I_{fd}^{lim} = E_q^{lim} \tag{3.49}$$

- **OXL using proportional control** (see Fig. 3.13). The linear AVR relationship (3.41) becomes:

$$v_{fd} = G\left[V_o - V - K_p(i_{fd} - I_{fd}^{lim})\right]$$

and the unsaturated emf E_q is given in per unit by:

$$E_q = G(V_o - V) - GK_p(E_q - E_q^{lim})$$

which can be rewritten as:

$$E_q = \frac{G}{1 + GK_p}(V_o - V) + \frac{GK_p}{1 + GK_p}E_q^{lim} \tag{3.50}$$

In practice typical values of K_p are around 10. Note incidentally that a large value of K_p corresponds to a takeover limiter, as defined in Section 3.3.1. For a large value of K_p, (3.50) can be approximated by (3.49). In the remaining of this section, we assume that the OXL is adequately represented in steady-state by (3.49).

Simplified case : unsaturated round-rotor machine

In the simple case of an unsaturated round-rotor machine, the phasor diagram of Fig. 3.15 still holds but with E_q held at the constant value (3.49). Basically, the machine behaves as a constant emf behind synchronous reactance. Therefrom the VQ relationship is easily shown to be:

$$Q = -\frac{V^2}{X} + \frac{1}{X}\sqrt{\left(VE_q^{lim}\right)^2 - (XP)^2} \tag{3.51}$$

General case : detailed modelling

As in the regulated case of section 3.4.1, there are four equations describing the machine but now (3.49) is substituted for (3.43). Again a numerical solution is needed to obtain Q as a function of V, as illustrated below.

Example. We carry on with our example of Fig. 3.16. Table 3.1 shows the values of I_{fB} (see Fig. 3.7) and I_{fd}^{lim}. One easily deduces:

$$E_q^{lim} = \frac{8300}{2671} = 3.107 \text{ pu}$$

The corresponding VQ characteristics are shown with solid lines in Fig. 3.17[5] for three levels of active power P. Similar examples may be found in [Ost93, PMS96].

---

[5] the other curves will be explained in the next section

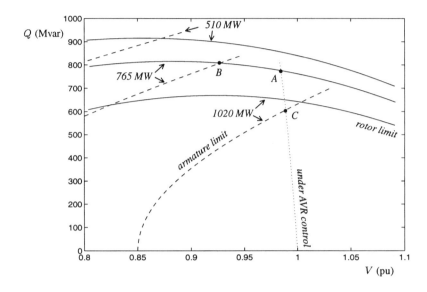

Figure 3.17   Generator VQ characteristics under resp. AVR control, field and armature
current limits

Equation (3.51) of the round-rotor unsaturated case can be used to approximate, at least
over some range of voltage values, the real VQ characteristic of a saturated round-rotor
machine under field current limit. However, this requires us to adjust the E_q^{lim} and X
parameters that appear in (3.51). The properly corrected reactance X_{cr} is significantly
smaller than X [BCR84]. Systematic investigations performed on Belgian generators
have shown that it lies in the range 0.5 – 1.5 pu on the machine base. Once X_{cr}
is known, the adjusted value of E_q^{lim} is easily computed by matching one point of
the real VQ characteristic. Applications of this corrected model to voltage stability
computations are illustrated in [BCR84, VC91a].

In contingency evaluation based on load flow calculations, it is common practice to
represent a field current limit by switching the machine bus to the so-called PQ type,
i.e. a constant reactive power generation is assumed. Figure 3.17 shows that: (i)
the reactive power is not exactly constant, and (ii) the reactive power limit must be
adjusted as a function of the active power generation.

3.4.3 Machine under armature current limit

The armature current limit does not depend on the machine modelling and is easily derived as follows.

Let I^{max} be the maximum armature current of concern. From the definition of apparent power:

$$S = \sqrt{P^2 + Q^2} = V I^{max} \tag{3.52}$$

one readily obtains:

$$Q = \sqrt{(V I^{max})^2 - P^2} \tag{3.53}$$

Furthermore, if I^{max} is the current that corresponds to operation under rated voltage and power, we have in per unit on the machine base:

$$S = 1 \text{ pu}, V = 1 \text{ pu} \quad \text{and hence} \quad I^{max} = 1 \text{ pu}$$

Example. In the previous example, the armature current limits were shown with dashed lines in Fig. 3.16, corresponding to the same three levels of active power considered for rotor limits. Here too, the larger P, the more constraining the armature limit.

3.4.4 Discussion

The three modes of operation of a machine are illustrated on Fig. 3.17, where we have added the voltage droop characteristics shown with dotted line and derived as explained in Section 3.4.1. In normal conditions the machine operates along this (almost vertical) line.

Consider first the case of a 765 MW generation and assume that external system conditions are imposed on the machine to produce more and more reactive power to maintain its terminal voltage. The operating point thus moves along the dotted line up to point A where the rotor limit is reached. This causes the OXL to operate, after some delay, as discussed previously. From there on, the machine voltage is no longer controlled and if external system conditions keep on deteriorating, the operating point moves along the solid line corresponding to $P = 765$ MW. It does so up to point B where the armature limit becomes the most constraining one. As can be seen, when the latter limit is enforced, the reactive power generation of the machine is drastically reduced.

In the case of a 1200 MW generation, the armature limit is the first constraint to be met for lowering voltage and its enforcement is even more stringent. However, if the voltage reference V_o of the AVR was set at a higher value, the characteristic under AVR control would be shifted to the right and the machine would again meet the rotor limit first.

Note finally that if the voltage continues to decrease, the machine may be tripped by a low voltage protection. The latter however should not be set too high, in order to avoid loosing the support of the machine at a time where it is precisely needed to face stressed system conditions. Several reported incidents have involved undue tripping of machines caused by undervoltage protections.

3.5 CAPABILITY CURVES

The operation of a generator can be characterized by three variables : P, Q and V. In the previous section we have characterized the generator operating limits through VQ relationships under constant P. In this section we derive these limits in terms of PQ relationships under constant V. The latter take on the form of the so-called *capability curves*, probably the best known graphical representation of the machine operating limits.

Remark. Note that for both the stator and the rotor, capability curves may refer to either the permanent admissible value of the current or the current after limiter action. The derivations of this section encompass any choice, as it is just a matter of choosing the value of the maximum current. The given example refer to the OXL limit I_{fd}^{lim} for the rotor and the permanent admissible current for the stator. Let us repeat here that the machine cooling conditions significantly influence the maximum admissible current and hence the capability curves.

3.5.1 Simplified case : unsaturated round-rotor machine

Again we start from the simple case of an unsaturated, round-rotor machine characterized by $X_d = X_q = X$.

Under constant V, the armature current limit is easily shown from (3.52) to be a circle centered at the origin, with a radius equal to $V\, I^{max}$. As regards the rotor limit, (3.45) and (3.49) give:

$$V^2 \left(E_q^{lim} \right)^2 = (V^2 + XQ)^2 + (XP)^2$$

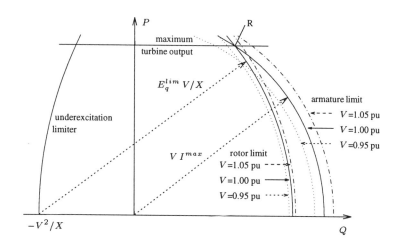

Figure 3.18 Capability curves (saliency and saturation neglected)

Again this corresponds to a circle with center at $(P = 0, Q = -V^2/X)$ and radius equal to $V E_q^{lim}/X$. The corresponding curves are shown in Fig. 3.18, together with the limits corresponding to maximum turbine output and underexcitation limiter operation.

Point R in this figure corresponds to operation at rated power. This is the intersection of the turbine and armature limits under rated voltage. In the case shown, E_q^{lim} has been chosen so that the field limit also passes through point R. In practice the three curves, although very close, may not intersect exactly.

The figure also shows the effect of the terminal voltage. A larger terminal voltage yields a larger armature limit and, for this simplified modelling, a slightly larger rotor limit as well.

Operating limits other than those considered above may also come into the picture, such as the minimum turbine output, bounds on the network terminal voltage and auxiliary bus voltage, etc. A comprehensive set of such limits is considered in [AM94].

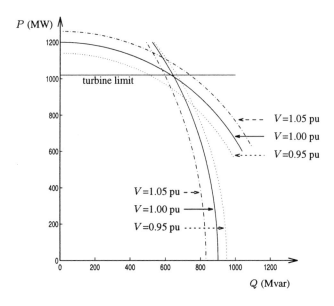

Figure 3.19 Capability curves (saliency and saturation taken into account)

3.5.2 General case : detailed modelling

Among the turbine, armature and rotor limits, only the last one is affected by saliency and saturation effects. As explained previously, the detailed model consists of the four equations (3.32a,3.32b,3.33,3.49) involving six variables ($P, Q, V, \delta - \theta, E_q$ and E_q^s). For a specified V, these equations can be solved numerically to obtain Q as a function of P.

Example. The capability curves of the example of Table 3.1 are shown in Fig. 3.19 for three levels of the terminal voltage.

These curves inspire the following comments:

■   unlike to the simplified case of Fig. 3.18, the above figure shows a decrease in the rotor limit when the voltage increases. This is due to the saturation effects on the machine emf and reactances. Higher voltages cause higher saturation;

■   at lower voltages and/or higher active power outputs, the armature limit may be more constraining than the rotor one;

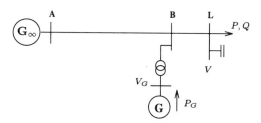

Figure 3.20 Two-generator system

- under any of the two limits, one way of increasing the machine reactive power capability is to decrease its active power output. However the benefit of this action should be balanced against the consequence of having the missing power picked up by other machines; if the latter are located too far from the load center(s) the network loading corresponding to this active power transfer may be detrimental to voltage stability.

3.6 EFFECT OF MACHINE LIMITATIONS ON DELIVERABLE POWER

We consider finally the role that generator limits play on voltage stability, more precisely on the maximum power that can be delivered to loads.

We have shown in the two-bus example of Chapter 2 that, as an operating point moves along the PV curve towards lower voltages, the reactive power produced by the generator increases. Expectedly, when the latter reaches a limit, the shape of the PV curve changes drastically.

We will illustrate this fact on the slightly more elaborate example of Fig. 3.20, inspired of that treated in [BB80] (see Problems 2.9 and 3.7 for the system data). In this system, the transmission line AB is assumed much longer than the local link BL and generator G plays the role of a local support to the load, while G_∞ is a remote, constant voltage source. Initially, the load active power is covered by generator G. It is then increased under constant power factor, with the corresponding active power provided by G_∞ and transmitted over ABL.

The corresponding network PV curves are shown in Fig. 3.21. As already quoted in Section 2.5.2, these curves represent steady-state (or equilibrium) conditions of the transmission and generation system. Hence, the curves have been computed with the generator G modelled as in the previous Sections 3.4 and 3.5.

The curve shown with solid line corresponds to generator G under AVR control, with

$$E_q = G(V_o - V)$$

irrespective of the field current value. The dotted curves correspond to the generator operating under field current limit, i.e.

$$E_q = E_q^{lim}$$

for various values of E_q^{lim}, chosen as follows. Consider the nominal operating point R in Fig. 3.18. Denoting by P_R (resp. Q_R) the corresponding active (resp. reactive) power generation, the *nominal Generator Power Factor* (GPF) is given by:

$$\text{GPF} = \frac{P_R}{\sqrt{P_R^2 + Q_R^2}}$$

The dotted curves of Fig. 3.21 correspond to the same maximum turbine output P_R but four values of GPF. The nominal apparent power thus changes with the value of GPF.

These curves inspire the following comments:

- the maximum load power is severely reduced when the field current of the local generator becomes limited;

- note also how the generator *reactive* power limit affects the load *active* power limit, showing once more that voltage instability involves a strong coupling between active and reactive power;

- higher nominal power factors correspond to smaller reactive reserves, as can be seen from Fig. 3.18. The curves illustrate that it is important to maintain adequate reactive reserves near the load centers. At the planning stage, the nominal power factor should be taken small enough. The cost associated with the higher apparent power is often reasonable;

- in many cases, the generator limits bring the maximum power point at higher voltages. The more stringent the reactive limitation, the more pronounced this effect.

Note also that at low generator voltage, an armature current limit may yield an even more stringent reduction in deliverable load power.

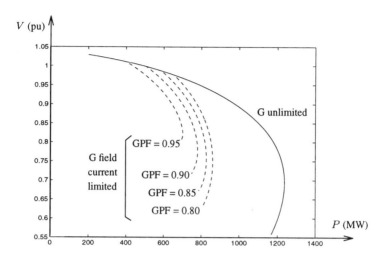

Figure 3.21 PV curves showing the effect of machine reactive limits

3.7 PROBLEMS

3.1. Derive the function g_{avr} of (3.40) for the excitation system models DC1, AC4 and ST1 of Ref. [IWG81].

3.2. Consider the steady-state operation of an unsaturated, round-rotor machine with an AVR described by (3.44). The slope of the corresponding VQ curve (see Fig. 3.16) can be characterized by the equivalent reactance X_{mQ} defined by:

$$\Delta V = X_{mQ}\,\Delta I_Q$$

where $I_Q = Q/V$ is the reactive current produced by the machine. Assuming that G is large enough and ΔV is small enough, show that X_{mQ} is given by:

$$X_{mQ} \simeq \frac{X}{G}\cos(\delta - \theta)$$

where $(\delta - \theta)$ is the machine internal (or load) angle, shown in Fig. 3.3.
Hint: start from (3.46) and assume a small variation of V and Q. See [CG84, BCR84, VC91a] for more details.

3.3. Consider a saturated machine under AVR control. Write down a program to solve Eqs. (3.32a,3.32b,3.33,3.44) for given P and V. Running this program for various values of V, obtain the VQ curves of Fig. 3.16. Repeat for several values of P.

Hint: use Newton method to solve these nonlinear equations.

3.4. Consider a saturated machine under field current limit. Write down a program to solve Eqs. (3.32a,3.32b,3.33,3.49) for given P and V. Running this program for various values of V, obtain one of the VQ curves of Fig. 3.17. Repeat for several values of P.

3.5. Consider a saturated machine under field current limit. Write down a program to solve Eqs. (3.32a,3.32b,3.33,3.49) for given P and V. Running this program for various values of P, obtain the field capability curves of Fig. 3.19. Repeat for several values of V.

3.6. A problem of practical interest is to check the value of the field current limit E_q^{lim} from the manufacturer's capability curves (Fig. 3.19). Given the values P, Q and V at one point of these curves, write down a program to solve Eqs. (3.32a,3.32b,3.33) and obtain the value of E_q^{lim}. Check this with Fig. 3.19 and the data given in Table 3.1.

3.7. Redo Problem 2.9 with a maximum reactive power output of 95 Mvar (0.95 pu) on generator G. The solution is the dotted curve GPF = 0.90 in Fig. 3.21, when using generator equilibrium equations instead of the PQ-bus approximation.

3.8. Redo Problem 2.10 with the above reactive limit on generator G.

4

LOAD ASPECTS

"Ex nihilo nihil"[1]
Lucretius

After dealing in the two previous chapters with transmission and generation aspects, we come now to the third factor of voltage instability: power system loads. Load dynamic response is a key mechanism of power system voltage stability driving the dynamic evolution of voltages and, in extreme cases, leading to voltage collapse.

Load modelling is a difficult problem because power system loads are aggregates of many different devices. The heart of the problem is the identification of the load composition at a given time and the modelling of the aggregate. However, in order to understand the nature of voltage stability and its close relationship with load dynamics it is necessary to start with the analysis of individual loads

In this chapter we first address the voltage dependence of loads, focusing on properties of the widely used exponential and polynomial loads models. Then, after introducing load dynamics from the viewpoint of load power restoration, three important components exhibiting power restoration are considered in detail: induction motors, load tap changers and thermostatic loads. We finally come back to aggregate loads and discuss some general models that have been proposed for their analysis.

---

[1] Nothing comes out of nothing

4.1 VOLTAGE DEPENDENCE OF LOADS

4.1.1 Load characteristics

As discussed in Section 2.4.1, a *load voltage characteristic*, or simply *load characteristic*, is an expression giving the active or reactive power consumed by the load as a function of voltage, as well as of an independent variable, which we will call the load *demand*. Noting the latter as z the general form of load characteristics is:

$$P = P(z, V) \tag{4.1a}$$
$$Q = Q(z, V) \tag{4.1b}$$

It is important to emphasize right away the clear distinction made between actually consumed load power (P, Q) and load demand z. This distinction is necessary for the understanding of a basic instability mechanism, by which increased demand may result in reduced consumption of power.

The frequency dependence of loads is not addressed in this book, since in voltage stability incidents the frequency excursions are not of primary concern.

Exponential load

A widely used load characteristic is the well-known *exponential* load, which has the general form:

$$P = zP_o \left(\frac{V}{V_o} \right)^\alpha \tag{4.2a}$$

$$Q = zQ_o \left(\frac{V}{V_o} \right)^\beta \tag{4.2b}$$

where z is a dimensionless demand variable, V_o is the reference voltage, and the exponents α and β depend on the type of load (motor, heating, lighting, etc.). Note that zP_o and zQ_o are the active and reactive powers consumed under a voltage V equal to the reference V_o and relate to the amount of connected equipment. These have been called *nominal* load powers [ITF95b, CTF93], in contrast to the *consumed* powers P, Q.

A graphical representation of the exponential load characteristic in the PV plane was given in Figs. 2.9 and 2.11 of Chapter 2.

Three particular cases of load exponents are noteworthy :

Table 4.1   A sample of fractional load exponents

load component	α	β
incandescent lamps	1.54	–
room air conditioner	0.50	2.5
furnace fan	0.08	1.6
battery charger	2.59	4.06
electronic compact fluorescent	0.95-1.03	0.31-0.46
conventional fluorescent	2.07	3.21

■ $\alpha = \beta = 2$: constant impedance load (often noted Z)

■ $\alpha = \beta = 1$: constant current load (often noted I)

■ $\alpha = \beta = 0$: constant power load (often noted P).

Fractional exponents have been attributed to certain load components, some of which are shown in Table 4.1 [PWM88, Tay94, HD97].

Care should be taken when using the exponential load at low voltage levels, because when voltage drops below a threshold value (e.g. $V < 0.6$) many loads may be disconnected, or may have their characteristics completely altered.

In the sequel we will demonstrate two important properties of the exponential load. To facilitate the presentation we will assume $z = 1$.

Due to the exponential relation, the reference voltage V_o and the corresponding P_o, Q_o can be specified arbitrarily without changing the characteristic. Consider for instance a voltage level V_1, for which the load power is:

$$P_1 = P_o \left(\frac{V_1}{V_o} \right)^{\alpha}$$

By using this relation to substitute P_o in (4.2a) we obtain:

$$P = P_1 \left(\frac{V}{V_1} \right)^{\alpha}$$

We have thus replaced V_o with V_1 and P_o with P_1 showing that any voltage level can be used as a reference, in order to initialize the exponential model.

The exponents α and β of the exponential load model determine the *sensitivity* of load power to voltage. Assuming any reference voltage V_o, for which the load active power is P_o, the sensitivity of active power with respect to voltage is calculated as:

$$\frac{dP}{dV} = \alpha P_o \left(\frac{V}{V_o}\right)^{\alpha-1} \frac{1}{V_o}$$

A similar relation holds for the reactive power. By rearranging the above expression and evaluating the sensitivity at $V = V_o$ we find:

$$\frac{dP/P_o}{dV/V_o} = \alpha \tag{4.3a}$$

$$\frac{dQ/Q_o}{dV/V_o} = \beta \tag{4.3b}$$

Thus, the normalized sensitivities of real and reactive load power are equal to the corresponding load exponents. Note that the normalized sensitivities are the same at any reference voltage.

Polynomial load

As seen in Table 4.1, different load components exhibit different voltage characteristics. Thus an alternative load representation is based on summing up load components which have the same (or almost the same) exponent. When the exponents are all integer, the load characteristic becomes a polynomial in V. A special case is the ZIP model, which is made up of three components: constant impedance, constant current and constant power. The real and reactive characteristics of the ZIP load model are given by the following quadratic expressions:

$$P = zP_o \left[a_P \left(\frac{V}{V_o}\right)^2 + b_P \frac{V}{V_o} + c_P \right] \tag{4.4a}$$

$$Q = zQ_o \left[a_Q \left(\frac{V}{V_o}\right)^2 + b_Q \frac{V}{V_o} + c_Q \right] \tag{4.4b}$$

where $a_P + b_P + c_P = a_Q + b_Q + c_Q = 1$, while zP_o and zQ_o are the load real and reactive powers consumed at the reference voltage V_o.

When polynomial load parameters are obtained from measurements, some of them, usually the one defining current contribution b_P (or b_Q) may assume negative values. Typical polynomial active load characteristics with b positive or negative are shown in Fig. 4.1a,b for various levels of demand. To emphasize that the ZIP model is unrealistic for low voltages, the lower part of the characteristics of Fig. 4.1a,b is plotted with dotted lines.

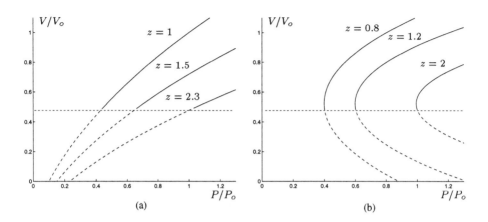

Figure 4.1 ZIP load characteristics: (a) $a = 0.4$, $b = 0.5$, $c = 0.1$, (b) $a = 2.2$, $b = -2.3$, $c = 1.1$

4.2 LOAD RESTORATION DYNAMICS

In the previous section we have seen that the power consumed by loads depends on their voltage characteristics. This dependence may be permanent, in which case the load is purely static, or it may change with time, in which case the load is dynamic. The dynamics of various load components and control mechanisms tend to restore load power, at least to a certain extent. We refer to this process as *load restoration*.

Before considering specific loads, let us show a concise way to describe load dynamics in general. Consider that the power consumed by the load at any time depends upon the instantaneous value of a load state variable, denoted as x:

$$P = P_t(z, V, x) \tag{4.5a}$$
$$Q = Q_t(z, V, x) \tag{4.5b}$$

where P_t, Q_t are smooth functions of demand, voltage, and load state and are called the *transient load characteristics*.

Consider also that the load dynamics are described by the smooth differential equation:

$$\dot{x} = f(z, V, x) \tag{4.6}$$

The steady state of load dynamics is characterized by the following algebraic equation:

$$f(z, V, x) = 0 \tag{4.7}$$

Table 4.2 Dynamic load state and demand variables

load component	state variable x	demand variable z
induction motor	rotor slip s	mechanical torque
load behind LTC	tap position	load demand
thermostatic	connected equipment	energy requirement

In general (i.e. when $\partial f/\partial x \neq 0$), (4.7) can be used to obtain the state variable x as a function of z and V:

$$x = h(z, V) \tag{4.8}$$

with h satisfying:

$$f(z, V, h(z, V)) = 0 \tag{4.9}$$

Substituting (4.8) into (4.5a,b) we obtain:

$$P = P_t(z, V, h(z, V)) = P_s(z, V) \tag{4.10a}$$
$$Q = Q_t(z, V, h(z, V)) = Q_s(z, V) \tag{4.10b}$$

where P_s, Q_s are the *steady-state load characteristics*. Note that the steady-state load characteristics do not depend on the load state variable.

The transition towards the steady-state load characteristics is driven by the load dynamics (4.6). Typical dynamic load components with the corresponding state and demand variables are shown in Table 4.2. The close similarities among these three types of load restoration mechanisms were pointed out and analyzed in [VC88a].

Usually, the transient load characteristic is more sensitive to voltage than the steady-state load characteristic, so that in the steady state the load power is restored closer to its pre-disturbance value. A typical example is the transition from constant impedance to constant power load. In the following sections we will discuss in detail the load components mentioned in Table 4.2. An extensive bibliography on load models can be found in [ITF95a], which includes both static and dynamic load components and models.

4.3 INDUCTION MOTORS

4.3.1 The significance of induction motors

Induction motor load is an important component in power system voltage stability assessment for the following reasons:
1. it is a fast restoring load in the time frame of a second;
2. it is a low power factor load with a high reactive power demand;
3. it is prone to stalling, when voltage is low and/or the mechanical load is increased.

There are various types of induction motors. In power system studies we usually assume aggregate motor models [RDM84, NKP87, SL97], i.e. one motor representing a large number of similar motors fed (through distribution lines) by the same substation. If the motors connected to the same bus are not similar, it may be necessary to use more than one aggregate motors to represent the load properly.

In terms of individual motor modelling, one has to distinguish between three-phase and single-phase motors, as well as motors having a constant rotor resistance and motors having double-cage, or deep bar rotor.

In this section we will discuss both three-phase and single-phase motors with constant rotor resistance. Motor models with saturation, as well as with variable rotor resistance can be found in [Kun94]. In all cases we use the per unit system on motor rating. The effect of variable frequency, as pointed out in the introduction to this chapter, is neglected for all load components.

4.3.2 Motor modelling

The stator of a three-phase induction machine is similar to that of a synchronous machine (Fig. 3.1). Using the Park transformation the three phase windings can be substituted by the two equivalent d- and q-axis windings, as in Fig. 3.2. The rotor of an induction motor may have a short-circuited, three-phase winding, or a squirrel-cage construction. In either case, the rotor can be analyzed also with two equivalent, short-circuited, d- and q-axis windings. In fact, the $d1$, $q1$ windings of Figs. 3.1 and 3.2 give rise to an asynchronous torque, typical of induction machines.

Induction machine dynamics

The following types of transients are present in an induction machine:

- Stator transients similar to those of synchronous machines. These will be neglected as was done for generator modelling in the previous chapter.

- Rotor electric transients involving the d- and q-axis equivalent rotor circuits. These are in the time frame of subtransient generator time constants (damper windings).

- Rotor mechanical motion characterized by the corresponding acceleration equation.

Usually it is assumed that the rotor electric transients are faster than the mechanical transients, so that the motor can be represented with only acceleration dynamics. In [AZT91] it is shown that this assumption is valid for small motors, whereas an alternative reduced order model can be used for large motors. In our analysis of induction motor behaviour we will neglect rotor electrical transients for the sake of simplicity. In detailed numerical simulation, however, the rotor electrical transients can be easily incorporated, provided that motor data is available.

Steady-state equivalent circuit

Assuming that the rotor electrical transients have died out, a three-phase induction machine with constant rotor resistance can be represented by the well-known equivalent circuit of Fig. 4.2 [FKU83], in which s is the motor slip defined as:

$$s = \frac{\omega_o - \dot{\theta}_r}{\omega_o} \tag{4.11}$$

where ω_o is the nominal angular frequency and $\dot{\theta}_r$ is the rotor speed in electrical radians per second. In the same figure, X_s, X_r are the stator and rotor *leakage* reactances, whereas X_m is the *magnetizing* reactance. All impedances are referred to the stator side.

The circuit of Fig. 4.2 is called the steady-state equivalent circuit, in the sense that all *electrical* transients have reached steady state. Note however that the motor is not necessarily at mechanical steady state, and therefore the motor slip can vary.

Typical equivalent circuit parameters taken from the literature are grouped in Table 4.3 for various types of induction motors. All values are in per unit at motor rating.

Two equivalent circuits derived from Fig. 4.2 are shown in Fig. 4.3a,b. In Fig. 4.3a the circuit to the left of BB' is replaced by a Thévenin equivalent with the following parameters:

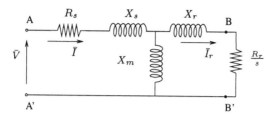

Figure 4.2 Steady-state equivalent circuit

Table 4.3 Typical induction motor parameters (pu on motor base)

Motor	R_s	X_s	X_m	R_r	X_r
Small industrial [ITF95b]	0.031	0.10	3.2	0.018	0.18
Large industrial [ITF95b]	0.013	0.067	3.8	0.009	0.17
Mean values for 11 kVA motors[FM94]	0.016	0.063	0.96	0.009	0.016
Small industrial [Kun94]	0.078	0.065	2.67	0.044	0.049
commercial+feeder [Kun94]	0.001	0.23	3.0	0.02	0.23
aggregate residential [ITF95b]	0.077	0.107	2.22	0.079	0.098
single phase [Tay94]	0.11	0.12	2.0	0.11	0.13

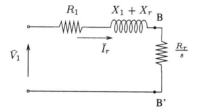

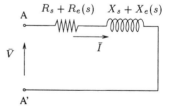

Figure 4.3a Thévenin equivalent seen from BB'

Figure 4.3b Equivalent circuit seen from AA'

$$V_1 = \frac{X_m V}{\sqrt{R_s^2 + (X_s + X_m)^2}} \qquad (4.12a)$$

$$R_1 + jX_1 = \frac{jX_m(R_s + jX_s)}{R_s + j(X_s + X_m)} \qquad (4.12b)$$

Note that when the motor is connected to a voltage supply through an impedance Z, the external impedance Z can be added to the stator one $R_s + jX_s$ and the same formulae apply, where V is the source voltage.

Figure 4.3b shows the equivalent impedance as seen from the stator terminals AA'. The stator current drawn by the motor is:

$$\bar{I} = \frac{\bar{V}}{(R_s + R_e) + j(X_s + X_e)} \tag{4.13}$$

where R_e and X_e depend on s and are given by:

$$R_e + jX_e = \frac{jX_m \left(\dfrac{R_r}{s} + jX_r \right)}{\dfrac{R_r}{s} + j(X_m + X_r)} \tag{4.14}$$

Torque and power

The per unit active power transferred from the stator to the rotor through the air gap, called the *air-gap power*, is readily calculated from the equivalent circuit as:

$$P_g = I_r^2 \frac{R_r}{s} \tag{4.15}$$

Subtracting the rotor losses $I_r^2 R_r$ from the above power P_g, one gets the power delivered by the motor through the electromagnetic torque:

$$P_e = I_r^2 \frac{R_r}{s}(1 - s) \tag{4.16}$$

The same power can be found multiplying the per unit electromagnetic torque T_e by the per unit rotor speed:

$$P_e = T_e \frac{\dot{\theta}_r}{\omega_o} = T_e(1 - s) \tag{4.17}$$

Comparison of (4.16) to (4.17) gives the expression of T_e:

$$T_e = I_r^2 \frac{R_r}{s} = P_g \tag{4.18}$$

Solving for I_r^2 from the Thévenin equivalent shown in Fig. 4.3a, we obtain the torque developed by the motor as a function of voltage V_1 and slip:

$$T_e(V_1, s) = \frac{V_1^2 \dfrac{R_r}{s}}{\left(R_1 + \dfrac{R_r}{s} \right)^2 + (X_1 + X_r)^2} \tag{4.19}$$

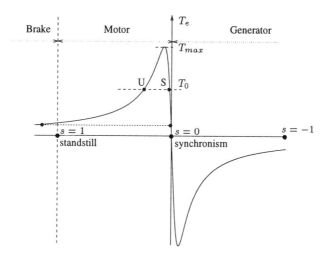

Figure 4.4 Slip-torque characteristic

or as a function of terminal voltage and slip:

$$T_e(V, s) = \frac{V^2 X_m^2 \dfrac{R_r}{s}}{\left[\left(R_1 + \dfrac{R_r}{s} \right)^2 + (X_1 + X_r)^2 \right] [R_s^2 + (X_s + X_m)^2]} \qquad (4.20)$$

The well-known slip-torque characteristic of an induction machine is shown in Fig. 4.4. The torque is considered positive for motor operation. As seen, the same model applies to generator operation, in which case the slip is negative (speed above synchronous). According to (4.19) a negative slip produces a negative electromagnetic torque. A third mode of operation corresponds to *braking*, where the direction of rotation is opposite to that of the rotating field ($s > 1$) and the motor absorbs both electrical and mechanical power.

The real and reactive power absorbed by the induction machine at steady state are given as functions of voltage and slip by:

$$P(V, s) = \frac{(R_s + R_e)V^2}{(R_s + R_e)^2 + (X_s + X_e)^2} \qquad (4.21a)$$

$$Q(V,s) = \frac{(X_s + X_e)V^2}{(R_s + R_e)^2 + (X_s + X_e)^2} \qquad (4.21b)$$

Mechanical motion dynamics

The differential equation of rotor motion can be written in terms of slip as:

$$2H\dot{s} = T_m(s) - T_e(V,s) \qquad (4.22)$$

where H is the inertia constant (s), and T_m the per unit mechanical torque (including the effect of mechanical losses), which depends in general upon the speed of rotation and therefore on the value of slip. Note that increasing load results in an increase of slip.

Just after a disturbance, the slip cannot change instantaneously due to mechanical inertia. With s fixed at its pre-disturbance value, Fig. 4.2 shows that the motor merely behaves as an impedance and thus the motor transient load characteristic is of constant impedance type. Thus, both real and reactive power, as well as the electric torque, decrease after a voltage drop. The torque reduction will cause eventually the rotor to decelerate according to (4.22), thus increasing the active power consumption until a new operating point is achieved.

The torque equilibrium condition is:

$$T_e(V,s) = T_m(s) \qquad (4.23)$$

Hence, the steady-state load characteristic is quite different from a constant impedance one, and it depends upon the mechanical torque characteristic, which we discuss in the following section.

4.3.3   Motor behaviour as affected by mechanical torque

In this section we consider the motor steady-state behaviour under different mechanical torque models, namely constant torque, quadratic torque and composite torque. Note that these are strict motor characteristics obtained by assuming a constant terminal voltage V. Examples of interaction of motor loads with the external system will be considered in later chapters.

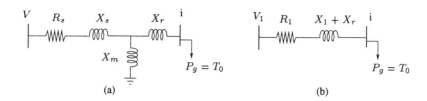

Figure 4.5 One-line diagram of constant torque motor and Thévenin equivalent

Constant torque model

The constant torque model is simply:

$$T_m(s) = T_0 \tag{4.24}$$

The mechanical torque characteristic is thus parallel to the s-axis in the slip-torque diagram and there are two intersection points with the electrical torque characteristic (shown as points S and U in Fig. 4.4), when $T_0 < T_{max}$. On the contrary, there are no points of intersection, when $T_0 > T_{max}$. In the latter case the motor *stalls*, i.e. decelerates to a complete stop. When T_0 is small, the second point of intersection may lie in the braking region, where $s > 1$ (dotted line in Fig. 4.4).

The stability of operating points S and U can be judged empirically as follows. At point S a small increase in slip produces a surplus of electric torque, which according to (4.22) will tend to reduce slip, thus bringing the operating point back to S. Similarly, a small decrease of slip will create a torque driving the motor back to point S. Thus it is concluded that S is a *stable* equilibrium. The reverse holds at point U: a small increase in slip results in a deficit of electrical torque, so that the motor will decelerate, further increasing the slip until standstill ($s = 1$). On the other hand, a small decrease of slip from operating point U will result in a surplus of electrical power, and the machine will accelerate up to the stable equilibrium point S. We thus conclude that point U is an *unstable* operating point.

Note that at steady state the constant torque motor model is equivalent to a constant power load behind the rotor leakage reactance X_r. Indeed from (4.18) the per unit electrical torque is equal to the per unit air gap power P_g, so when the former becomes equal to the constant mechanical torque, the latter is restored. This leads to the one-line diagram representation of an induction motor shown in Fig. 4.5. This diagram corresponds to the steady state of both electrical and mechanical transients and is valid only for constant mechanical torque.

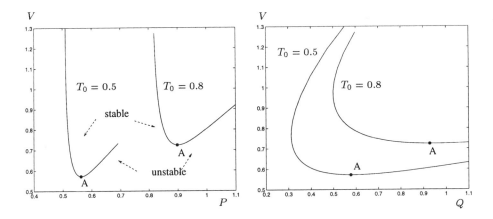

Figure 4.6 Motor steady-state characteristic – Constant torque

To summarize, the induction motor dynamics are such that the load characteristic at the internal node "i" (see Fig. 4.5) changes from constant impedance (transient characteristic for constant slip) to constant power $P_g = T_0$ (steady-state characteristic). Note finally that the loadability limit for an induction motor with constant mechanical torque and constant stator voltage corresponds to the maximum torque condition, for which the variable resistance R_r/s matches exactly the transfer impedance, $R_1 + j(X_1 + X_r)$:

$$\frac{R_r}{s_{maxT}} = \sqrt{R_1^2 + (X_1 + X_r)^2} \qquad (4.25)$$

This can be seen as a direct application of the derivation for maximum power under constant power factor in section 2.2.2.

The P and Q steady-state characteristics of a motor operating with constant torque, and with the terminal voltage V considered as an independent parameter, are drawn in Fig. 4.6. Note that the active power absorbed is almost constant for voltages above 0.8 pu, with just a slight negative slope. The increased consumption for lowered voltage is due to the increased stator losses due to higher current. The slightly negative slope becomes more pronounced as voltage is lowered until the stalling point A is reached. If voltage is lowered below this point there will be no steady-state solutions as the motor will stall. The operating points lying to the right of the stalling point A are unstable as explained above.

The reactive power characteristic is quite different. At high voltage levels the slope is positive, meaning that for increased stator voltage the increased magnetizing reactance consumption (V^2/X_m) dominates over the reduced leakage reactive losses (XI^2). However, at a certain voltage level, which can be quite close to nominal for a heavily loaded motor, the slope reverses sign and becomes negative. Thus the reactive consumption of the motor increases considerably as the stalling point A is approached.

Quadratic torque model

Constant torque is the simplest mechanical load model, but not necessarily the most realistic. Many loads, such as circulation pumps exhibit a quadratic mechanical torque characteristic:

$$T_m(s) = T_2(1 - s)^2 \qquad (4.26)$$

One important property of this load is that the number of intersection points of mechanical and electrical torque is either one, or three, as shown in Fig. 4.7, where a few quadratic mechanical load characteristics are plotted as functions of slip, together with an electrical torque characteristic. When T_2 is small, the only intersection point is close to the synchronous speed. As T_2 increases the points of intersection become three. For further load increase there is again a single intersection point, but it is now far from the synchronous speed and close to standstill. Empirical reasoning shows that when there is only one operating point, this operating point is stable. In the case of three intersection points, the operating point U in the middle is unstable, whereas the other two (S_1 and S_2) are both stable.

The practical loadability limit is point B, which is a local maximum corresponding to the maximum loading for which there is a normal, high speed (low slip) operating point. Point A is a local minimum and represents the minimum loading for which a stable low speed (high slip) operating point exists. The practical consequence of the limit point A is that for a loading parameter $T_2 > T_{2A}$ the motor cannot start properly as will be discussed later.

As in the constant torque case, the P and Q steady-state characteristics of a motor with quadratic mechanical torque are drawn in Fig. 4.8. Note that now the slope of the active power characteristic is slightly positive at normal voltage levels, since the mechanical load and thus the air-gap power increases with voltage. The reactive power characteristic has a similar form to that drawn for the constant torque model. The stalling point is marked as point B and corresponds to loadability limit B in Fig. 4.7.

Finally, in Fig. 4.9 the real and reactive power absorbed by an induction motor following a sudden, 15% voltage drop (at time $t = 0.1$ s) are plotted. The mechanical torque is quadratic. As seen, both powers drop instantaneously, but their recovery is very fast, taking less than a second. The active power recovers almost to the pre-disturbance value, as expected from the corresponding steady-state characteristic, whereas the

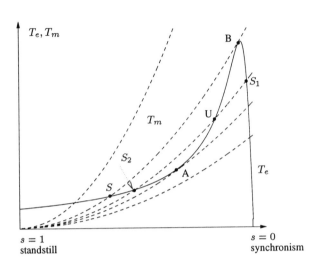

Figure 4.7   Quadratic mechanical torque model

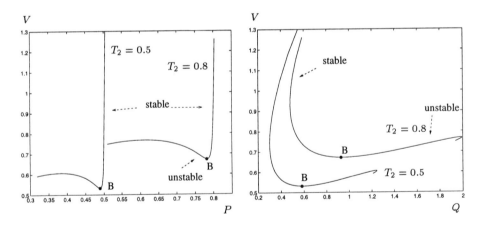

Figure 4.8   Motor steady-state characteristic – Quadratic torque

reactive power drops at steady state, meaning that the operating point is on the upper
part of the reactive power characteristic of Fig. 4.8.

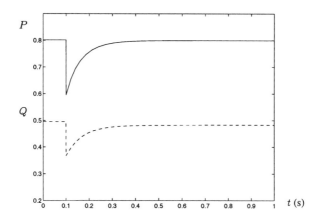

Figure 4.9   Motor load recovery after voltage drop – Quadratic torque

Composite torque model

Constant torque and quadratic torque models can be combined to form composite
mechanical loads. A third component of mechanical torque that is important during
starting is the *static* torque, which is due to friction. When the motor is at standstill the
static torque impedes its rotation, but as the rotor starts to rotate this torque component
is reduced. Thus, the static torque can be modelled as being proportional to slip:

$$T_m(s) = T_s s \tag{4.27}$$

Incorporating the static torque, a composite mechanical torque model takes the following general form:

$$T_m(s) = T_0 + T_s s + T_2(1 - s)^2 \tag{4.28}$$

In Fig. 4.10 a composite mechanical torque characteristic of the form (4.28) is shown.

Motor starting and stalling

At starting, the motor is originally at standstill ($s = 1$). If the electromagnetic torque
$T_e(V, 1)$ exceeds the mechanical torque $T_m(1)$ (for $s = 1$) the motor will start to
accelerate. Note from the equivalent circuit that the motor impedance for $s = 1$ is
significantly smaller than that of normal operation, where s is very small. Thus the
starting current of the motor is several times larger than its rating. In the case of
constant power torque, starting under load is not possible, unless the starting torque

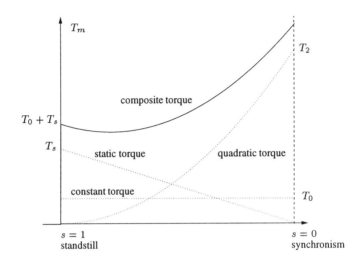

Figure 4.10   Static torque and composite mechanical torque characteristic

is higher than the load torque T_0. This means that the unstable operating point U of Fig. 4.4 must be in the braking region for the motor to be able to start under load without employing special starting schemes.

In general, when the mechanical torque model (4.28) has a nonzero value at standstill (i.e. $T_m(1) \neq 0$), this has to be interpreted as follows:

■   if the starting electrical torque exceeds the mechanical $(T_e(V, 1) > T_m(1))$, then (4.22) applies and the rotor accelerates,

■   if the electrical torque at standstill is insufficient for starting $(T_e(V, 1) \leq T_m(1))$, the motor will not decelerate into the braking region, as seems to be implied by (4.22), but will remain at standstill.

Let us now investigate the motor operation at stalling.   Looking again at Fig. 4.6, which shows the steady-state motor characteristics for constant mechanical torque, it can be seen that close to the stalling point A, the stator current for a mechanical torque $T_0 = 0.8$ pu is close to 1.6 pu. Motor protection will not trip for this current, which is still 3–4 times less than the motor starting current. If the motor stalls, it will eventually come to a complete stop and its current will become equal to the starting current. The motor protection will not trip for this value either. Thus, in order to avoid the severe consequences of motor stalling, undervoltage protection is installed on many industrial motors tripping the motor if the voltage is below a threshold [Tay94]. Small motors have no undervoltage protection, but only thermal overload protection which still has to allow starting current for several seconds. Thus a stalled motor can remain on-line absorbing its highly reactive starting current.

In the case of quadratic mechanical torque it should be noted that when the motor stalls, it will not stop completely, but it will be operating at an abnormal, low-speed operating point, shown as point S in Fig. 4.7. At this operating condition the current drawn by the motor is close to the value corresponding to starting. As discussed above, motor protection will allow this abnormal operation (at least for some time), unless undervoltage protection is installed.

4.3.4 Single-phase induction motors

A significant proportion of residential load consists of a large number of low-rating, single-phase motors. A simple way to represent this load component is by considering a single-phase motor of appropriate rated power connected to each phase, so as to maintain balanced operation.

The single-phase motor can be analyzed by resolving the pulsating, stationary magnetic field produced in the air gap, into two traveling waves, rotating in opposite directions: one forward, and one backward [FKU83]. For the forward traveling wave a steady-state equivalent circuit similar to that of Fig. 4.2 applies. The same is true for the backward traveling wave, but with respect to this the slip is:

$$\frac{-\omega_o - \dot{\theta}_r}{-\omega_o} = 1 + \frac{\dot{\theta}_r}{\omega_o} = 2 - s$$

This leads to the equivalent circuit of Fig. 4.11, where both the forward and the backward rotating fields are represented.

The torque produced by the forward rotating field is in the positive direction, whereas the torque generated by the backward field is negative. Thus, the per unit electromagnetic torque developed by the single-phase motor is given as:

$$T_e = T_f - T_b = I^2(R_f - R_b) \tag{4.29}$$

where:

$$R_f = \frac{0.5 \frac{R_r}{s} X_m^2}{(\frac{R_r}{s})^2 + (X_r + X_m)^2} \tag{4.30a}$$

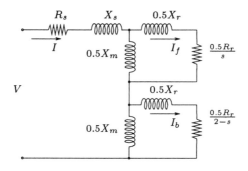

Figure 4.11   Single-phase motor equivalent circuit

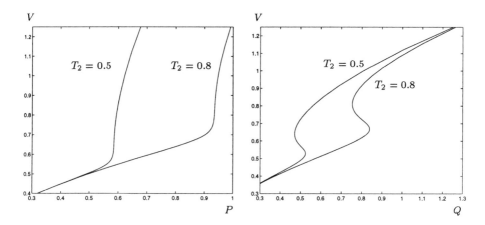

Figure 4.12   Single-phase motor steady-state characteristics

$$R_b = \frac{0.5 \dfrac{R_r}{2-s} X_m^2}{(\dfrac{R_r}{2-s})^2 + (X_r + X_m)^2} \tag{4.30b}$$

and the stator current I can be calculated from the equivalent circuit of Fig. 4.11.

Note that at $s = 1$ (standstill) $T_f = T_b$, so the single-phase motor has no starting torque. A description of auxiliary mechanisms and devices for starting a single-phase motor is outside the scope of this book.

The PV and QV characteristics of a single-phase motor are plotted in Fig. 4.12 assuming a quadratic mechanical torque model. At low voltage, the active and reactive powers absorbed are independent of loading. The same holds for the reactive power at high voltages.

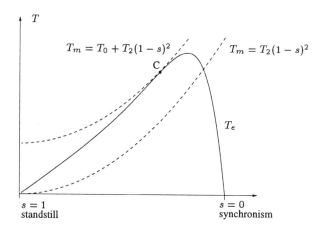

Figure 4.13 Single-phase motor slip-torque characteristics

Note that when supply voltage drops, the motor speed is reduced gradually, unlike the three-phase motor with the same quadratic torque model (Fig. 4.8), which presents a discontinuity at the stalling point B. This property of the single-phase motor is associated with the absence of starting torque as becomes evident by inspecting the slip-torque characteristic of Fig. 4.13. Of course the single-phase motor has a stalling point (similar to that of the three-phase one) when the mechanical load has a constant torque component, as is also shown in Fig. 4.13 (point C).

4.4 LOAD TAP CHANGERS

4.4.1 Description

One of the key mechanisms in load restoration is the voltage regulation performed automatically by the tap changing devices of main power delivery transformers. Figure 4.14 shows a typical one-line diagram. The tap changer controls the voltage of the distribution, Medium Voltage (MV) side V_2 by changing the transformer ratio r. In most cases the variable tap is on the high voltage (HV) side. One reason for this is that

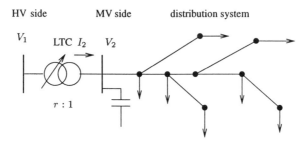

Figure 4.14   One-line diagram of bulk power delivery transformer

the current is lower at this side making commutation easier. Another reason is that more turns are available at the HV side making regulation more precise. Exceptions to this rule can be found, mostly in the case of autotransformers.

Similarly to generator AVRs (section 3.2), the voltage error may be compensated with a current measurement, so that the reference voltage is effectively increased at high loading levels. This is known as line drop compensation [Cal84]. Effectively this is the same as regulating voltage further away from the transformer secondary.

Various acronyms have been suggested for the transformer tap changer mechanisms: on-load tap changers, under load tap changers [Kun94], tap changers under load [Cal84, SP94]. In this book we will adopt the simpler term *Load Tap Changers* (LTC), as in [CTF93, Tay94].

The LTCs are slowly acting, discrete (discontinuous) devices changing the tap by one step at a time, if the voltage error remains outside a deadband longer than a specified time delay. The minimum time required for the tap changer to complete one tap movement is usually close to 5 seconds. We will call this the *mechanical time delay* and denote it by T_m. Various intentional time delays (ranging from several seconds to a couple of minutes) are usually added to the mechanical time delay to avoid frequent or unnecessary tap movements, which are a cause of wear to equipment. The intentional time delays can be either constant, or variable. In the latter case an inverse-time characteristic is often used. The essence of the inverse-time characteristic is that the time delay becomes shorter for larger voltage errors.

One important constraint in LTC operation is that the variable tap ratio has a limited regulation range:

$$r^{min} \leq r \leq r^{max}$$

Typical values of the lower limit are from 0.85–0.90 pu and for the upper limit 1.10–1.15 pu.

The size of a tap step is usually in the range of 0.5%–1.5%. A typical value used in North America is 0.00625 (5/8%). For obvious reasons the deadband must be larger than the tap step size. Typically the deadband is twice the tap step.

Many LTC systems accept a tap blocking signal, which disables the automatic regulation of secondary voltage. This feature aims at tackling the unstable LTC response, which we will analyze later in this book. Other strategies used in emergency conditions consist in reducing the voltage setpoint, or moving the tap to a prespecified position.

4.4.2 LTC modelling

In this section we introduce two types of LTC modelling: discrete models representing discontinuous, step by step tap change, and an approximate continuous model.

For simplicity we will assume here a transformer with negligible resistance and magnetizing reactance, having a constant leakage reactance X_t. Of course, more detailed models including resistance and magnetizing reactance, and possibly the effect of variable tap on transformer impedance, are used in computer simulations. The equivalent one-line diagram of the LTC transformer is shown in Fig. 4.15 using an ideal transformer with an $r : 1$ ratio.

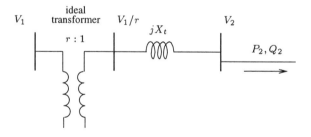

Figure 4.15   Equivalent circuit of an off-nominal tap transformer

Discrete LTC models

The discrete LTC models assume that when the LTC is activated it will raise or lower the transformer ratio by one tap step instantaneously. We denote the size of each tap step by Δr. The LTC can operate at discrete time instants denoted by $t_k, k = 0, 1, \ldots$ and given by the recursive formula:

$$t_{k+1} = t_k + \Delta T_k \qquad (4.31)$$

Note that unlike usual discrete-time systems, t_k is not an independent variable, and ΔT_k is not necessarily constant, since it depends in general upon the device characteristics and the voltage error. The integer counter will advance from k to $k + 1$, when the time elapsed since t_k becomes equal to (or exceeds) ΔT_k.

A universal formula for ΔT_k including fixed and inverse-time delays is the following:

$$\Delta T_k = T_d \frac{d}{|V_2 - V_2^0|} + T_f + T_m \qquad (4.32)$$

where V_2 is the controlled voltage, V_2^0 is the reference voltage, d is half the LTC deadband (defined below), T_d is the maximum time delay of the inverse-time characteristic, T_f is the fixed intentional time delay, and T_m is the mechanical time necessary to perform the tap change, as discussed above.

The tap changing logic at time instant t_k is the following:

$$r_{k+1} = \begin{cases} r_k + \Delta r & \text{if} & V_2 > V_2^0 + d & \text{and} & r_k < r^{max} \\ r_k - \Delta r & \text{if} & V_2 < V_2^0 - d & \text{and} & r_k > r^{min} \\ r_k & \text{otherwise} \end{cases} \qquad (4.33)$$

where r^{max}, r^{min} are the upper and lower tap limits.

The LTC is activated and the counter k is set to zero at each time t_o the voltage error increases beyond the deadband limits:

$$k = 0 \quad \text{if} \quad |V_2(t_o^+) - V_2^0| > d + \epsilon \quad \text{and} \quad |V_2(t_o^-) - V_2^0| \leq d + \epsilon$$

where ϵ is an (optional) hysteresis term. Using the hysteresis term, the effective deadband is larger for the first tap movement, thus the LTC becomes more "reluctant" to initiate a sequence of tap changes. Note that the hysteresis term ϵ does not enter into expression (4.33), so the LTC will bring the error back within the proper deadband (if limits are not met first).

We can distinguish between two modes of LTC operation depending on whether each tap movement is considered independently or in sequence [SP94]. The *sequential* mode of operation consists of a sequence of tap changes starting after an initial time delay ΔT_o (either fixed, or constant) and continuing at constant time intervals until the error is brought back inside the deadband, or until the tap limits are reached.

Table 4.4 Examples of LTC settings

Utility	Mode	T_m (s)	T_{do} (s)	T_{fo} (s)	T_d (s)	T_f (s)	d (%)
EU1	seq.	5	0	25 (EHV/HV)	0	5	1.5
				55 (HV/MV)			2
EU2	non-seq.	5	–	–	140	10	1
NA1	seq.	4	50	0	0	0	5/8
NA2	seq.	8	0	20	0	0	5/8

The general formula for the first tap time delay is similar to (4.32):

$$\Delta T_o = T_{do} \frac{d}{|V_2 - V_2^0|} + T_{fo} + T_m \qquad (4.34)$$

where T_{do}, T_{fo} are the values of the inverse-time and the fixed intentional delay respectively, for the first tap step. Subsequent taps are performed at constant time intervals corresponding to (4.32) with $T_d = 0$

In the *non-sequential mode* of operations the LTC makes no distinction between first and subsequent taps. Time starts to count when either the error exceeds the deadband limits (plus the optional hysteresis), or after a tap movement is performed. Thus all time delays are given by the same formula (4.32).

In Table 4.4 we give typical examples for LTC settings as used by European (EU1, EU2) and North American (NA1, NA2) utilities. Note that utility EU1 has two levels of LTCs, namely at Extra High Voltage - High Voltage (EHV/HV) and High Voltage - Medium Voltage (HV/MV) transformers. The reason for choosing longer time delay for the downstream HV/MV level will be discussed in Section 4.4.4.

Continuous LTC model

The continuous LTC model is based on the assumption of a continuously changing tap $r(t)$, which can take all real values between r^{min} and r^{max}. Usually the effect of the deadband is neglected in a continuous LTC model, so that the following differential equation results:

$$T_c \dot{r} = V_2 - V_2^0 \qquad r^{min} \leq r \leq r^{max} \qquad (4.35)$$

Note that when using (4.35) the LTC is modelled as an *integral controller*. The continuous LTC model is less accurate than the discrete ones, but is a useful approximation, particularly convenient for analytical purposes. Its use in time simulation is limited.

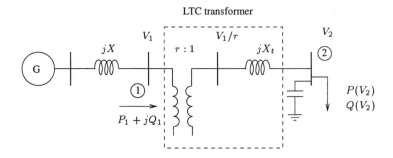

Figure 4.16   Generator-line-LTC system

The model (4.35) is a continuous approximation of a non-sequential LTC with $T_f = T_m = 0$, in which case the time constant T_c has been derived in [SP94] as:

$$T_c = \frac{T_d\, d}{\Delta r} \tag{4.36}$$

In practice one could assume a slightly larger time constant to compensate for the inevitable mechanical time delay T_m.

4.4.3   Load restoration through LTC

The load restoration performed by LTCs is indirect: when the LTC succeeds to restore the distribution side voltage V_2 close to its reference value V_2^0, the load power, which in general depends on bus voltage, is also restored.

The analysis of LTC dynamics is facilitated by the fact that the LTC is a slowly acting device. We may, therefore, substitute generators and induction motors by their steady-state equations. In this way, the LTC dynamics are the only ones to be considered.

We will illustrate the load restoration through LTC operation using the simple system shown in Fig. 4.16, which consists of a generator feeding an LTC transformer through a transmission line. We have shown the LTC in this figure as an ideal transformer in series with a leakage reactance. The copper losses in the transformer windings have been neglected for simplicity. Two network $P_1 V_1$ characteristics relating the transmission side voltage V_1 to the power P_1 absorbed by the transformer are drawn with a solid line in Fig. 4.17. The two characteristics are drawn for different values of the line impedance X. Note that the network characteristic has to be drawn for the P_1, Q_1 pairs corresponding to the load connected to bus 2.

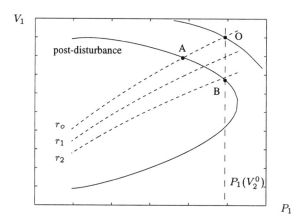

Figure 4.17 PV curves of generator-line-LTC system

Consider now the *load* characteristic as seen by the LTC primary side. We assume a general load-voltage relationship giving the load power as a function of load voltage:

$$P = P(V_2)$$
$$Q = Q(V_2)$$

The load-side voltage V_2 is linked to V_1 through the following equation:

$$\left(\frac{V_1}{r}\right)^2 = \left[V_2 + \left(\frac{Q(V_2)}{V_2} - BV_2\right)X_t\right]^2 + \left[\frac{P(V_2)X_t}{V_2}\right]^2 \qquad (4.37)$$

The active and reactive power P_1, Q_1 absorbed by the ideal transformer of Fig. 4.16 is made up of load power, plus reactive losses in the transformer leakage reactance X_t, minus reactive compensation and is also a function of V_2:

$$P_1 = P(V_2) \qquad (4.38a)$$
$$Q_1 = Q(V_2) + \frac{P(V_2)^2 + Q(V_2)^2}{V_2^2}X_t - BV_2^2 \qquad (4.38b)$$

Thus, we can in principle eliminate V_2 using (4.37) to obtain P_1, Q_1 as functions of only V_1/r:

$$P_1 = P(V_1/r)$$
$$Q_1 = Q(V_1/r)$$

This gives the *transient* load characteristic (as seen from the LTC) corresponding to a particular value of the tap ratio r. If r changes, so does the transient load characteristic. Three transient load characteristics for different values of the tap ratio r ($r_o > r_1 > r_2$) are plotted in Fig. 4.17 with dotted lines.

A different load characteristic can be derived when V_2 is restored to its reference value V_2^0, in which case the load will consume a constant amount of real and reactive power given by (4.38a,b) with V_2 substituted by its setpoint V_2^0. Since this value is independent of V_1, the load power is shown in the graph of Fig. 4.17 as a vertical dashed line. This is the *steady-state* load characteristic as seen by the LTC primary side. Note that both real and reactive powers are constant in the steady-state load characteristic, whereas the primary voltage V_1 changes with the variable tap r, so as to restore the secondary voltage.

To illustrate the LTC operation, let us assume that the system is initially at point O in Fig. 4.17, when a disturbance (e.g. an impedance increase) forces the network characteristic to the post-disturbance one. The primary voltage V_1 will initially drop along the transient LTC load characteristic for the given value of $r = r_o$ from point O to point A. At this point the power consumed by the load is less than that corresponding to V_2^0, meaning that $V_2 < V_2^0$. Since V_2 is lower than the reference, the LTC will react by decreasing the tap ratio according to (4.33), so as to boost the secondary side voltage. This will change the transient load characteristic, and the operating point will move along the post-disturbance network characteristic, until a new operating condition is reached close to point B, where the steady-state load characteristic intersects the network characteristic. Note that during this operation the LTC is restoring both secondary voltage and load power. This constitutes the *stable* LTC operation. Cases of unstable LTC operation will be discussed in Chapters 7 and 8.

4.4.4   Multiple LTC levels

As discussed in Section 2.7, in many power systems the transformers connecting different transmission voltage levels are equipped with LTC mechanisms. It was shown that this practice helps to increase the transfer capability of the transmission system by breaking up the total impedance between generation and load.

In this section we investigate the dynamic interactions between cascaded levels of tap-changing transformers. A typical case is the interaction between LTCs at the EHV/HV interconnection and those at the bulk distribution (HV/MV) substations. Consider the simple radial system of Fig. 4.18. When the upstream EHV/HV transformer ratio r_1 is reduced, both the controlled high voltage V_H and the medium voltage V_M are normally

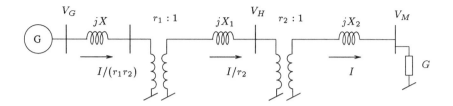

Figure 4.18 Radial system with two LTC transformers

Table 4.5 Parameters for the two LTC system

Case	Fast upstream tapping		Oscillatory response	
level	EHV/HV	HV/MV	EHV/HV	HV/MV
$T_{fo} + T_m$ (s)	20	50	30	45
$T_f + T_m$ (s)	5	10	10	10
tap step (%)	0.625	0.625	0.625	1.25

raised. However, when the downstream HV/MV transformer ratio r_2 is reduced in order to boost V_M, the HV transmission voltage V_H is also reduced.

Due to the above reasons the usual practice in coordinating the two LTC levels is to make the EHV/HV level faster than the HV/MV one [Lac79]. This will result in general in more effective and fewer tap operations. Intentional time delays for the first step of the HV/MV level are typically 20–40 seconds larger than those of the EHV/HV level. The same rule applies in case of several downstream LTC levels: they should become slower as they approach the end consumer.

One particular dynamic aspect of cascaded LTC levels is that they tend to be oscillatory [Lac79, HH93]. In general any cascade of load restoration mechanisms can produce an oscillatory response when they operate in the same time frame [VVC95]. In the particular case of the two LTC levels of Fig. 4.18 oscillations may arise, when the EHV/HV level fails to bring the load side (MV) voltage within its deadband before the HV/MV LTC starts to act.

We illustrate the above principle with an example based on the system of Fig. 4.18. We assume first that we conform to the aforementioned rule of making the EHV/HV LTC level faster by choosing the settings shown in the second column of Table 4.5 marked "Fast upstream tapping". The disturbance simulated is a 20% increase of the

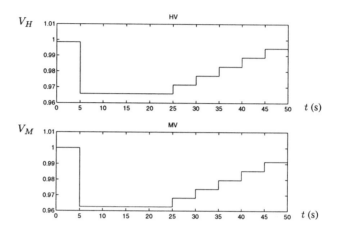

Figure 4.19   Response of the two LTC system with fast r_1

load conductance G at time $t = 5$ sec. As seen in Fig. 4.19, the EHV/HV LTC is able to restore both voltages inside their corresponding deadbands (assumed here between 0.99 and 1.01 pu) with only 5 taps, i.e. before the downstream LTC is activated.

Oscillatory behaviour in the form of voltage overshoot is experienced when using the settings of the third column of Table 4.5 marked "Oscillatory response". Note that due to the larger tap step the speed of the downstream LTC, after the initial time-delay, is effectively double that of the upstream.

As seen in Fig. 4.20, the upstream LTC is no longer able to restore voltages before the downstream LTC is activated at time $t = 50$ sec. Thus the LTCs are racing each other from this point on until the MV side voltage V_M is brought inside its deadband at time $t = 60$ sec. After that, r_1 takes another 3 steps before the HV side voltage V_H is back in its deadband. However, at that time, V_M is brought above its own deadband, so r_2 has to tap back reducing V_M (and increasing V_H) as seen in Fig. 4.20.

The overall settling time for the same disturbance is almost double that of the previous, well-tuned case. Also more tap changes were required: 6 for r_1 (all in one direction, reducing tap) and 3 for r_2 (two down and one up). Finally, the overshoot shown in the response of V_M is undesirable in power system operation. Note that the amount of overshoot depends on the disturbance initiating the LTC operation. Therefore, for large disturbances even well-tuned cascaded LTCs could exhibit oscillatory behaviour.

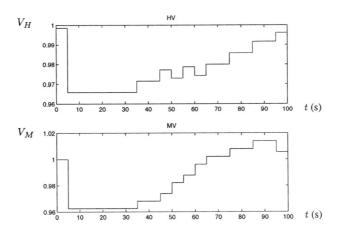

Figure 4.20   Response of the two LTC system with slow r_1

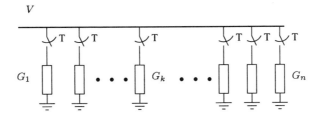

Figure 4.21   Thermostat controlled devices in parallel

4.5   THERMOSTATIC LOAD RECOVERY

As discussed in Section 4.2, the load dependence on voltage may be varying with time exhibiting a power restoration tendency. One typical category of such self-restoring load are the devices controlled by thermostats, which are used in all types of heating, i.e. water heating, space heating, industrial process heating, etc. The significance of thermostatic load recovery for voltage stability was brought forward by some relatively early publications [AFI82, Cla87, Gra88].

Let us consider n unity power factor, constant conductance devices, controlled by individual thermostats and connected in parallel, as in Fig. 4.21. Each switch determines the load cycle of the device, typically lasting for several minutes, such that the mean

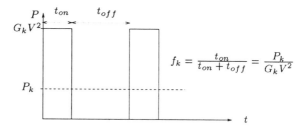

Figure 4.22   Thermostat duty cycle

power consumed during each cycle is equal to that necessary for keeping the required temperature under the given weather conditions. Denoting the required power of the kth component by P_k we can write:

$$P_k = f_k G_k V^2 \tag{4.39}$$

where $0 < f_k < 1$ is the *duty cycle* parameter, i.e. the percentage of on-time of the device during a cycle. The duty cycle is illustrated in the graph of Fig. 4.22.

The above equation (4.39) applies when the required power does not exceed that which can be achieved by the device under voltage V, i.e. under the condition:

$$P_k \leq G_k V^2 \tag{4.40}$$

It is possible that the above condition is violated after a significant voltage drop, in which case the device remains on-line continuously giving $f_k = 1$ (full duty cycle).

Note that the duty cycle parameter f_k defined above gives also the *probability* that the kth device is on-line at any specific point in time. Thus the average power consumed by all devices at time t is given by the transient load voltage characteristic:

$$P_t(V) = \sum_{k=1}^{n} f_k(t) G_k V^2 \tag{4.41}$$

Assuming an initial voltage V_o, such that (4.40) holds for all k, the total power drawn by all devices is:

$$P_o = \sum_{k=1}^{n} P_k = \sum_{k=1}^{n} f_{ko} G_k V_o^2 \tag{4.42}$$

where $f_{ko} < 1$ for all k.

A step reduction in supply voltage from V_o to V at time $t = 0$ will result in an immediate decrease of the power consumed by the devices that were on-line at the time of the disturbance. As time passes, however, the individual devices will stay on-line longer, as the duty cycle variables $f_k(t)$ will tend to increase until they satisfy (4.39) or, alternatively, until they reach their maximum value of 1. Consider the set:

$$\mathcal{L}(V) = \{k, \text{such that} \quad P_k > G_k V^2\}$$

This set contains the indices of the devices that are unable to restore their power demand for a supply voltage equal to V. The total power consumed on the average by all devices at steady state, i.e. after the duty cycles of all devices have been adapted to the new voltage, is:

$$P_s(V) = \sum_{k=1}^{n} f_k(\infty) G_k V^2 = \sum_{k \notin \mathcal{L}} P_k + V^2 \sum_{k \in \mathcal{L}} G_k \qquad (4.43)$$

Clearly this power is less than P_o if at least one device is unable to restore its power requirement. It is expected that as the voltage becomes lower, more devices are likely to remain on-line for all time, until for a large enough voltage drop all devices will remain on-line continuously thus giving a purely impedance load characteristic even in the steady state.

The dynamic response of aggregate thermostatic load can be formulated as an equivalent time varying conductance:

$$T_L \dot{G} = P_o/V^2 - G \qquad (4.44)$$

where T_L is the thermostatic load recovery time constant (usually in the order of magnitude of several minutes). To the above differential equation we have to add the limitation:

$$G \leq G_{max}(V)$$

where:

$$G_{max}(V) = \frac{P_s(V)}{V^2} = \sum_{k=1}^{n} f_k(\infty) G_k$$

depends on the amount of equipment on full duty cycle and thus depends on the voltage level V. The extreme value of $G_{max}(V)$ is achieved when all equipment is on full duty cycle, in which case G_{max} is the sum of all conductances. In general:

$$G_{max}(V) \leq \sum_{k=1}^{n} G_k$$

In the existing literature the dependence of G_{max} on V is usually neglected due to the lack of data. The total load power at any point of time is given by:

$$P = GV^2 \qquad\qquad (4.45)$$

A typical response of aggregate thermostatic load derived by simulation of 10,000 individual thermostatic loads is shown in [Gra88]. A more detailed model of thermostatic load dynamics involving two time constants can be found in [Kun94].

[Kar94] reports on a study showing that the supply voltage influences the behaviour of thermostat controllers. Indeed, older bimetallic thermostats are equipped with acceleration and compensation elements, which anticipate the changes in the room temperature. The heat produced by these elements is affected by the variations in supply voltage and hence the duty cycle of the thermostat is also influenced. For instance, it is observed that under reduced voltage, the heating devices are left longer under service which makes load restoration faster. This explains why observed time constants are smaller than what could be expected from equipment thermal inertia. Modern, electronic thermostats are less affected by such changes in the supply voltage.

4.6   GENERIC AGGREGATE LOAD MODELS

4.6.1   Load aggregation

The total load seen by a bulk power delivery transformer, like the one shown in Fig. 4.14, is a composition of a large number of individual loads (including customer wiring, transformers and capacitors) fed through medium voltage and low voltage distribution lines (primary and secondary feeders), voltage regulating transformers, distribution (MV/LV) transformers, switched capacitors, etc. [ITF93]. It also consists of components without restoration dynamics, as well as of components with load restoration at various time scales.

Although statistically the load of the same substation, at the same time of day, at the same season of the year, and for similar weather conditions, tends to be quite consistent, large variations are observed in load behaviour of different substations. Even in the same substation, the load behaves differently in different seasons. Load response may also vary with the weather conditions, between week-days and week-ends, or between day and night.

Thus the problem of modelling the aggregate load as such, is not easy to address. In this section we will review models that approximate the static and dynamic behaviour of aggregate loads. As in the rest of this chapter, only the voltage characteristics of loads are considered.

One technique often used for the construction of an aggregate load model is based on the assumption that the load of a particular substation consists of a mixture of components, which have more or less a given set of characteristics. In most cases the load of the substation is divided into percentages of commercial, residential, industrial, and agricultural load. The popularity of this method is due partly to the fact that it is relatively easy to obtain the required information from billing data, usually available with power utilities. One can then apply typical characteristics for each load class from measurements in other substations, or from the literature.

One note of caution is needed at this point, since the general characteristics of a residential load may be quite different from place to place, even in the same area, much more so in different countries. Also, the nature of the industrial, or commercial, use of electricity may be critical for the characteristic of the overall industrial, or commercial load. However, in the area of a single utility, and concentrating on the more sensitive buses, it is possible to specify quite accurately the characteristics of each of the above load classes. This was achieved for instance in [XVM97].

Specialized computer programs (such as LOADSYN [PWM88]) can be used to provide load parameters (including equivalent circuit parameters for aggregate induction motors) for a specific load bus, when the load composition, as well as the total real and reactive load at the bus are specified.

4.6.2   Generic models of self-restoring load

In the previous section we have seen that the aggregate behaviour of a number of thermostat controlled conductance loads has a transient voltage characteristic with an exponent of 2 (constant impedance), whereas after some time the steady-state voltage characteristic is closer to constant power.

Thermostat control is just one form of load self-restoration following a disturbance. Other controllers operating in the same time frame of one or more minutes have similar effects on aggregate load response. For instance distribution feeders with voltage regulators and voltage controlled capacitors, will have a load restoration effect. Also, consumer reaction following a disturbance may have a similar effect by manually switching on more devices to compensate for the reduced power supplied.

This behaviour can be captured by the so called *generic* models of self-restoring load, as has been suggested in the literature [Hil93, KH94, XM94]. The generic load models are usually associated with an exponential type of voltage characteristic. We will adopt this convention, although a polynomial or any other type of voltage characteristic is also applicable.

Two variants of the generic load model can be identified: the *multiplicative* model, in which the load state variable multiplies the transient load characteristic, and the *additive* model, in which the load state variable is added to the transient characteristic. In both cases the transient voltage characteristic is exponential with exponents α_t, β_t.

Multiplicative generic load model

The power consumed by the multiplicative generic load model is given by:

$$P \;=\; z_P P_o \left(\frac{V}{V_o}\right)^{\alpha_t} \tag{4.46a}$$

$$Q \;=\; z_Q Q_o \left(\frac{V}{V_o}\right)^{\beta_t} \tag{4.46b}$$

where z_P and z_Q are dimensionless state variables associated with load dynamics.

In *steady state* the voltage characteristic of the generic load model becomes:

$$P_s \;=\; P_o \left(\frac{V}{V_o}\right)^{\alpha_s} \tag{4.47a}$$

$$Q_s \;=\; Q_o \left(\frac{V}{V_o}\right)^{\beta_s} \tag{4.47b}$$

Usually the transient load exponents α_t, β_t have larger values than the steady-state ones α_s, β_s, so that the transient characteristic is more voltage sensitive. The generic model is thus restoring load power.

The load dynamics of the multiplicative model are given by the following differential equations:

$$T_P \dot{z}_P \;=\; \left(\frac{V}{V_o}\right)^{\alpha_s} - z_P \left(\frac{V}{V_o}\right)^{\alpha_t} \tag{4.48a}$$

$$T_Q \dot{z}_Q \;=\; \left(\frac{V}{V_o}\right)^{\beta_s} - z_Q \left(\frac{V}{V_o}\right)^{\beta_t} \tag{4.48b}$$

which force the transient characteristic towards the steady-state one with a time constant T_P (resp. T_Q) for the active (resp. reactive) load.

The multiplicative model is initialized with $z_P = z_Q = 1$, so as to achieve a steady state for $V = V_o$.

As in the case of the thermostatic load, limits should be imposed on the load state variables:

$$z_P^{min} \leq z_P \leq z_P^{max} \tag{4.49a}$$
$$z_Q^{min} \leq z_Q \leq z_Q^{max} \tag{4.49b}$$

When a voltage drop is experienced on the load bus, the load will initially respond with its transient characteristics (4.46a,b) and the power consumed will drop instantaneously. Following this the state variables z_P, z_Q will start to increase according to (4.48a,b) causing both real and reactive power to recover to their steady-state characteristics (4.47a,b). This process will end when either the steady-state characteristics are achieved, or when the state variable limits (4.49a,b) are encountered.

Additive generic load model

In the *additive* load model the transient load characteristic is written as:

$$P = P_o \left[\left(\frac{V}{V_o} \right)^{\alpha_t} + z_P \right] \tag{4.50a}$$

$$Q = Q_o \left[\left(\frac{V}{V_o} \right)^{\beta_t} + z_Q \right] \tag{4.50b}$$

where again z_p, z_Q are dimensionless load state variables.

Note that the additive load model introduces a constant power term in the transient load characteristic. As we will see in Chapter 7, such loads may introduce nonphysical singularity problems in the response of the system. In contrast, the transient characteristic of the multiplicative load model does not include a constant power component, and is thus more realistic. Reference [AAH98] shows a numerical simulation that cannot proceed when loads are represented with the additive model, while it provides a perfectly interpretable response when the multiplicative load model is used instead.

The steady-state load characteristics (4.47a,b) apply also to the additive load model, whereas the load dynamics for this model are described by:

$$T_P \dot{z}_P = -z_P + \left(\frac{V}{V_o} \right)^{\alpha_s} - \left(\frac{V}{V_o} \right)^{\alpha_t} \tag{4.51a}$$

$$T_Q \dot{z}_Q = -z_Q + \left(\frac{V}{V_o} \right)^{\beta_s} - \left(\frac{V}{V_o} \right)^{\beta_t} \tag{4.51b}$$

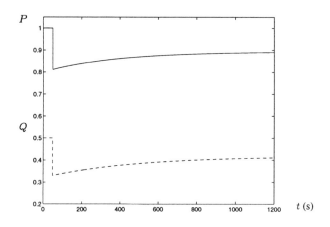

Figure 4.23   Generic load response

To initialize the additive load model for $V = V_o$, z_P and z_Q should be set to zero. As in the case of the multiplicative model, upper and lower limits on the state variables (4.49a,b) are in general imposed.

Figure 4.23 shows a typical response for a 10% voltage drop imposed on a load model with $\alpha_t = 1, \beta_t = 2, \alpha_s = 0.5, \beta_s = 0.8$ and time constants $T_P = T_Q = 300$ s. Note that the load restoration process described by the generic models presented in this section refers to the load behind the bulk power delivery transformers and is therefore in cascade with that performed by the LTCs. The steady-state load characteristics however incorporate the effect of downstream LTCs and other voltage regulating devices, as discussed earlier.

The transient and steady state load exponents, as well as the time constants T_P, T_Q can be identified from field tests performed at the main HV/MV substations. The process involves either an abrupt switching of one of two parallel transformers, or a more gradual voltage drop performed by changing the transformer tap [SSH77]. The exponents are determined as the normalized real and reactive power sensitivity to voltage, using formulae (4.3a,b). To give just one example, measurements performed by Hydro-Québec gave a transient real power exponent $\alpha_t = 1.4$ restoring to a steady-state one $\alpha_s = 0.6$. It should be pointed out, however, that measurements performed recently in another Canadian utility (BC Hydro) have found no load recovery, i.e. $\alpha_t = \alpha_s$ [XVM97].

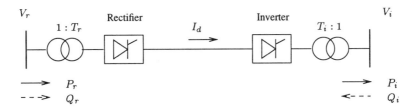

Figure 4.24 One-line diagram of a two-terminal HVDC link

4.7 HVDC LINKS

An HVDC link may connect two areas of a power system providing a parallel path for power transfer, or it may connect asynchronously two separate AC power systems. From the point of view of voltage stability analysis HVDC converter terminals can be regarded as loads of a special type [PSH92]. In this section we briefly discuss the behaviour of an HVDC link considering its active and reactive power characteristics as seen from the AC system.

A one-line diagram of a two-terminal HVDC link is shown in Fig. 4.24. By the subscript r we specify the rectifier end, where the active power flows from the AC to the DC system, and by the subscript i we denote the inverter end. Note that reactive power is absorbed at both ends of the DC interconnection. This power is provided by capacitors, harmonic filters (which are capacitive at fundamental frequency) and/or synchronous condensers.

A detailed description of converter dynamics, their modelling and the corresponding controls can be found in [Kun94]. The available control means include the rectifier and inverter gate control (which can be considered instantaneous) and the converter transformer tap changers, which usually operate in multiples of 5 seconds. A slow-acting power loop may also be present, adjusting the current setpoint. The effects of DC links on voltage stability are analyzed in [Tay94].

In normal operation the rectifier adjusts the firing angle so that the DC current I_d follows the setpoint I_{ord}, while the inverter controls the extinction angle (usually to its minimum value). The tap changer at the rectifier end is used to maintain the firing angle within the regulating range, whereas the tap changer at the inverter end is used for DC voltage control.

Under these conditions the active power at both ends has initially an almost constant current characteristic. Using an exponential model to represent this load, this means an exponent $\alpha_t \simeq 1$. At steady state, due to the DC voltage control and the outer loop power control, the active power restores to a constant value ($\alpha_s = 0$). A generic load model would thus be for the active power:

$$P \;=\; z\,P_o\,\left(\frac{V}{V_o}\right) \tag{4.52a}$$

$$T\,\dot{z} \;=\; 1 - z\,\left(\frac{V}{V_o}\right) \tag{4.52b}$$

where T is in the order of 5 to 15 seconds. Note that at the inverter terminal P_o is negative.

The response of reactive power is quite complex, because the power factor depends on both the AC voltage magnitude and the direct current. For instance, the power factor at the inverter end is given by:

$$\cos\phi_i = \cos\gamma - \frac{k\,I_d}{V_i}$$

where γ is the extinction angle, normally held constant to its minimum value in order to decrease the reactive power consumption. Note that the power factor decreases for reduced AC voltage and also for increased direct current.

Thus the reactive power Q can be represented as a nonlinear function of both P and V, which increases with active power and decreases with voltage.

To avoid the negative impact of increased reactive power consumption for reduced voltage, in voltage stability limited systems the HVDC links have preventive controls, such as a Voltage Dependent Current Order Limiter (VDCOL). This system decreases the current setpoint I_{ord} when it senses an AC voltage magnitude below some threshold (typically 0.9 to 0.95 pu).

The exact modelling of HVDC depends on the particular implementation and control design adopted. Apart from the above mentioned schemes there are various other possibilities of control coordination [CAD92]. Also, various auxiliary controls, such as damping signals or power modulation to increase transient stability, may interact with voltage stability and should be taken into account in detailed studies.

4.8 PROBLEMS

4.1. Draw the PV and QV steady-state characteristics for the motor whose parameters are listed in the first row of Table 4.3 for constant mechanical torque, and then for quadratic mechanical torque.

Hint: Since T_m is a known function of slip, you can use (4.19) to solve for V (for given slip). Then P, Q can be derived (for the same slip) using (4.21a,b)

4.2. Investigate the loadability limit of a three-phase induction motor with quadratic mechanical torque in the following way: substituting in the equilibrium condition (4.23) the electrical and the mechanical torque from (4.19) and (4.26) respectively derive an expression for the independent loading parameter T_2. A loadability limit is an extremum value of T_2, the necessary condition for which is:

$$\frac{dT_2}{ds} = 0$$

Compute the loadability limits using a symbolic processor to differentiate the expression for T_2 you have derived. Take any of the motors in the first 4 rows of Table 4.3 as a numerical example

4.3. In the system of Fig. 4.16 the load at the secondary transformer bus is made up of 50% constant impedance (for $V_2 = 1$), the other half being a 3-phase induction motor with a quadratic mechanical torque characteristic ($T_2 = 0.8$) and the parameters of the first row of Table 4.3. The reactive power of the impedance load is supposed to be compensated by a matching capacitor. Thus, the reactive power of the aggregate load is the one consumed by the motor. The generator operates under constant excitation voltage of 1.4 pu and the total reactance between the EMF and the LTC is 1.2 pu. The leakage reactance of the transformer is 0.15 pu on the transformer base. Draw the $P_1 V_1$ network and load characteristics for various values of the tap ratio r.

Hint: Use the motor slip as an independent variable. Then calculate the motor voltage from the equilibrium equation, the active load power, and V_1/r. The network characteristic has to be drawn with the P, Q pairs of the composite load derived as indicated.

4.4. Use a polynomial fit to produce approximate polynomial load characteristic for an induction motor, at the normal operating region and during starting or stalling. Apply to all motor models given in Figs. 4.6, 4.8, and 4.12 using data from Table 4.3.

PART II

INSTABILITY MECHANISMS AND ANALYSIS METHODS

5

MATHEMATICAL BACKGROUND

"There is no royal road in geometry"
Euclid

This chapter provides an overview of some mathematical concepts not always covered in electrical engineering curricula. Our main goal is to give the readers an overview of nonlinear system dynamics, a perspective that will prove useful when we embark on a more detailed analysis of complex power system voltage stability problems. In our exposition we assume some prior knowledge of linear algebra and linear systems theory. Readers who are already familiar with the theory of nonlinear systems may choose to skip this chapter. Those interested in a detailed mathematical analysis of nonlinear dynamics are urged to consult specialized textbooks, such as [HK92, GH83, Wig90, Sey88].

5.1 DIFFERENTIAL EQUATIONS (QUALITATIVE THEORY)

For the most part of this chapter we discuss properties of systems described by nonlinear Ordinary Differential Equations (ODEs). It is common engineering knowledge that most practical systems of this type are difficult, impractical, or even impossible to solve analytically. On the other hand, in most cases, systems of nonlinear ODEs are easily integrated numerically, once an initial condition is provided. The *qualitative* or *geometric* theory of ODEs introduced in 1881 by Henri Poincaré [Poi81] investigates the general properties of ODE solutions without resorting to an explicit integration.

5.1.1 Existence and uniqueness of solutions

Most engineering systems involving dynamics can be analyzed using a set of n ordinary differential equations, which are usually written in the compact form:

$$\dot{\mathbf{x}} = \mathbf{f}(\mathbf{x}) \tag{5.1}$$

where $\mathbf{x}$ is a $n \times 1$ vector and each f_i, $(i = 1, \ldots, n)$ is in general a nonlinear function of all x_i, $(i = 1, 2, \ldots, n)$.

We assume that the reader is familiar with the concept of the "state" of a system, which is defined by the n *state variables* forming the *state vector* $\mathbf{x}$. The state vector defines a point in *state space*, which is the space having the state variables as its coordinates.

The time response of a physical system is linked to a *solution* of the ODEs (5.1) for an *initial condition* giving the state vector at time $t = 0$:

$$\mathbf{x}(0) = \mathbf{x}_o \tag{5.2}$$

The initial condition (5.2) and the ODEs (5.1) constitute an *initial value problem*.

The solution $\mathbf{x}(t)$ for a given initial condition $\mathbf{x}_o$ can be pictured as a curve in the state space. We will call this curve a *trajectory* of the system passing through $\mathbf{x}_o$. The solution $\mathbf{x}(t)$ for $t > 0$ is called the *forward* trajectory and for $t < 0$ the *backward* trajectory.

Intuitively we expect that for each initial condition there exists a solution of the nonlinear system (5.1). The conditions under which this assertion is correct are outlined in the theorem of *existence and uniqueness* of solutions. Let us suppose that $\mathbf{f}(\mathbf{x})$ is defined in a domain U, which is a subset of R^n. According to the aforementioned theorem [HK92]:

1. If $\mathbf{f}$ is continuous in U, there exists a solution $\mathbf{x}(t)$ for all initial conditions $\mathbf{x}_o$ in U. Each solution is defined on a *maximal interval of existence* $I_{\mathbf{x}_o}$, which depends upon the initial condition:

$$I_{\mathbf{x}_o} : \alpha_{\mathbf{x}_o} < t < \beta_{\mathbf{x}_o} \tag{5.3}$$

Of course, $\alpha_{\mathbf{x}_o}$, or $\beta_{\mathbf{x}_o}$, or both may be infinite, in which case a solution exists for all positive and/or negative values of time.

2. When $\mathbf{f}$ is k times differentiable ($k \geq 1$), then the solution through $\mathbf{x}_o$ is *unique* and has k continuous derivatives. Therefore, the sufficient condition for the uniqueness of solutions is that $\mathbf{f}$ be *smooth*.

3. When the maximal interval of existence is finite, the limit points of the solution $\mathbf{x}(t)$ for $t \rightarrow \beta^-$ or $t \rightarrow \alpha^+$ either belong to the boundary of U, when U is bounded, or are infinite when U is unbounded.

In the rest of this chapter we will consider that $\mathbf{f}$ is smooth in U, so that we have a unique solution for all initial conditions. Note from (5.3) that the popular notion of a solution existing "forever" can be erroneous even when $\mathbf{f}$ is smooth.

Example

To illustrate the meaning of the maximal interval of existence, consider the differential equation, taken (with a sign modification) from [HK92]:

$$\dot{x} = -x^2 \tag{5.4}$$

The system defined in (5.4) has the rare advantage of possessing an explicit time solution, given by:

$$x = 0 \qquad \text{if} \quad x_o = 0 \tag{5.5}$$
$$x = \frac{1}{t + 1/x_o} \qquad \text{if} \quad x_o \neq 0 \tag{5.6}$$

where x_o is the initial condition for $t = 0$. Typical responses for positive and negative initial conditions, are plotted in Fig. 5.1. As can be seen, when the initial condition x_o is positive, the state variable x will eventually reduce to zero. However, for a negative initial condition, the system will "collapse" before reaching the critical time β_{x_o}, which is given by:

$$\beta_{x_o} = -1/x_o \qquad x_o < 0 \tag{5.7}$$

As seen in Fig. 5.1, the "lifetime" of an ODE solution can be limited. Each solution starting from a negative initial condition has a finite upper bound of its interval of existence depending on the initial condition.

On the other hand, when the initial condition is positive, the solutions exist for all time $t > 0$. Note that in this case the backward trajectory (i.e. for $t < 0$) has a finite lower bound α_{x_o} of its interval of existence given by:

$$\alpha_{x_o} = -1/x_o \qquad x_o > 0 \tag{5.8}$$

In more dramatic terms, the trajectories with $x_o < 0$ are doomed to collapse at some specific future point in time, whereas those with $x_o > 0$ originate from a "big bang" that occurred at some point of time in the past.

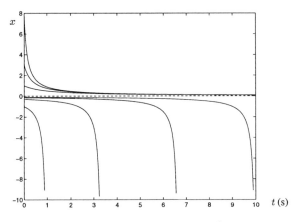

Figure 5.1   Time solutions of $\dot{x} = -x^2$

5.1.2   Equilibria and their stability

In this section we discuss the stability properties of *equilibrium points* (or simply *equilibria*) of the system defined by equation (5.1). The equilibrium points (if any) are given by the solutions $\mathbf{x}^*$ of the algebraic equations:

$$\mathbf{f}(\mathbf{x}) = 0 \tag{5.9}$$

An equilibrium point $\mathbf{x}^*$ is a particular solution of the ODE (5.1), since for $\mathbf{x}_o = \mathbf{x}^*$ one gets $\mathbf{x}(t) = \mathbf{x}^*$ for all time.

Stability

An equilibrium point $\mathbf{x}^*$ is called *stable* if all solutions with an initial condition close to $\mathbf{x}^*$ remain near $\mathbf{x}^*$ for all time. The following definition of stability is due to Liapunov [Lia66] and is illustrated graphically in Fig. 5.2:

> An equilibrium point $\mathbf{x}^*$ is *stable* if for every neighbourhood V of $\mathbf{x}^*$ we can find a neighborhood V_1 of $\mathbf{x}^*$ such that for all $\mathbf{x}_o \in V_1$ the solution $\mathbf{x}(t)$ exists and lies in V for all time $t > 0$.

Furthermore, we say that an equilibrium is *asymptotically stable* when all trajectories with $\mathbf{x}_o \in V_1$ approach $\mathbf{x}^*$ as $t \to \infty$. An equilibrium that is *not* stable is called *unstable*.

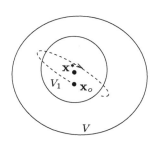

Figure 5.2   Definition of stability

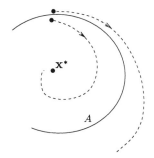

Figure 5.3   Region of attraction

As seen in the above definition, stability refers to an equilibrium point and is, therefore, a *local* property. If one is interested in the *global* behavior of (5.1), i.e. its solutions for all initial conditions, one question to be answered is which trajectories will be attracted by a stable equilibrium and which trajectories will diverge away from it. The largest set A, such that all trajectories with initial conditions $x_o \in A$ will approach eventually an asymptotically stable equilibrium x^* is known as the *region of attraction* (also called *domain*, or *basin* of attraction) of x^*. The region of attraction of an asymptotically stable equilibrium is illustrated graphically in Fig. 5.3.

Let us consider now a linear system of ODEs:

$$\dot{x} = Ax \tag{5.10}$$

It is well known that the linear system (5.10) has only one equilibrium point ($x^* = 0$), the stability of which is determined by the eigenvalues of the *state matrix* A. If all eigenvalues have negative real parts, the equilibrium is asymptotically stable. If at least one eigenvalue has a positive real part, the equilibrium is unstable. Note that in linear systems the region of attraction of an asymptotically stable equilibrium is the whole state space: all initial conditions give trajectories approaching the origin.

In contrast to linear systems, in a nonlinear system one should always be aware of the following facts:

- the number of equilibria varies. A system may have one, more than one, or no equilibria;

- the region of attraction of a stable equilibrium may be limited, therefore the existence of a stable equilibrium is not sufficient to guarantee stability.

In most cases we can determine the stability of a nonlinear system equilibrium by examining the linearized system *around* an equilibrium point x^*. Defining:

$$\Delta x = x - x^* \tag{5.11}$$

and keeping only the first-order term of the Taylor series expansion of f around x^*, one arrives at the following linear system:

$$\Delta \dot{x} = A \Delta x \tag{5.12}$$

where the state matrix A is defined as the *Jacobian* of f with respect to x evaluated at x^*:

$$A = \left. \frac{\partial f}{\partial x} \right|_{x=x^*} = f_x(x^*) \tag{5.13}$$

A is also called the *state Jacobian*. The stability of the equilibrium point x^* is determined by that of the linearized system (5.12). More specifically:

- if all the eigenvalues of f_x have negative real parts, the equilibrium x^* is asymptotically stable;

- if at least one eigenvalue of the linearized system state matrix has a positive real part, the equilibrium x^* is unstable.

Types of equilibria

Asymptotically stable equilibria are called *sinks*, or *stable nodes*. If all eigenvalues have positive real parts the unstable equilibrium is called *source* or *unstable node*. If some eigenvalues have positive real parts and all others have negative real parts, the unstable equilibrium is called a *saddle*.

Note that the linearization provides no information on the stability of an equilibrium, for which the corresponding Jacobian f_x has one or more eigenvalues with zero real parts. For instance, consider again the first-order system (5.4) which we introduced in the previous section. This system has only one equilibrium point at the origin ($x^* = 0$), around which the linearization results in a zero Jacobian:

$$\Delta \dot{x} = -2x^* \Delta x = 0 \tag{5.14}$$

The stability of this equilibrium cannot be determined by linearization, so we have to rely on the definition of stability: since in any neighbourhood of $x^* = 0$ there are initial conditions ($x_o < 0$), for which the forward trajectory is unbounded, the equilibrium is unstable. This particular type of equilibrium point is called a *saddle-node*.

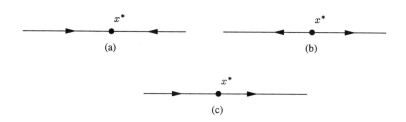

Figure 5.4 a) stable node, b) unstable node, c) saddle-node

For a first-order system, the three types of equilibrium points are shown in Fig. 5.4. These are the stable node, the unstable node and the saddle-node. The arrows in the figure show the direction of trajectories.

In multivariable systems a saddle-node is characterized by a state Jacobian, which has a zero eigenvalue. Near the saddle-node there exists a direction in state space, along which trajectories behave as shown in Fig. 5.4c, approaching the equilibrium from the one side, and diverging on the other. Note that when $\mathbf{f_x}$ has a zero eigenvalue its determinant is also zero.

Another, special type of equilibrium is the *center*. A center is an equilibrium point, the linearized model of which is characterized by a pair of complex conjugate eigenvalues with zero real part. We will give an example of a center in the next subsection.

If one's goal were to study a system for just one set of conditions, parameter values, etc., then it would be extremely unlikely to encounter strange equilibria, such as saddle-nodes, or centers. However, in power systems engineering one has often to examine a whole range of operating points, some of which may be stable, while others are unstable. It is therefore inevitable that a special case of an equilibrium, such as the saddle-node, exists in between stable and unstable cases. We will return to this notion when discussing bifurcations in Section 5.3.

5.1.3 Invariant Manifolds

We have seen in the previous section that the linearized system (5.12) can provide useful information concerning the stability of the equilibrium points of the nonlinear system (5.1). We will now show how the eigenvectors of the linearized system can be generalized in the nonlinear case.

The invariance property of eigenvectors

It is well known that the response of the linear system (5.10) for any initial condition can be expressed in terms of the right ($\mathbf{v}$) and left ($\mathbf{w}$) eigenvectors of the state matrix $\mathbf{A}$. These eigenvectors satisfy the following relations:

$$\mathbf{A}\mathbf{v}_i \;=\; \lambda_i\mathbf{v}_i \qquad i = 1,\dots,n \tag{5.15}$$

$$\mathbf{w}_i^T\mathbf{A} \;=\; \lambda_i\mathbf{w}_i^T \qquad i = 1,\dots,n \tag{5.16}$$

For simplicity, we will assume that the linearized system has n distinct eigenvalues, in which case the left and right eigenvectors for different eigenvalues are orthogonal to each other:

$$\mathbf{w}_i^T\mathbf{v}_j = 0 \qquad i \neq j \tag{5.17}$$

The response of the linear system for an initial condition $\mathbf{x}_o$ is given by:

$$\mathbf{x}(t) = \sum_{i=1}^{n} e^{\lambda_i t}\mathbf{v}_i\mathbf{w}_i^T\mathbf{x}_o \tag{5.18}$$

Consider now an initial condition $\mathbf{x}_o$ collinear with a right eigenvector $\mathbf{v}_i$, i.e.

$$\mathbf{x}_o = a\mathbf{v}_i \tag{5.19}$$

Substituting in (5.18) and making use of (5.17) we get:

$$\mathbf{x}(t) = ae^{\lambda_i t}(\mathbf{w}_i^T\mathbf{v}_i)\mathbf{v}_i = b\mathbf{v}_i \tag{5.20}$$

Equation (5.20) demonstrates the invariance property of a right eigenvector: "once on an eigenvector, always on the eigenvector". Moreover, if λ_i has a negative real part, the trajectory $\mathbf{x}(t)$ given by (5.20) approaches the equilibrium point at the origin, since $e^{\lambda_i t}$ tends to zero as $t \to \infty$. Similarly, if λ_i has a *positive* real part, the backward trajectory $\mathbf{x}(t)$ originates from the equilibrium point (for $t \to -\infty$).

Note that all subspaces spanned by a number of (right) eigenvectors have the same invariance property. For instance consider an initial point on the plane spanned by two eigenvectors $\mathbf{v}_i, \mathbf{v}_j$:

$$\mathbf{x}_o = a\mathbf{v}_i + b\mathbf{v}_j \tag{5.21}$$

The trajectory starting at $\mathbf{x}_o$ is:

$$\mathbf{x}(t) = \left[ae^{\lambda_i t}(\mathbf{w}_i^T\mathbf{v}_i)\right]\mathbf{v}_i + \left[be^{\lambda_j t}(\mathbf{w}_j^T\mathbf{v}_j)\right]\mathbf{v}_j = c\mathbf{v}_i + d\mathbf{v}_j \tag{5.22}$$

Therefore, it lies entirely on the plane defined by $\mathbf{v}_i, \mathbf{v}_j$.

In particular, the subspace spanned by all the eigenvectors corresponding to eigen-values with *negative* real parts forms the *stable eigenspace* of the linearized system. All trajectories on the stable eigenspace approach the origin as $t \rightarrow \infty$. The sub-space spanned by all the eigenvectors corresponding to eigenvalues with *positive* real parts forms the *unstable eigenspace*. All trajectories on the unstable eigenspace orig-inate from the equilibrium point. Finally, the subspace spanned by the eigenvectors corresponding to eigenvalues with *zero* real parts forms the *center eigenspace* [GH83].

From eigenspaces to manifolds

In a nonlinear system the invariance property of eigenvectors and eigenspaces is transferred to *invariant manifolds* of the same dimensions. By the term "manifold" we mean a smooth curved line, surface, or hypersurface, without self-intersections, or other singular points. An invariant manifold is a subset of state space, such that any trajectory starting on the manifold will remain on it for all time.

The stable, unstable and center eigenspaces of the linearized system around an equilib-rium point $\mathbf{x}^*$ correspond to the *stable*, *unstable* and *center* manifolds of the nonlinear system. Just as in the case of the stable eigenspace, all trajectories with initial con-ditions on the stable manifold remain on the manifold for all time, and approach the equilibrium point $\mathbf{x}^*$ as $t \rightarrow \infty$. Similarly, all backward trajectories on the *unstable* manifold originate from $\mathbf{x}^*$, i.e. they approach $\mathbf{x}^*$ as $t \rightarrow -\infty$. Near the equilibrium point the *local* stable, unstable and center manifolds are tangent to the corresponding eigenspaces of the linearized system.

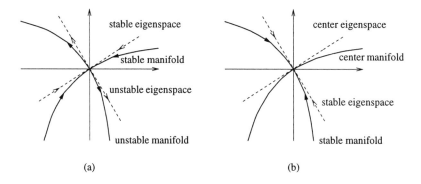

Figure 5.5   Local stable, unstable, and center manifolds

In Fig. 5.5a,b, we illustrate the stable and unstable manifolds (with solid curves) and the corresponding right eigenvectors (with dashed lines) of two second-order systems. Both systems have an equilibrium point at the origin. The equilibrium in Fig. 5.5a is a saddle with a Jacobian having one positive and one negative real eigenvalue, and the system in Fig. 5.5b has a Jacobian with one zero eigenvalue and a negative eigenvalue. Note that we are not allowed to assign the direction of trajectories on the center manifold without knowing the nonlinear characteristics of the system.

Figure 5.5 provides only a local view of stable, unstable and center manifolds and eigenspaces. A global perspective of stable and unstable manifolds may reveal further interesting properties, as the following example illustrates.

Example: The "fish"

Consider the second-order system (which could correspond to the classical model of a single generator - infinite bus system with zero damping):

$$\dot{x}_1 = a^2 x_2 \qquad a > 0 \qquad (5.23a)$$
$$\dot{x}_2 = 0.5 - \sin x_1 \qquad (5.23b)$$

This system has two equilibria:

$$\mathbf{x}^{(1)} = [\pi/6 \quad 0]^T$$
$$\mathbf{x}^{(2)} = [5\pi/6 \quad 0]^T$$

The Jacobian of this system is:

$$\mathbf{f_x} = \begin{bmatrix} 0 & a^2 \\ -\cos x_1^* & 0 \end{bmatrix}$$

At $\mathbf{x}^{(1)}$ the eigenvalues of the linearized system are purely imaginary:

$$\lambda_{1,2}^{(1)} = \pm j\, 0.931a$$

therefore this equilibrium point is a center and its stability cannot be decided by linearization. The other equilibrium point at $\mathbf{x}^{(2)}$ is a saddle with one positive and one negative eigenvalue:

$$\lambda_{1,2}^{(2)} = \pm 0.931a$$

In Fig. 5.6 we have plotted the trajectories of the system (5.23a,b) for a few initial conditions. Such a set of trajectories is usually called a *phase portrait*. It is easy to check the orientation of trajectories shown in Fig. 5.6: due to (5.23a) the trajectories at

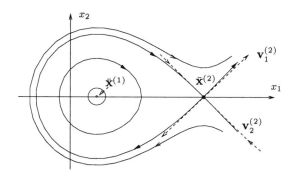

Figure 5.6 Phase portrait of "the fish"

the upper half plane move to the right and at the lower half plane they move to the left. Following [HK92] the system exhibiting the phase portrait of Fig. 5.6 will be referred to as "the fish".

It is easy to verify that all trajectories near the center $\mathbf{x}^{(1)}$ are periodic oscillations with constant amplitude. Going back to the definition of stability that was given in the previous section we can verify easily that this equilibrium is stable but not asymptotically stable.

Let us return now to the unstable equilibrium point $\mathbf{x}^{(2)}$. The eigenvector of $\lambda_1^{(2)} = +0.931a$ forms the unstable eigenspace of the saddle point and is given by:

$$\mathbf{v}_1^{(2)} = [\ a \quad 0.931\ ]^T$$

As seen in Fig. 5.6, $\mathbf{v}_1^{(2)}$ is tangent to the local unstable manifold. The stable eigenspace is the eigenvector of the stable eigenvalue $\lambda_2^{(2)} = -0.931a$, which is:

$$\mathbf{v}_2^{(2)} = [\ a \quad -0.931\ ]^T$$

and it is tangent to the local stable manifold.

One important feature of the undamped system (5.23a,b), shown in Fig. 5.6 is that one branch of the stable manifold of the unstable equilibrium coincides with one branch of the unstable manifold of the same equilibrium forming what is known as a *homoclinic loop*. This property is just one example of the various exotic behaviours that can be encountered when analyzing special cases of nonlinear systems.

5.1.4 Limit cycles and their stability

In this subsection we discuss another aspect of nonlinear systems: the existence of periodic solutions of (5.1). A periodic solution is a function $\mathbf{x}(t)$ satisfying (5.1) and having the property:

$$\mathbf{x}(t + T) = \mathbf{x}(t) \tag{5.24}$$

for all t. The *period* of the periodic solution is the smallest number T for which (5.24) holds.

Consider a trajectory starting at a point $\mathbf{x}_o$ lying on a periodic solution. According to (5.24) at time $t = T$ the trajectory will pass again through $\mathbf{x}_o$ and will retrace its course from there on. Therefore, a periodic solution forms a closed curve in n-dimensional space.

A limit cycle is an *isolated* periodic solution of the system (5.1). Note that the periodic solutions shown in Fig. 5.6 surrounding the stable center $\mathbf{x}^{(1)}$ are not limit cycles, because they are not isolated: there is an infinite number of periodic solutions near each one of them.

The stability definition for a limit cycle is analogous to that of an equilibrium:

- a limit cycle is asymptotically stable, if all trajectories starting near the limit cycle approach the limit cycle with increasing time;

- an unstable limit cycle is one, for which there are trajectories starting near the limit cycle that diverge away.

We will now give two examples of nonlinear systems exhibiting limit cycles.

Example: stable limit cycle

This example is a slight modification of the Van der Pol oscillator [GH83]. It is a second-order system, whose equations are:

$$\dot{x}_1 = x_2 \tag{5.25a}$$
$$\dot{x}_2 = 10(-x_1 + x_2 - x_2^3) \tag{5.25b}$$

The only equilibrium point of this system is the origin $\mathbf{x}^* = 0$. Linearizing around this equilibrium we obtain:

$$\Delta\dot{x}_1 = \Delta x_2 \tag{5.26a}$$

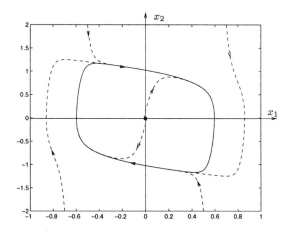

Figure 5.7 Stable Van der Pol oscillator

$$\Delta \dot{x}_2 = -10\Delta x_1 + 10\Delta x_2 \qquad (5.26b)$$

The eigenvalues of the state matrix are:

$$\lambda_1 = +8.87 \qquad \lambda_2 = +1.13$$

and therefore the equilibrium is an unstable node.

It can be easily verified using numerical simulation that all trajectories starting near the unstable equilibrium eventually end up in a periodic oscillation, shown in the state space of Fig. 5.7 as a closed curve. Since the periodic solution is isolated, it is a limit cycle. As seen in the figure all trajectories originating either inside, or outside the limit cycle approach the periodic solution. Thus the limit cycle is *asymptotically stable*.

Example: unstable limit cycle

Let us now modify the equations (5.25a,b) of the stable oscillator as follows:

$$\dot{x}_1 = 10x_2 \qquad (5.27a)$$
$$\dot{x}_2 = -x_1 - x_2 + x_2^3 \qquad (5.27b)$$

The system (5.27a,b) has again a single equilibrium point at the origin with the following eigenvalues:

$$\lambda_{1,2} = -0.5 \pm j3.12$$

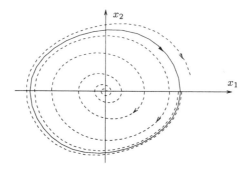

Figure 5.8   Unstable limit cycle

As seen from the eigenvalue real parts, this equilibrium point is now *stable*. The trajectories starting from initial conditions on the positive x_1 axis, up to a critical point $\tilde{x}_1 = 3.685$, spiral inward towards the stable equilibrium, while those starting beyond this critical point diverge. Two trajectories starting near the critical point, one converging to the equilibrium and one diverging are shown in Fig. 5.8 with dashed lines. In between these two trajectories we can identify an *unstable* limit cycle shown in Fig. 5.8 with a solid line.

Since this is a second-order system, the limit cycle separates the state space into two components. Note that in this case, the region of attraction of the stable equilibrium is the interior of the limit cycle. The boundary of the region of attraction is the limit cycle itself.

Similar to the equilibrium points, limit cycles have invariant manifolds. These however, are difficult to visualize in higher dimensions. The stable manifold of a limit cycle is the largest set, such that all initial conditions in the set result in trajectories approaching the limit cycle as $t \to \infty$, and the unstable manifold of a limit cycle is the largest set, such that all backward trajectories passing through its points originate from the limit cycle as $t \to -\infty$.

It is easy to see that for second-order systems the stable and unstable manifolds of limit cycles are trivial: the stable manifold of the stable limit cycle of Fig. 5.7 is the whole state space; the stable manifold of the unstable limit cycle of Fig. 5.8 is the limit cycle itself and all the remaining state space is its unstable manifold.

5.1.5 Region of Attraction

In Fig. 5.3 we have introduced the concept of the region of attraction of an asymptotically stable equilibrium $\mathbf{x}^*$. The region of attraction is the largest subset $A(\mathbf{x}^*)$ of state space for which:

$$\mathbf{x}_o \in A(\mathbf{x}^*) \iff \lim_{t \to \infty} \mathbf{x}(t) = \mathbf{x}^* \tag{5.28}$$

It follows directly from the above definition that if a point belongs to the region of attraction A, its backward trajectory is also part of A.

The region of attraction is not necessarily bounded. When it exists, the boundary of the region of attraction consists of parts of the stable manifolds of unstable equilibrium points and unstable limit cycles. We illustrate the significance of stable manifolds of unstable equilibria and limit cycles using two of our previous examples of second-order systems.

In the case of the "fish" (Fig. 5.6) the stable equilibrium point $\mathbf{x}^{(1)}$ is not asymptotically stable. It is easy to see that the branch of the stable manifold of the unstable equilibrium $\mathbf{x}^{(2)}$ encircling $\mathbf{x}^{(1)}$ (homoclinic loop) is separating the state space into two regions: one region of oscillatory stability inside the homoclinic loop, and one region of diverging trajectories outside the loop.

Let us now modify this system by adding some damping in (5.23b):

$$\dot{x}_1 = a^2 x_2 \tag{5.29a}$$
$$\dot{x}_2 = 0.5 - 0.111 x_2 - \sin x_1 \tag{5.29b}$$

The equilibrium points remain the same, since at equilibrium $x_2 = 0$.

The state Jacobian of the damped system (5.29a,b) is:

$$\mathbf{f_x} = \begin{bmatrix} 0 & a^2 \\ -\cos x_1^* & -0.111 \end{bmatrix}$$

Choosing for example $a^2 = 314.16$, at the equilibrium $x_1^{(1)} = \pi/6$ the eigenvalues of the linearized system are:

$$\lambda_{1,2}^{(1)} = -0.0555 \pm j16.5$$

Therefore, this is an asymptotically stable equilibrium. The other equilibrium point at $x_1^{(2)} = 5\pi/6$ is again a saddle, with one positive and one negative eigenvalue:

$$\lambda_1^{(2)} = +16.44 \quad \lambda_2^{(2)} = -16.55$$

The stable and unstable manifolds of the unstable equilibrium point of the damped system (5.29a,b) are shown in Fig. 5.9.

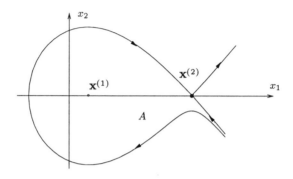

Figure 5.9   Region of attraction of damped system

The homoclinic loop is now broken. Only one branch of the local unstable manifold of the equilibrium $\mathbf{x}^{(2)}$ is shown, the other branch consisting of a spiral converging slowly to the stable equilibrium. The boundary of the region of attraction A of the asymptotically stable equilibrium $\mathbf{x}^{(1)}$ is formed by both branches of the stable manifold of $\mathbf{x}^{(2)}$. Note that along the direction, where the two branches of the stable manifold approach each other asymptotically, the region of attraction is unbounded. Inside the region of attraction A the trajectories spiral inwards towards $\mathbf{x}^{(1)}$.

Let us now consider again the system (5.27a,b), whose phase portrait was shown in Fig. 5.8. As discussed before, the region of attraction of the only equilibrium point $\mathbf{x}^* = 0$ is bounded by the unstable limit cycle. In a higher dimensional system with a similar structure, the boundary of the region of attraction consists of the *stable manifold* of the unstable limit cycle close to the equilibrium. This is hard to illustrate with a two dimensional picture, therefore we have to rely upon the reader's imagination.

One note is due concerning the dimensions of stable manifolds in n-dimensional state space. Unstable equilibrium points with only one positive eigenvalue (often called *type 1*) have a stable manifold whose dimension is $n - 1$. Locally the stable manifold is tangent to the eigenspace spanned by the eigenvectors of the stable $n - 1$ eigenvalues. Similarly, a type-1 unstable limit cycle has a stable manifold of dimension $n - 1$. Note that due to the uniqueness of solutions of a smooth system, stable manifolds of distinct equilibria or limit cycles cannot intersect.

5.2 BIFURCATIONS

In this section we explain in simple terms the concept of bifurcation and its importance for nonlinear systems, trying to answer questions legitimately raised by many
practicing engineers. As a starting point, let us state that bifurcation theory deals with
one key aspect of nonlinear systems: the emergence of sudden changes in system
response arising from smooth, continuous parameter variations [Arn86]. Admittedly,
this description brings to mind a typical scenario of voltage collapse.

5.2.1 What is a bifurcation?

In this section we will consider families of smooth ODEs of the form:

$$\dot{x} = f(x, p) \tag{5.30}$$

where x is a $n \times 1$ state vector and p is a $k \times 1$ parameter vector. For every value of
p the equilibrium points of the system (5.30) are given by the solutions of:

$$f(x^*, p) = 0 \tag{5.31}$$

The above equation defines the k-dimensional *equilibrium manifold* in the $(n + k)$-
dimensional space of states and parameters. Consider an equilibrium point $x^{(1)}$
corresponding to the parameter values p_o, and assume that the Jacobian of f with
respect to x is nonsingular at this point:

$$\det f_x(x^{(1)}, p_o) \neq 0 \tag{5.32}$$

By the implicit function theorem, there exists a unique smooth function:

$$x^* = g^{(1)}(p) \tag{5.33}$$

with $x^{(1)} = g^{(1)}(p_o)$, giving a branch of equilibrium points of (5.30) as a function of
p.

Consider now that for the same value p_o there is another equilibrium point $x^{(2)}$, i.e. a
second solution of (5.31), for which also the Jacobian $f_x(x^{(2)}, p_o)$ is nonsingular. By
the implicit function theorem, we have a second function:

$$x^* = g^{(2)}(p) \tag{5.34}$$

with $x^{(2)} = g^{(2)}(p_o)$, giving another branch of equilibrium points of (5.30) as a
function of p.

The term "bifurcation" originates from the concept of different branches of equilibrium points intersecting each other, and thus "bifurcating". In such bifurcation points the Jacobian $\mathbf{f_x}$ is singular, and consequently the implicit function theorem cannot be applied.

Let us illustrate the above concept with an example. Consider the first-order system:

$$\dot{x} = x^2 - 2x + 1.1 - \mu \tag{5.35}$$

where μ is a scalar parameter ($\mathbf{p} = \mu$ and $k = 1$). In Fig. 5.10 we plot the two branches of equilibrium points $g^{(1)}(\mu)$ and $g^{(2)}(\mu)$ of this system in the state-parameter space, i.e. on the plane (μ, x). The two branches intersect at the bifurcation point B ($\mu = 0.1, x^* = 1$), where $\partial f / \partial x = 0$.

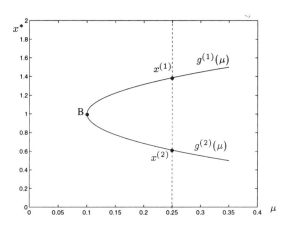

Figure 5.10   Bifurcation of equilibria

Generalizing the concept of bifurcating equilibrium branches discussed above, we say that a *bifurcation* occurs at any point in parameter space, for which the *qualitative structure* of the system (5.30) changes for a small variation of the parameter vector $\mathbf{p}$. We will not attempt a rigorous definition of the qualitative structure of a system, which is linked to concepts such as structural stability [PSL95a, PSL95b]. We will just give below a nonexclusive list of properties, a change in any of which constitutes a bifurcation:

1. number of equilibrium points;
2. number of limit cycles;

3. stability of equilibrium points or limit cycles;
4. period of periodic solutions.

In this book we will discuss only the first 3 types of qualitative structure changes and the corresponding bifurcations. Since these deal with local properties such as the stability of equilibria and periodic orbits, these bifurcations are *local* ones. The far more complex behaviour encountered in global bifurcations will not be analyzed in this book, although such phenomena have been reported in the power system literature [LA93].

Furthermore, we will restrict ourselves to the bifurcations that are expected to be encountered in single parameter families of ODEs. These are the Saddle-Node Bifurcation (SNB) and the Poincaré-Andronov-Hopf bifurcation which is customarily referred to by the last name, i.e. as Hopf Bifurcation (HB).

Other, more complex, bifurcations are not *generic* in single-parameter families, meaning that they can occur only as isolated exceptions. However, bifurcations more complex than those discussed here become generic in the multiparameter case.

5.2.2 Saddle-node bifurcation

Consider a single-parameter family of ODEs:

$$\dot{\mathbf{x}} = \mathbf{f}(\mathbf{x}, \mu) \tag{5.36}$$

with the following equilibrium condition:

$$\mathbf{f}(\mathbf{x}^*, \mu) = 0 \tag{5.37}$$

An SNB is a point where two branches of equilibria meet, such as point B in the example of Fig. 5.10. At the bifurcation the equilibrium becomes a saddle-node, hence the name of the bifurcation. As discussed above, at this point the state Jacobian $\mathbf{f_x}$ has to be singular. Therefore, the necessary conditions for an SNB are given by the equilibrium equations (5.37) together with the following singularity condition:

$$\det \mathbf{f_x}(\mathbf{x}^*) = 0 \tag{5.38}$$

These are $n + 1$ equations in $n + 1$ variables ($\mathbf{x}$ and μ).

Not all points satisfying these necessary conditions are SNB points. To illustrate the nature of sufficient conditions we consider a scalar system. For such a system the sufficient SNB conditions are:

$$f(x^*, \mu) = 0 \tag{5.39a}$$

$$\frac{\partial f}{\partial x} = 0 \tag{5.39b}$$

$$\frac{\partial f}{\partial \mu} \neq 0 \tag{5.39c}$$

$$\frac{\partial^2 f}{\partial x^2} \neq 0 \tag{5.39d}$$

Conditions (5.39c,d) are usually called *transversality* conditions. The first one (5.39c) guarantees that there exists a smooth local function $\mu = h(x)$ at the bifurcation point (μ_o, x_o^*). In geometrical terms this means that the equilibrium manifold (5.39a) intersects the line $x = x_o^*$ *transversally*. The last condition (5.39d) implies that the equilibrium manifold remains locally on the one side of the line $\mu = \mu_o$.

Note that (5.39a-d) are also the sufficient conditions for an extremum (either maximum, or minimum) of μ subject to the constraint (5.39a).

Figure 5.11a,b,c illustrates three cases of first-order, single parameter systems, for which the condition $\partial f/\partial x = 0$ is satisfied at $\mu_o = x_o^* = 0$:

$$\dot{x} = x^2 + \mu \tag{5.40a}$$

$$\dot{x} = x^3 + \mu \tag{5.40b}$$

$$\dot{x} = x^2 + \mu^3 \tag{5.40c}$$

Of these systems only the first one, shown in Fig. 5.11a, has a saddle-node bifurcation at the origin. The second system, shown in Fig. 5.11b, violates the second derivative condition (5.39d), whereas the third system, shown in Fig. 5.11c, violates the first transversality condition (5.39c). Note that the violation of either one of conditions (5.39c,d) requires one additional equation (total of $n+2$ equations in $n+1$ unknowns). Hence, in single parameter families, the equilibria satisfying the necessary condition (5.39b) will in general satisfy also (5.39c,d) and will thus be SNB points.

Let us now examine more closely the system (5.40a), the one that does have an SNB for $\mu = 0$. As seen in Fig. 5.11a, this system has two equilibrium points for $\mu < 0$, a single equilibrium for $\mu = 0$ and no equilibrium points for $\mu > 0$. The equilibrium at $\mu = 0$ is a saddle-node similar to the one we have analyzed in Fig. 5.1, but with a different direction of time, i.e. the trajectories for negative initial conditions will converge to the equilibrium point $x^* = 0$, whereas the trajectories with positive initial conditions x_o will explode in finite time $\beta_{x_o} = 1/x_o$.

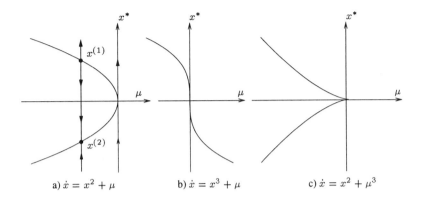

Figure 5.11 a) Sufficient conditions for SNB; b) violation of (5.39b); c) violation of (5.39a)

Consider now the equilibrium points for $\mu < 0$. Setting $\mu = -a^2$, with $a > 0$ the equilibrium points are:

$$x^{(1)} = a \qquad x^{(2)} = -a$$

The Jacobian of (5.40a) is:

$$\frac{\partial f}{\partial x} = 2x^*$$

Therefore $x^{(1)}$ is an unstable equilibrium and $x^{(2)}$ is stable. This illustrates that for a first-order system:

> at a saddle-node bifurcation two equilibrium points, one stable and one unstable, coalesce and disappear.

From another viewpoint, if the parameter μ is larger than the bifurcation value μ_o and decreases slowly, at the SNB point two equilibria, one stable and one unstable, emerge simultaneously.

This property is generalized for multivariable systems as follows:

> at a saddle-node bifurcation two equilibrium points coalesce and disappear (or emerge simultaneously). One of the equilibrium points has a real positive and the other a real negative eigenvalue, both becoming zero at the bifurcation.

If all other eigenvalues of the multivariable system (except the one becoming zero at the bifurcation) have negative real parts, one of the equilibria coalescing at the SNB is stable and the other one unstable.

The direction of trajectories approaching the stable equilibrium $x^{(2)}$, or departing from the unstable one $x^{(1)}$ is shown with arrows in Fig. 5.11a. Trajectories with initial conditions $x_o < a$ will converge to the stable equilibrium point $x^{(2)} = -a$. Trajectories with initial conditions $x_o > a$ will explode in finite time. Therefore, the region of attraction of the stable equilibrium point is bounded by the unstable equilibrium (its stable manifold in the case of multivariable systems). Note that the region of attraction shrinks as the SNB is approached.

Consider now the equilibrium point at the origin, which is a saddle-node. The stable manifold of the saddle-node is the negative x-axis, and its unstable manifold is the positive x-axis. The whole x-axis forms the center manifold of the saddle-node, which in this scalar system is the whole state-space.

Finally, it can be verified, either analytically, or by numerical integration that for $\mu > 0$ all trajectories, whatever the initial condition explode in finite time. No equilibrium points exist in this case.

As mentioned before, in single parameter families of ODEs the necessary conditions (5.37,5.38) will in general yield SNB points. In the multiparameter case, where $\mathbf{p}$ is a $k \times 1$ vector, the points in the $(n + k)$-dimensional state and parameter space which satisfy the necessary conditions for an SNB form a $(k - 1)$-dimensional manifold. The points violating the sufficient conditions form $(k - 2)$-dimensional submanifolds lying on this manifold. Note that when the parameter vector moves along a given curve, e.g. when the k parameters depend upon a scalar μ:

$$\mathbf{p} = \mathbf{p}(\mu)$$

the problem reduces to a single parameter one and the points satisfying the necessary conditions are in general saddle-node bifurcations.

Geometrically, at a SNB point the manifold of equilibrium points "folds" with respect to the parameter space, as the curve of Fig. 5.11a folds with respect to the μ axis. The projection of the SNB points onto the k-dimensional parameter space is a hypersurface of dimension $k - 1$, which we call *bifurcation surface*. The bifurcation surface forms a boundary of the *feasibility region* [VSZ91], i.e. the region in parameter space for which equilibrium points exist. Going through a saddle-node bifurcation the number of projection points of the equilibrium manifold on the parameter space changes by two. In Fig. 5.11a the feasibility region is the negative μ axis, bounded by the point $\mu = 0$.

In multivariable systems, a saddle-node bifurcation is an equilibrium with a simple zero eigenvalue satisfying transversality conditions [Dob94]. The center manifold of the saddle-node is a curve in the n-dimensional state space tangent to the right eigenvector of the zero eigenvalue. The center manifold is made up of a stable and an unstable manifold separated by the equilibrium point, as in Fig.5.4c.

5.2.3 Hopf bifurcation

We have seen up to now that an SNB is characterized by a zero eigenvalue and that the response of the system at a saddle-node is monotonic. We will now discuss the emergence of oscillatory instability. It is well known that a stable equilibrium point can become unstable following a parameter variation that forces a pair of complex eigenvalues to cross the imaginary axis in the complex plane. This type of oscillatory instability is associated in nonlinear systems with the Hopf Bifurcation (HB) mentioned in subsection 5.2.1.

In a saddle-node bifurcation the region of attraction of a stable equilibrium shrinks due to an approaching unstable equilibrium and stability is eventually lost when the two equilibria coalesce and disappear. At a Hopf bifurcation the stability of an equilibrium is lost through its interaction with a limit cycle. There are two types of Hopf bifurcation depending on the nature of this interaction:

subcritical HB: an unstable limit cycle, existing prior to the bifurcation, shrinks and eventually disappears as it coalesces with a stable equilibrium point at the bifurcation. After the bifurcation, the equilibrium point becomes unstable resulting in growing oscillations;

supercritical HB: a stable limit cycle is generated at the bifurcation, and a stable equilibrium point becomes unstable with increasing amplitude oscillations, which are eventually attracted by the stable limit cycle.

The necessary condition for a Hopf bifurcation is the existence of an equilibrium with purely imaginary eigenvalues. This condition is not as easy to establish as the zero eigenvalue condition, which is simply that of a vanishing determinant. Most of the equilibria with purely imaginary eigenvalues will be Hopf bifurcation points, but, similarly to the SNB case, there may exist exceptional cases, for which the real part of the critical eigenvalue pair does not change sign after going to zero: these points are not HB points.

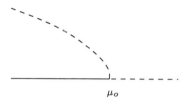

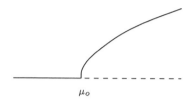

μ_o

μ_o

Figure 5.12a   Subcritical Hopf bifurcation   **Figure 5.12b**   Supercritical HB

In Fig. 5.12a,b the two types of Hopf bifurcation are illustrated graphically. The straight line corresponds to an equilibrium point and the curved line represents the amplitude of the limit cycle. Solid lines correspond to stable equilibria or limit cycles, whereas the dashed lines indicate unstable equilibria or limit cycles. In each figure the abscissa is the parameter value μ, and μ_o is the bifurcation value. The ordinate represents the amplitude of the limit cycle

In Figs. 5.12a the amplitude of the unstable limit cycle is seen to diminish as the parameter approaches the bifurcation value. At the bifurcation point the limit cycle disappears and the equilibrium point becomes unstable. The region of attraction of the stable equilibrium prior to the bifurcation is bounded by the stable manifold of the unstable limit cycle. Trajectories after the bifurcation are unbounded, with oscillations of increasing amplitude. This is sometimes called "hard loss of stability" [Arn86]. The bifurcation is called subcritical, because the branch of limit cycles emanating at the bifurcation is directed to the left, i.e. it exists for smaller values of the parameter.

The evolution of the system response is quite different in the case of the supercritical Hopf bifurcation illustrated in Fig. 5.12b. Before the bifurcation there is no limit cycle bounding the region of attraction of the stable equilibrium. A limit cycle is generated at the bifurcation point and it is stable. After the bifurcation the trajectories starting near the unstable equilibrium are attracted by the stable limit cycle and the oscillations are bounded. This has been called a "soft loss of stability" [Arn86]. From a global viewpoint, this bifurcation makes little change initially. Indeed, prior to the bifurcation the trajectories converge to the stable equilibrium with lightly damped, sustained oscillations; immediately after the bifurcation the same trajectories are attracted by a stable, small amplitude limit cycle. Note that from an engineering viewpoint both cases (just before, or just after the bifurcation) are usually unacceptable.

5.3 DIFFERENTIAL-ALGEBRAIC SYSTEMS

In this section we analyze systems described by a set of differential equations, which include a number of algebraic variables and are subject to a set of algebraic constraints. Of particular interest to the subject of this book are the differential-algebraic systems that result from the time-scale decomposition that will be introduced in the next section.

5.3.1   Equilibrium points and stability

The Differential-Algebraic (D-A) systems are described by of a set of n differential and m algebraic equations, which are assumed smooth throughout this analysis:

$$\dot{\mathbf{x}} = \mathbf{f}(\mathbf{x}, \mathbf{y}, \mathbf{p}) \tag{5.41a}$$
$$0 = \mathbf{g}(\mathbf{x}, \mathbf{y}, \mathbf{p}) \tag{5.41b}$$

where, $\mathbf{x}$ is the vector of n state variables, $\mathbf{y}$ is the vector of m algebraic variables, and $\mathbf{p}$ are the k parameter variables. The m algebraic equations (5.41b) define a manifold of dimension $n+k$, called *constraint manifold* [HH91], in the $(n+m+k)$-dimensional space of $\mathbf{x}, \mathbf{y}, \mathbf{p}$.

Systems in the form of (5.41a,b) have been considered "theoretically problematic" [HH89], since the system of m nonlinear algebraic equations (5.41b) may have singular points where it cannot be solved for the m dependent algebraic variables $\mathbf{y}$. At these points the response of the system cannot be defined, which contradicts engineering intuition claiming that "any physical system must have a time solution" [PSL95b].

Differential-algebraic systems are analyzed using the implicit function theorem. Consider a point $\mathbf{x}, \mathbf{y}, \mathbf{p}$ for which the algebraic Jacobian $\mathbf{g_y}(\mathbf{x}, \mathbf{y}, \mathbf{p})$ is nonsingular. According to the implicit function theorem, there exists a locally unique, smooth function $\mathbf{F}$ of the form:

$$\dot{\mathbf{x}} = \mathbf{F}(\mathbf{x}, \mathbf{p}) \tag{5.42}$$

from which the algebraic variables have been eliminated. Since $\mathbf{F}$ can be defined and is smooth at all points where $\mathbf{g_y}$ is nonsingular, we know from the existence theorem of Section 5.1.1 that there exists a unique time solution of the differential-algebraic system (5.41a,b) for all these points. We denote the domain of $\mathbf{F}$ in state space for a given value of the parameter vector $\mathbf{p}$ as $U_{\mathbf{p}}$. The domain $U_{\mathbf{p}}$ is bounded by the points satisfying the singularity condition of $\mathbf{g_y}$. It may also be bounded by hard limits imposed on the state variables [JVS96].

For a fixed value of $\mathbf{p}$ the equilibrium points of (5.41a,b) are solutions of the system:

$$\mathbf{f}(\mathbf{x}, \mathbf{y}, \mathbf{p}) = 0 \qquad (5.43a)$$
$$\mathbf{g}(\mathbf{x}, \mathbf{y}, \mathbf{p}) = 0 \qquad (5.43b)$$

The stability of equilibrium points can be determined by linearizing (5.41a,b) around the equilibrium:

$$\begin{bmatrix} \Delta \dot{\mathbf{x}} \\ 0 \end{bmatrix} = \mathbf{J} \begin{bmatrix} \Delta \mathbf{x} \\ \Delta \mathbf{y} \end{bmatrix} \qquad (5.44)$$

where $\mathbf{J}$ is the *unreduced* Jacobian of the D-A system:

$$\mathbf{J} = \begin{bmatrix} \mathbf{f_x} & \mathbf{f_y} \\ \mathbf{g_x} & \mathbf{g_y} \end{bmatrix} \qquad (5.45)$$

Assuming $\mathbf{g_y}$ is nonsingular we can eliminate $\Delta \mathbf{y}$ from (5.44):

$$\Delta \dot{\mathbf{x}} = [\mathbf{f_x} - \mathbf{f_y} \mathbf{g_y}^{-1} \mathbf{g_x}] \Delta \mathbf{x} \qquad (5.46)$$

We can thus obtain the state matrix $\mathbf{A}$ of the linearized system:

$$\mathbf{A} = \mathbf{F_x} = [\mathbf{f_x} - \mathbf{f_y} \mathbf{g_y}^{-1} \mathbf{g_x}] \qquad (5.47)$$

which is the Schur complement [Cot74] of the algebraic equation Jacobian $\mathbf{g_y}$ in the unreduced Jacobian $\mathbf{J}$. For this reason, in the power system literature, $\mathbf{A}$ is often called the *reduced* Jacobian as opposed to the unreduced one.

The stability of an equilibrium point of the D-A system for a given value of $\mathbf{p}$ depends upon the eigenvalues of the state matrix $\mathbf{A}$. The stable and unstable manifolds of the equilibrium point lie on the constraint manifold, when seen in the space of state and algebraic variables. As $\mathbf{p}$ varies, the D-A system may experience bifurcations like a simple ODE system. The case of a SNB deserves special attention. Schur's well known determinant formula [Sch17] yields (for nonsingular $\mathbf{g_y}$):

$$\det \mathbf{J} = \det \mathbf{g_y} \det[\mathbf{f_x} - \mathbf{f_y} \mathbf{g_y}^{-1} \mathbf{g_x}] = \det \mathbf{g_y} \det \mathbf{A} \qquad (5.48)$$

We will use (5.48) at various places in later chapters. Its main implication is that the state matrix $\mathbf{A}$ becomes singular together with the unreduced Jacobian $\mathbf{J}$, when $\mathbf{g_y}$ is nonsingular. Therefore, a necessary condition for a saddle-node bifurcation of the D-A system is the singularity of the unreduced Jacobian $\mathbf{J}$.

5.3.2   Investigating algebraic singularities

As pointed out in the previous subsection, one complexity introduced by D-A systems is the existence of algebraic singularities. The implications of this are discussed below.

Impasse and singularity surfaces

The points on the constraint manifold, for which the algebraic equation Jacobian is singular are given by the simultaneous solution of the m algebraic constraint equations (5.41b) and the following scalar equation:

$$\det \mathbf{g_y}(\mathbf{x}, \mathbf{y}, \mathbf{p}) = 0 \qquad (5.49)$$

Equations (5.41b) and (5.49) form in general an $(n+k-1)$-dimensional surface, lying on the constraint manifold, since one more algebraic equation is added to the m initial ones. This hypersurface is called *impasse surface* [HH89, HH91], because it cannot be crossed by the trajectories of the system. The constraint manifold is decomposed by the impasse surface into different components called *causality regions*. Causality is linked to the existence of a local solution of (5.41b) for $\mathbf{y}$ [KPB86].

For given parameters $\mathbf{p}$, the projection of the impasse surface onto the state space, gives an $(n-1)$-dimensional surface S_p, consisting of algebraic equation singularity points. According to the above discussion this surface (if it exists) bounds the domain U_p, on which a differential equation in the form (5.42) can be defined for the given value of $\mathbf{p}$. The region $U_\mathbf{p}$ can shrink or expand depending on the variation of $\mathbf{p}$.

Geometrically, the impasse surface can be viewed as the collection of points, on which the constraint manifold "folds" with respect to the state space having two projections before $S_\mathbf{p}$ and none after. We will not discuss here higher order singularities, such as 'cusps' [Arn86].

Let us consider an initial point $(\mathbf{x}_o, \mathbf{y}_o, \mathbf{p}_o)$ on the constraint manifold, whose projection on the state space is $\mathbf{x}_o \in U_{\mathbf{p}_o}$. Applying the existence theorem of Section 5.1.1, it can be seen that the solutions of (5.42) having a finite upper bound on their interval of existence, either diverge to infinity, or end up on the boundary of $U_\mathbf{p}$, part of which is the singularity surface. Similarly, the trajectories with a a finite lower bound α on their interval of existence, are either unbounded as $t \rightarrow \alpha$, or emerge from the singularity surface. Thus the singularity surface contains points from which trajectories emerge, and points on which trajectories vanish at finite time.

Singularity induced bifurcation

A special type of bifurcation, existing only in D-A systems, occurs when the k-dimensional equilibrium manifold, which lies on the constraint manifold of a D-A system, intersects the $(n+k-1)$-dimensional impasse surface. The points in question satisfy both singularity (5.49), and equilibrium (5.43a,b) conditions. Note that such points exist in general, even in the single parameter case $(k = 1)$. Technically, these points are not equilibria, because the system cannot be defined on the impasse surface. However, equilibrium points may exist arbitrarily close to both sides of such a singularity.

Consider a family of equilibrium points approaching the impasse surface under a slow parameter variation. The determinant of the algebraic equation Jacobian g_y calculated at these points becomes progressively very small as the impasse surface is approached, and consequently the determinant of the reduced Jacobian A gets very large due to (5.48). Therefore, at least one of the state matrix eigenvalues tends to infinity. Similarly, on the other side of the impasse surface, the equilibrium points have also an eigenvalue tending to infinity, but with an opposite sign. We have therefore a change of the stability properties of the system, on the two sides of the singularity and this constitutes a bifurcation, that has been called *singularity induced bifurcation* [VSZ91, VSZ93].

Note that in D-A systems the determinant of the state matrix A can change sign, either when it becomes singular, having a zero eigenvalue, or when it has an infinite eigenvalue going from minus infinity to plus infinity.

Example

Let us illustrate the concept of the singularity induced bifurcation with an example. Consider the system:

$$\dot{x} = z - y\sin x \qquad\qquad (5.50a)$$
$$0 = -0.5z - y^2 + y\cos x \qquad\qquad (5.50b)$$

where z is a scalar parameter and $y > 0$. In this system $n = m = k = 1$. The state and parameter space has dimension $n + m + k = 3$. In this space the constraint manifold is a surface of dimension $n + k = 2$, the "impasse surface" is a curve $(n + k - 1 = 1)$ and the equilibrium manifold is another curve $(k = 1)$ lying also on the constraint manifold.

Figure 5.13 shows a view of the 3-dimensional x, y, z space, with z corresponding to the elevation. The 2-dimensional constraint manifold, defined by (5.50b), is represented through a collection of contours shown with solid lines for different values of z.

The impasse curve is defined by the singularity condition :

$$\frac{\partial g}{\partial y} = -2y + \cos x = 0 \qquad\qquad (5.51)$$

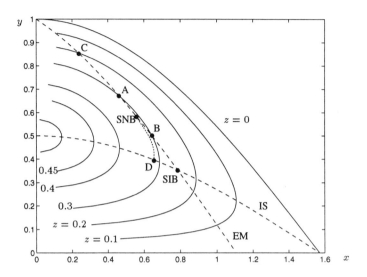

Figure 5.13   Singularity induced bifurcation

together with the constraint (5.50b). It is shown as a dashed curve marked IS in Fig. 5.13. The latter divides the constraint manifold in two causality regions. Note that each (solid) curve of the constraint manifold folds with respect to the x axis at its crossing point with the impasse curve.

The equilibrium manifold satisfies the condition:

$$z - y \sin x = 0 \tag{5.52}$$

together with (5.50b). Eliminating z from these two relationships, we obtain the curve:

$$y = \cos x - 0.5 \sin x \tag{5.53}$$

marked EM in Fig. 5.13. As seen in this figure, the equilibrium manifold intersects the impasse surface IS at the point SIB, which is a singularity induced bifurcation, with coordinates given by the simultaneous solution of equations, (5.50b,5.51,5.52):

$$x = \pi/4 \quad y = \sqrt{2}/4 \quad z = 0.25$$

Let us now follow the equilibrium curve as the parameter z varies slowly starting from 0. For $z = 0$ we have a single equilibrium $x = 0, y = 1$. This equilibrium is stable and globally attracting. As the parameter z increases slowly we move uphill along the dashed equilibrium curve, until we reach a point (for z slightly larger than 0.3), after which there is no equilibrium for increasing z. This is a saddle-node bifurcation point, at which a stable equilibrium (such as A) coalesces with an unstable equilibrium (such as B), which bounds its region of attraction. The descending path in the direction of B, is made up of unstable equilibria becoming more and more unstable with an eigenvalue approaching $+\infty$ at the SIB point,

In this case the singularity induced bifurcation is "protected" by a path of unstable equilibrium points. If the system reaches the saddle-node bifurcation by a gradual increase of z, like the one described above, the trajectory will depart from the equilibrium surface along the unstable manifold of the SNB and it will end up on an algebraic singularity at point D (which is not an equilibrium).

5.4   MULTIPLE TIME SCALES

5.4.1   Singular perturbation

Many systems have dynamics evolving in different time scales. Some are fast, others slow. In most cases it is not practical to handle all these dynamics combined in a single model. In deriving fast component models we usually consider that the slow states are practically constant during fast transients and in deriving slow component models we assume that the fast transients are not excited during slow changes. Such time-scale separation is many times performed intuitively: for instance, when deriving generator models in Chapter 3, it was considered that electromagnetic transients die out very fast after a disturbance, so that they need not be included in the model developed.

When a multi-time-scale model is available, one can derive accurate, reduced-order models suitable for each time scale. This process is called time-scale decomposition. The proper way to do this is based on the analysis known as *singular perturbation* [O'M74, KKO86].

A singularly perturbed system is one for which a small parameter ϵ multiplies one or more state derivatives. Therefore for $\epsilon = 0$ the order of the system changes. In its standard form a singularly perturbed system is written as:

$$\dot{\mathbf{x}} = \mathbf{f}(\mathbf{x}, \mathbf{y}, \epsilon) \tag{5.54a}$$

$$\epsilon\dot{\mathbf{y}} = \mathbf{g}(\mathbf{x}, \mathbf{y}, \epsilon) \tag{5.54b}$$

where $\mathbf{x}$ is an $n \times 1$ vector and $\mathbf{y}$ an $m \times 1$ vector. Although $\mathbf{f}$ and $\mathbf{g}$ may depend on ϵ we will not show this explicitly in the sequel to simplify notation. For a physical system there may be more than one way to derive a model in the standard form (5.54a,b).

The time-scale decomposition consists in deriving two reduced order subsystems, such that one describes the slow dynamics and the other the fast dynamics of (5.54a,b). We denote with $\mathbf{x}_s, \mathbf{y}_s$ the slow and by $\mathbf{x}_f, \mathbf{y}_f$ the fast components of the state variables, such that:

$$\mathbf{x} = \mathbf{x}_s + \mathbf{x}_f \qquad (5.55)$$

$$\mathbf{y} = \mathbf{y}_s + \mathbf{y}_f \qquad (5.56)$$

Due to the ϵ term, the dynamics of $\mathbf{y}$ are faster than those of $\mathbf{x}$. Thus an often acceptable, intuitively appealing approximation of the slow dynamics of (5.54a,b) consists of taking $\epsilon = 0$ in (5.54b). This defines the *quasi-steady-state* (QSS) approximation of the slow subsystem:

$$\dot{\mathbf{x}}_s = \mathbf{f}(\mathbf{x}_s, \mathbf{y}_s) \qquad (5.57a)$$

$$0 = \mathbf{g}(\mathbf{x}_s, \mathbf{y}_s) \qquad (5.57b)$$

For given $\mathbf{x}_s$ (5.57b) is obviously the equilibrium condition for $\mathbf{y}$, but since this is achieved by setting $\epsilon = 0$ we can still have $\dot{\mathbf{y}}_s \neq 0$, and $\mathbf{y}_s$ is allowed to change so that (5.57b) is satisfied as $\mathbf{x}_s$ varies during slow transients. The rate of change of $\mathbf{y}$ during slow, quasi-steady-state transients can in fact be calculated by differentiating (5.57b) using the chain rule and (5.54a).

The QSS slow subsystem (5.57a,b) is a differential-algebraic system which can be analyzed as described in the previous section. In particular, the stability of this system is characterized be the state (or reduced) matrix $\mathbf{A}$ given by (5.47). As pointed out before, the singularity of this matrix coincides with the singularity of the unreduced Jacobian (5.45).

We will discuss the fast subsystem after introducing the concept of the slow or integral manifold.

5.4.2 Slow manifold

The QSS approximation can be improved to any degree of accuracy by introducing the concept of the *integral manifold*, also called *slow manifold*.

We start with an example of a linear second-order system:

$$\dot{x} = y \qquad (5.58a)$$

$$\epsilon \dot{y} \;=\; -x - 1.1y \qquad\qquad (5.58b)$$

When ϵ is small this system has two negative eigenvalues corresponding to slow (λ_1), and fast (λ_2) dynamics, i.e.:

$$\lambda_2 \ll \lambda_1 < 0$$

Consider the right eigenvector $\mathbf{v}_1$ corresponding to the slow eigenvalue. Due to the invariance property of eigenvectors (5.20), only the slow transients are excited for initial conditions lying on $\mathbf{v}_1$. The trajectories starting outside this eigenvector will converge onto it with a fast transient and then they will slowly approach the equilibrium point remaining on $\mathbf{v}_1$. This is shown in Fig. 5.14a where various trajectories are plotted with dashed lines, and the two eigenvectors with solid lines. The value of ϵ used in this figure is 0.05, for which the eigenvalues are $\lambda_2 = -20.05$, $\lambda_1 = -0.95$.

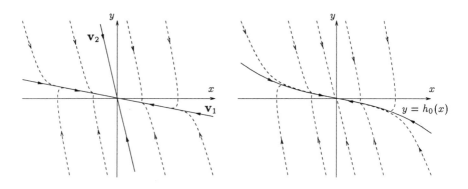

Figure 5.14a   Slow and fast eigenvectors   **Figure 5.14b**   Approximate slow manifold

In Section 5.1.3 we have seen that in a nonlinear system, eigenspaces are replaced by invariant manifolds. Thus, in a nonlinear system the slow eigenvector is replaced by a *slow manifold*. The latter is defined as an invariant manifold on which the fast dynamics are not excited, so that $\mathbf{x} = \mathbf{x}_s$, $\mathbf{y} = \mathbf{y}_s$. The slow manifold is n-dimensional (corresponding to the n slow variables) and can be defined by the m equations:

$$\mathbf{y}_s = \mathbf{h}(\mathbf{x}_s) \qquad\qquad (5.59)$$

For $\mathbf{x}_s, \mathbf{y}_s$ to be solutions of the original system (5.54a,b), the functions $\mathbf{h}$ defining the slow manifold must satisfy the following condition, which is obtained by substituting (5.59) and (5.54a) in (5.54b):

$$\epsilon \mathbf{h}_x \mathbf{f}(\mathbf{x}_s, \mathbf{h}) = \mathbf{g}(\mathbf{x}_s, \mathbf{h}) \qquad\qquad (5.60)$$

where $\mathbf{h_x}$ is the Jacobian of $\mathbf{h}$ with respect to $\mathbf{x}$. These partial differential equations cannot be solved analytically, except in some very special cases [SAZ84]. Thus, the slow manifold is usually approximated using a series expansion in ϵ.

For $\epsilon = 0$ we obtain the first term $\mathbf{h_0}$ of this expansion, for which (5.60) amounts to:

$$\mathbf{g}(\mathbf{x_s}, \mathbf{h_0}) = 0$$

That is, for $\epsilon \rightarrow 0$ the slow manifold is given by its QSS approximation (5.57b). Since the error made by this approximation tends to zero as $\epsilon \rightarrow 0$, we say that the approximation error is "order ϵ". In general we write $e = O(\epsilon)$, when there are constants α and A such that: $e \leq A\epsilon$ for $\epsilon < \alpha$.

We illustrate now the slow manifold concept and its approximation with an example. Consider the second-order nonlinear system:

$$\dot{x} = y \tag{5.61a}$$
$$\epsilon \dot{y} = -x - x^3 - 1.1y \tag{5.61b}$$

This system has only one equilibrium at the origin. The linearized system around this equilibrium is the one we discussed above (5.58a,b). The responses of the nonlinear system for various initial conditions are plotted with dashed lines in Fig. 5.14b. We also plot in this figure the approximate slow manifold corresponding to $\epsilon = 0$ in (5.61b):

$$y_s = h_0(x_s) = -(x_s + x_s^3)/1.1$$

As seen, this approximates closely the actual slow dynamics of the nonlinear, two-time-scale system. The accuracy of the approximation can be improved by using more terms of the Taylor series expansion of $\mathbf{h}$. Using this technique it is straightforward to calculate components of the slow manifold up to any power of ϵ using (5.60) [KKO86]. In this book we will use only the QSS approximation.

5.4.3 Slow and fast dynamics

Once the slow manifold $\mathbf{h}(\mathbf{x_s})$ is calculated to any desired accuracy, the slow dynamics are given by the reduced-order slow subsystem:

$$\dot{\mathbf{x}}_s = \mathbf{f}(\mathbf{x_s}, \mathbf{h}(\mathbf{x_s})) \tag{5.62}$$

and the slow components $\mathbf{y}_s$ of $\mathbf{y}$ are given by (5.59).

In order to approximate the fast subsystem we consider that $\mathbf{x}$ is predominately slow ($\mathbf{x} \simeq \mathbf{x_s}$). Thus, the state variables of the fast subsystem are the fast components $\mathbf{y}_f$

already introduced, which are also called *off-manifold variables*:

$$\mathbf{y}_f = \mathbf{y} - \mathbf{y}_s = \mathbf{y} - \mathbf{h}(\mathbf{x}_s) \tag{5.63}$$

Substituting (5.63) in (5.54b) we obtain the approximate fast subsystem:

$$\epsilon \dot{\mathbf{y}}_f = \epsilon \dot{\mathbf{y}} - \epsilon \dot{\mathbf{y}}_s \simeq \mathbf{g}(\mathbf{x}_s, \mathbf{y}_f + \mathbf{h}(\mathbf{x_s})) \tag{5.64}$$

Note that (5.64) defines a system with equilibrium point on the slow manifold (where $\mathbf{y}_f = 0$), and that the slow variables $\mathbf{x}_s$ are parameters for the fast subsystem.

Linearizing (5.64) on a point $(\mathbf{x}_s, \mathbf{h}(\mathbf{x}_s))$ lying on the slow manifold we obtain:

$$\epsilon \Delta \dot{\mathbf{y}}_f = \mathbf{g}_\mathbf{y} \Delta \mathbf{y}_f \tag{5.65}$$

Therefore, the stability of the off-manifold dynamics is determined by the Jacobian $\mathbf{g_y}$.

Improved accuracy is needed when large disturbances are imposed on the fast variables. This is clear in Fig. 5.14a,b where it is seen that the slow variable cannot be considered constant during fast transients, when the initial conditions are not close to the slow eigenvector (manifold). The corresponding component $\mathbf{x}_f$ dies out with the fast transient. The accuracy of (5.64) can be increased to any order of ϵ by calculating the component $\mathbf{x}_f$ with the asymptotic expansion method described in [O'M74].

Asymptotic expansion theorems [O'M74, KKO86] guarantee the validity of the time scale decomposition outlined above. In particular it has been proved that for $t > t_1$ (i.e. excluding the first initial transient when starting outside the slow manifold) the QSS slow subsystem (5.57a,b) approximates the slow dynamics of the original two-time-scale system (5.54a,b) with an order ϵ error:

$$\mathbf{x} = \mathbf{x}_s + O(\epsilon) \tag{5.66}$$
$$\mathbf{y} = \mathbf{y}_s + O(\epsilon) \tag{5.67}$$

under the following conditions:

1. The Jacobian $\mathbf{g_y}$ is nonsingular.

2. The off-manifold dynamics are stable, i.e. $\mathbf{g_y}$ has eigenvalues with strictly negative real parts.

3. All disturbances remain in the region of attraction of the stable equilibrium of the off-manifold dynamics.

The first assumption is implied by the second one, but it is stated separately because of its importance for the definition of a slow manifold $\mathbf{h}(\mathbf{x})$.

As stated before, during slow transients the slow variables act as parameters of the fast subsystem describing the off-manifold dynamics. Therefore, as a slow transient evolves, the fast dynamics subsystem (5.64) may experience a bifurcation, for instance a saddle-node bifurcation, or a Hopf bifurcation. At such points the time-scale decomposition breaks down, since assumptions 1 and 2 (in the SNB case), or assumption 2 (in the HB case) are violated. In practice, the time scale decomposition will break down and the trajectory will depart from the slow manifold, even before the bifurcation point, due for instance to the shrinking region of attraction of the off-manifold dynamics equilibrium that may cause a violation of assumption 3.

Note that a singularity induced bifurcation (see Section 5.3.2) of the QSS system (5.57a,b) is equivalent to a saddle-node bifurcation of the fast subsystem, both bifurcations being characterized by the singularity of $\mathbf{g_y}$. This provides a way to remove the singularity of a D-A system by relaxing algebraic constraints into differential equations, as we will see in Section 5.4.5.

5.4.4 Example: decomposed oscillator

We demonstrate now the time-scale decomposition and the limits of its validity using the stable oscillator introduced in (5.25a,b) rewritten as:

$$\dot{x} = y \qquad (5.68a)$$
$$\epsilon\dot{y} = -x + y - y^3 \qquad (5.68b)$$

The QSS approximation of the slow manifold is found by taking $\epsilon = 0$:

$$-x_s + y_s - y_s^3 = 0 \qquad (5.69)$$

and is shown as the S-shaped curve in Fig. 5.15. All trajectories starting outside this curve converge on it with a fast transient. As stated above, points on the slow manifold are equilibria of the off-manifold subsystem (5.64), which in our case is:

$$\epsilon\dot{y}_f = (1 - 3y_s^2)y_f - 3y_s y_f^2 - y_f^3$$

where y_s is the slow component of y given by (5.69), and $y_f = y - y_s$ is the off-manifold variable. As seen, y_s is a parameter of the off-manifold system depending on x_s. As x_s varies, two SNB points can be identified on the slow manifold at the

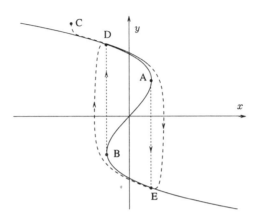

Figure 5.15   Slow manifold and limit cycle

values $y_s = \pm 1/\sqrt{3}$ where the linear part of the off-manifold system vanishes. These
are shown as points A and B in Fig. 5.15.

Consider a trajectory starting from point C in Fig. 5.15. The trajectory will first
converge on the slow manifold with a fast transient. From there on, the system is
driven by slow dynamics along the slow manifold. Due to (5.68a), the slow dynamics
take the system to the right on the upper half plane and to the left on the lower half
plane. By so doing the SNB at point A will be approached. Assuming that ϵ is infinitely
small, the system will come very close to point A. where the trajectory will depart from
the slow manifold with a fast transient, as shown by the dotted line. This transient will
converge on the lower side of the slow manifold at point E, after which a slow transient
will drive the system to the left, until the second SNB point B is encountered. At this
point another fast transient will take the system to point D, where the same sequence
is repeated all over again. The system will thus exhibit a limit cycle.

Comparing the so obtained limit cycle to the exact one for $\epsilon = 0.1$ (shown in Fig. 5.15
with a dashed line) we see that the two are qualitatively similar. Note, however, that
due to the not-so-small value of ϵ the exact limit cycle departs from the approximate
slow manifold before the SNB points A or B are met.

5.4.5 Singularity in two-time-scale systems

We will now discuss one way to deal with algebraic singularity in D-A systems. This consists in considering the algebraic equations as the equilibrium conditions of fast, non-modelled dynamics. In this way the D-A system is transformed into a two-time-scale system with $\epsilon \rightarrow 0$ [DB84, DO90].

When ϵ is nonzero, the actual trajectory of a two-time-scale system departs from the slow manifold before the singularity point, as seen in Fig. 5.15. This discrepancy is reduced as ϵ decreases and at the limit, as $\epsilon \rightarrow 0$ the "jump" from the slow manifold resembles a *discrete transition*. This transition may be catastrophic (in which case we can call it a collapse), or it may end up on another stable equilibrium, as in the case of Fig. 5.15. Thus the response of the D-A system can be predicted as the limit for $\epsilon \rightarrow 0$ of that of the two-time-scale system, derived by relaxing algebraic constraints into differential equations.

We finish this discussion by remarking that when the D-A system is derived from a time scale decomposition of a physical system, the fast dynamics forcing y on the constraint manifold cannot be stable on both sides of the impasse surface. We can revisit in this light the example of Fig. 5.13. In this example we assumed that the normal operating region is the upper part of the figure, above the impasse curve IS. This means that the off-manifold dynamics are stable in this causality region. Following our analysis, the points on the impasse surface are SNBs of the off-manifold dynamics. Therefore, in the causality region below the impasse curve IS, the off-manifold dynamics are unstable and a time-scale decomposition not valid.

6

MODELLING : SYSTEM PERSPECTIVE

"All science is dominated by the idea of approximation"
Bertrand Russell

After dealing with component modelling in Chapters 2 to 4, we devote this chapter to deriving and discussing models of the whole system.

First, we outline a dynamic model, staying at a rather high level of generality and referring where appropriate to Chapters 2 to 4. Next, we consider in more detail the network model, which is present in even the simplest voltage stability analysis tool. At this point, we propose a detailed illustration on a small but representative system, which will serve as an example in Chapter 8.

Following this, we introduce the important concept of time-scale decomposition which will provide in later chapters a framework for classifying instability mechanisms and deriving appropriate analytical methods.

The remaining of the chapter is devoted to deriving equilibrium models, which play an important rôle in voltage stability analysis. We come back to the simple example for illustrating these concepts.

6.1 OUTLINE OF A GENERAL DYNAMIC MODEL

In this section we outline the derivation of a general dynamic model of the power system, of the type typically used in power system dynamic simulation. The model is described in compact, vector form, going from faster to slower dynamic phenomena.

6.1.1   Instantaneous response : the network

An *instantaneous* response is assumed for the network. Indeed, as pointed out in Section 3.1.2, the corresponding transients, of electromagnetic type, are very fast compared to the time interval of interest in voltage stability studies. Consequently, the network is described by a set of algebraic equations:

$$0 = g(x, y, z_c, z_d) \qquad (6.1)$$

where g are smooth functions and y is the vector of bus voltages. Vectors g and y have the same dimension. Vectors x, z_c and z_d are defined in the sequel. These equations play an important rôle in voltage stability analysis and Section 6.2 is devoted to their detailed derivation.

6.1.2   Short-term dynamics

The *short-term* time scale is the time scale of synchronous generators and their regulators (AVRs and governors), induction motors, HVDC components and SVCs. The corresponding dynamics last typically for several seconds following a disturbance and have been also referred to as the *transient* dynamics. The term transient is in accordance with transient angle instability, which involves the same components and evolves in the same time frame. However, for reasons explained in Chapter 1, we adopt the short-term terminology throughout this book.

The short-term dynamics are captured by the following differential equations:

$$\dot{x} = f(x, y, z_c, z_d) \qquad (6.2)$$

where f are smooth functions and x is the corresponding state vector. Examples of the above equations have been given in Sections 3.1.2, 3.1.3 and 3.1.4 for the synchronous generator and Section 4.3.2 for the induction motor, respectively.

6.1.3   Long-term dynamics

The *long-term* time scale is the time scale of phenomena, controllers, and protecting devices that act typically over several minutes following a disturbance. In the case of controllers and protecting devices, the components are generally designed to act after the short-term transients have died out, to avoid unnecessary actions or even unstable interactions with the short-term dynamics. Table 6.1 lists the components most relevant to voltage stability, with reference to sections where they have been described.

Table 6.1   Important components of long-term dynamics

type	component	description
phenomena	thermostatic load recovery	§4.5
	aggregate load recovery	§4.6.2
controllers	secondary voltage control	§3.2.3
	load-frequency control	§3.2.1
	transformer Load Tap Changers (LTCs)	§4.4
	shunt capacitor/reactor switching	§2.6.2
protecting devices	OvereXcitation Limiters (OXLs)	§3.3
	armature current limiters	§3.3

The long-term dynamics are represented by both continuous and discrete-time equations:

$$\dot{z}_c \;=\; h_c(x, y, z_c, z_d) \tag{6.3}$$

$$z_d(k+1) \;=\; h_d(x, y, z_c, z_d(k)) \tag{6.4}$$

where z_c (resp. z_d) are the continuous (resp. discrete) long-term state vectors.

The following are devices involving either continuous, or discrete long-term dynamics:

■   Shunt compensation switching and LTC operation are typical discrete events captured by (6.4); the z_d variable is the shunt susceptance and the transformer ratio, respectively.

■   In present-day practice, secondary voltage and frequency controllers are digital and transmit to generators discrete changes in voltage and power setpoints, which are thus components of z_d. On the other hand, their internal control laws may involve continuous-time state equations of the type (6.3). A typical example is the PID control law used in the above controllers.

■   Overexcitation limitation can be reasonably considered as a discrete event, as regards the delayed decision to switch the field current under limit. Referring to Figs. 3.12 and 3.13, x_t and I_{ref} are components of z_c. The limit enforcement itself falls in the same short-term category as the synchronous generator.

■   The thermostatic and aggregate load recovery models (4.48a,b) or (4.51a,b) are in the form of the differential equations (6.3), as well as the continuous-time approximation (4.35) of the LTC dynamics.

Formally, a small enough time step ΔT should be assumed for the difference equations
(6.4), so that all discrete transitions take place at multiples of ΔT. In other words, the
discrete variables change values from $\mathbf{z}_d(k)$ to $\mathbf{z}_d(k+1)$ at times

$$t_k = k.\Delta T \qquad (k = 0, 1, 2, \ldots)$$

Note that with the above definition of ΔT, it is likely that most discrete long-term
components will remain unchanged for many time steps; it is even quite common to
have many time steps without any transition, especially during the short-term period.

6.2   NETWORK MODELLING

6.2.1   Network model in vector form

The network modelling relies on two basic assumptions:

- the quasi-sinusoidal, nominal frequency assumption recalled in Section 3.1.2; this
  allows us to represent voltages and currents through time-varying phasors and to
  use network impedances (or admittances);

- the definition of a single, common reference for all these phasors; such a reference
  was defined in Section 3.1.6 and takes on the form of two orthogonal axes rotating
  at the synchronous speed, and denoted by x and y respectively (see Fig. 3.4).

Under the above assumptions, the voltage-current relationships relative to an N-bus
system may be written in vector form as:

$$\bar{\mathbf{I}} - \mathbf{Y}\bar{\mathbf{V}} = 0 \qquad\qquad (6.5)$$

where $\bar{\mathbf{I}}$ is the N-dimensional vector of complex injected currents
       $\bar{\mathbf{V}}$ is the N-dimensional vector of complex bus voltages
       $\mathbf{Y}$ is the $N \times N$ bus admittance matrix of the network (with the ground
       node taken as the voltage reference).

Note that the injected currents $\bar{\mathbf{I}}$ are nonlinear functions of voltages $\bar{\mathbf{V}}$ and some
previously defined state variables $\mathbf{x}$, $\mathbf{z}_c$ and $\mathbf{z}_d$. Hence, Eq. (6.5) is in fact nonlinear.

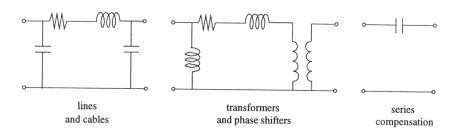

lines
and cables

transformers
and phase shifters

series
compensation

Figure 6.1 Two-port model of usual network components

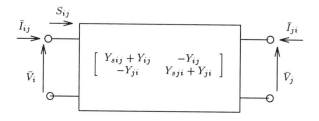

Figure 6.2 Two-port notation

6.2.2 Two-port modelling

The usual network components are mainly transmission lines, cables, transformers and possibly series capacitors and phase shifters. The standard models of these components are the two-ports shown in Fig. 6.1. A derivation of their parameters is available in many textbooks (e.g. [Elg71, Ber86, Kun94]).

Under the above assumptions, each two-port can be described by its nodal admittance matrix. For a two-port connecting buses i and j, as in Fig. 6.2, the following voltage-current relationship holds[1]:

$$\begin{bmatrix} \bar{I}_{ij} \\ \bar{I}_{ji} \end{bmatrix} = \begin{bmatrix} Y_{ij} + Y_{sij} & -Y_{ij} \\ -Y_{ji} & Y_{ji} + Y_{sji} \end{bmatrix} \begin{bmatrix} \bar{V}_i \\ \bar{V}_j \end{bmatrix} \tag{6.6}$$

---

[1] Note that Y_{ij} is *not* the (i,j) element of matrix $\mathbf{Y}$. The latter has the opposite sign and accounts for other branches connecting the same two buses

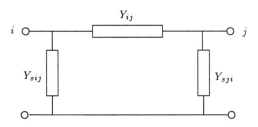

Figure 6.3 Pi-equivalent of reciprocal two-ports

Note that the two-port models of all the above mentioned components, *except the phase shifter*, are reciprocal [CDK87], which implies that : (i) their admittance matrix is symmetric, i.e. $Y_{ij} = Y_{ji}$; (ii) they can be represented by a pi-equivalent as shown in Fig. 6.3. Nevertheless the derivation which follows is general and is not restricted to reciprocal two-ports.

Note also that for transformers equipped with LTCs, the discrete changes in ratios, captured by (6.3), causes corresponding changes in the pi-equivalent parameters.

6.2.3   Complex current formulation

The network model is obtained by assembling the various two-ports according to the system topology. Let N be the number of system buses and let $\mathcal{N}(i)$ be the set of buses directly connected to bus i ($i = 1, \ldots, N$), as shown in Fig. 6.4. Using the Kirchhoff's current law, we have:

$$\bar{I}_i - Y_{si}\bar{V}_i - \sum_{j \in \mathcal{N}(i)} \bar{I}_{ij} = 0$$

and, using (6.6):

$$\bar{I}_i - Y_{si}\bar{V}_i - \sum_{j \in \mathcal{N}(i)} Y_{sij}\bar{V}_i - \sum_{j \in \mathcal{N}(i)} Y_{ij}(\bar{V}_i - \bar{V}_j) = 0 \qquad (6.7)$$

where Y_{si} is the admittance of all shunt elements present at bus i and $\bar{I}_i$ is the current *injected* by any generator, load and/or compensator connected to this bus (see Fig. 6.4).

In the next two sections we derive network models using $2N$ real equations instead of the N complex ones (6.7). To this purpose, we first project all phasors on the x and y axes of the synchronous frame of reference defined in Section 3.1.6 (see Fig. 3.4).

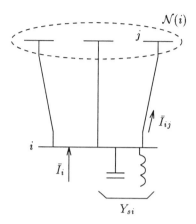

Figure 6.4 Definition of nodal quantities

This formulation is commonly used in power system dynamic simulation. Next we derive an alternative network model which is more convenient when dealing with the power system at equilibrium. In this model, each injected current is split into its active and reactive components, as shown in Fig. 3.15 and the voltage phasors are expressed in polar form.

The two types of phasor projection are summarized in Fig. 6.5

6.2.4 Real and imaginary current formulation

Let us define, according to Fig. 6.5:

$$\bar{I}_i = i_{xi} + j\, i_{yi} \qquad (6.8a)$$
$$\bar{V}_i = v_{xi} + j\, v_{yi} \qquad (6.8b)$$

as well as the following conductances and susceptances:

$$Y_{si} = jB_{si} \qquad (6.9a)$$
$$Y_{sij} = G_{sij} + jB_{sij} \qquad (6.9b)$$
$$Y_{ij} = G_{ij} + jB_{ij} \qquad (6.9c)$$

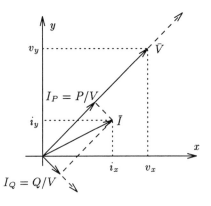

Figure 6.5   The two types of phasor projections

Introducing (6.8a,b) and (6.9a,b,c) into (6.7) yields:

$$i_{xi} \quad +B_{si}v_{yi} - \sum_{j \in \mathcal{N}(i)} (G_{sij}v_{xi} - B_{sij}v_{yi})$$

$$- \sum_{j \in \mathcal{N}(i)} [G_{ij}(v_{xi} - v_{xj}) - B_{ij}(v_{yi} - v_{yj})] = 0 \qquad (6.10a)$$

$$i_{yi} \quad -B_{si}v_{xi} - \sum_{j \in \mathcal{N}(i)} (B_{sij}v_{xi} + G_{sij}v_{yi})$$

$$- \sum_{j \in \mathcal{N}(i)} [B_{ij}(v_{xi} - v_{xj}) + G_{ij}(v_{yi} - v_{yj})] = 0 \qquad (6.10b)$$

As already mentioned, the currents i_x and i_y are functions of voltages v_x and v_y, as well as some short- and long-term state variables. Examples of such functions will be given in Section 6.3 for the synchronous generator and the induction motor. Other components like HVDC and FACTS devices (including SVCs) [CIE95, CTF96], although not detailed here, are easily included in the model through appropriate expressions of the i_x and i_y currents. For such devices, the current injected at one bus generally involves the voltage at another bus in addition to the device state variables.

Defining the network variables **y** as:

$$\mathbf{y} = \begin{bmatrix} \mathbf{v}_x \\ \mathbf{v}_y \end{bmatrix} \qquad (6.11)$$

the above equations (6.10a,b) take on the compact form (6.1).

Let us point out the *sparse* character of the above equations. Sparsity results from the fact that (i) a bus has a small number of direct neighbours; (ii) each i_x or i_y current involves only the state variables of the component connected to the corresponding bus. For the above quoted devices, it also involves the voltage at another bus but this cross-coupling affects the model sparsity only marginally.

6.2.5 Active and reactive current formulation

We consider now the alternative network model. The latter uses as network variables the bus voltage magnitudes and phase angles, related to complex voltages through:

$$\bar{V}_i = V_i \, e^{j\theta_i}$$

Taking the complex conjugate of (6.7) and multiplying by $\bar{V}_i$ yields:

$$\bar{V}_i \bar{I}_i^* - Y_{si}^* V_i^2 - \sum_{j \in \mathcal{N}(i)} Y_{sij}^* V_i^2 - \sum_{j \in \mathcal{N}(i)} Y_{ij}^* (V_i^2 - \bar{V}_i \bar{V}_j^*) = 0$$

Using (6.9a,b,c) and replacing the complex power $\bar{V}_i \bar{I}_i^*$ by its active and reactive components, the above equation becomes:

$$P_i + jQ_i \quad + \quad jB_{si} V_i^2 - \sum_{j \in \mathcal{N}(i)} (G_{sij} - jB_{sij}) V_i^2$$

$$- \quad \sum_{j \in \mathcal{N}(i)} (G_{ij} - jB_{ij}) \left[V_i^2 - V_i V_j \left(\cos(\theta_i - \theta_j) + j\sin(\theta_i - \theta_j) \right) \right] = 0$$

which decomposes into:

$$P_i \quad - \quad V_i^2 \sum_{j \in \mathcal{N}(i)} (G_{sij} + G_{ij})$$

$$+ \quad V_i \sum_{j \in \mathcal{N}(i)} V_j \left[G_{ij} \cos(\theta_i - \theta_j) + B_{ij} \sin(\theta_i - \theta_j) \right] = 0 \qquad (6.12a)$$

$$Q_i \quad + \quad V_i^2 \sum_{j \in \mathcal{N}(i)} (B_{sij} + B_{ij}) + B_{si} V_i^2$$

$$- \quad V_i \sum_{j \in \mathcal{N}(i)} V_j \left[B_{ij} \cos(\theta_i - \theta_j) - G_{ij} \sin(\theta_i - \theta_j) \right] = 0 \qquad (6.12b)$$

These equations make up a network model in terms of powers, which is used very frequently in the literature. Equivalently, a division of (6.12a,b) by V_i yields a formulation in terms of active and reactive currents, as defined in Fig. 6.5:

$$I_{Pi} \quad - \quad V_i \sum_{j \in \mathcal{N}(i)} (G_{sij} + G_{ij})$$

$$+ \quad \sum_{j \in \mathcal{N}(i)} V_j [G_{ij} \cos(\theta_i - \theta_j) + B_{ij} \sin(\theta_i - \theta_j)] = 0 \qquad (6.13a)$$

$$I_{Qi} \quad + \quad V_i \sum_{j \in \mathcal{N}(i)} (B_{sij} + B_{ij}) + B_{si} V_i$$

$$- \quad \sum_{j \in \mathcal{N}(i)} V_j [B_{ij} \cos(\theta_i - \theta_j) + G_{ij} \sin(\theta_i - \theta_j)] = 0 \qquad (6.13b)$$

With respect to the power equations (6.12a,b), the advantage of the current equations (6.13a,b) lies in the fact that they are slightly less nonlinear, and unless otherwise specified, we will use them in the remaining of this book.

As mentioned in the previous section, the injected currents I_{Pi} and I_{Qi} in (6.13a,b) must be expressed in terms of bus voltage magnitudes and phase angles as well as some dynamic state variables. We will detail these expressions in Section 6.5.1.

With the appropriate expressions substituted for the currents, the $2N$ equations (6.13a,b) make up another network model, equivalent to (6.10a,b). For this new model, we re-define the vector **y** of algebraic variables as:

$$\mathbf{y} = \begin{bmatrix} \theta \\ \mathbf{V} \end{bmatrix} \qquad (6.14)$$

with

$$\theta = [\theta_1 \ \dots \ \theta_N]^T \qquad \mathbf{V} = [V_1 \ \dots \ V_N]^T \qquad (6.15)$$

which allows to write (6.13a,b) in the same compact form (6.1).

6.3 A DETAILED EXAMPLE

In this section we present the detailed model of a simple system. The objective is twofold: (i) first, to illustrate the general modelling approach described in the previous section; (ii) second, to introduce a simple but representative test system that will be used in Chapter 8 to illustrate voltage instability mechanisms. This example is an improved version of the one detailed in [VC91b].

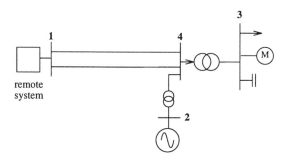

Figure 6.6 One-line diagram of the example system

6.3.1 Overall description

The system one-line diagram is shown in Fig. 6.6. The load at bus 3 is fed through an LTC transformer. Most of the power is provided by a remote system (bus 1) through a rather long, double-circuit transmission line, while the remaining is supplied by the local generator at bus 2.

The disturbance to be analyzed in Chapter 8 is the tripping of one circuit between buses 1 and 4. This drastically decreases the maximum power that can be delivered to the load while the latter tends to restore to constant power due to transformer LTC effects. The voltage support provided by the local generator is withdrawn some time after the disturbance, due to the limited overload capability of the field winding.

6.3.2 Main modelling assumptions

In order to keep the example as simple as possible, while retaining the essential of voltage instability phenomena, the following assumptions are made:

1. the system connected to bus 1 keeps the frequency constant. It is characterized by its (finite) short-circuit level at bus 1 and is represented by a Thévenin equivalent;

2. line 1-4 is modelled by its series reactance only;

3. each transformer is modelled by its leakage reactance in series with an ideal, off-nominal transformer;

4. the generator step-up transformer (between buses 2 and 4) has a fixed turn ratio;

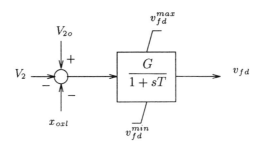

Figure 6.7   AVR model

5. the load transformer is equipped with an LTC aimed at keeping the voltage at bus 3 within a deadband;

6. the generator at bus 2 is represented by the simple model of Section 3.1.4, accounting for the field winding only. Saturation is neglected for simplicity, which is probably the most questionable simplification;

7. the AVR of this generator is represented by the simple first-order transfer function shown in Fig. 6.7, with non-windup limits on the field voltage, corresponding to minimum and ceiling field voltages, respectively. In addition, the machine is protected by an OXL with inverse-time characteristic and integral action, modelled as shown in Fig. 3.12 (variant shown with solid lines);

8. no governor is considered, assuming that the frequency is held constant by the system connected to bus 1. The mechanical power of the generator is thus considered constant;

9. the load is made up of : (i) one part represented by an exponential model; (ii) another part represented by an equivalent induction motor, and (iii) a shunt capacitor, for compensation purposes. A first-order model is used for the motor, corresponding to the equivalent circuit of Fig. 4.2, where stator losses are neglected and a constant mechanical torque is assumed.

Assumptions 1 to 3 above lead to the circuit representation shown in Fig. 6.8. In the Thévenin equivalent (bus 1), X_{th} is the reciprocal of the short-circuit level (in per unit) and E_{th} and θ_{th} are constant.

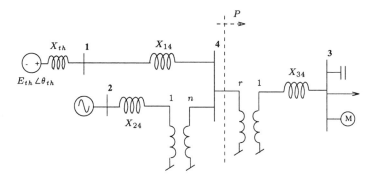

Figure 6.8   Circuit representation of the example system

Table 6.2   Pi-equivalent parameters

branch	B_{ij}	B_{sij}	B_{sji}
1 - 4	$B_{14} = -\dfrac{1}{X_{14}}$	$B_{s14} = 0$	$B_{s41} = 0$
2 - 4	$B_{24} = -\dfrac{1}{nX_{24}}$	$B_{s24} = \dfrac{1-n}{n}\dfrac{1}{X_{24}}$	$B_{s42} = \dfrac{n-1}{n^2}\dfrac{1}{X_{24}}$
3 - 4	$B_{34} = -\dfrac{1}{rX_{34}}$	$B_{s34} = \dfrac{1-r}{r}\dfrac{1}{X_{34}}$	$B_{s43} = \dfrac{r-1}{r^2}\dfrac{1}{X_{34}}$

6.3.3   Network modelling

All lines and transformers are represented by pi-equivalents as shown in Fig. 6.3. In these equivalents, all conductances are zero (due to assumptions 2 and 3 above) while the susceptances have the values shown in Table 6.2.

The network equations in real and imaginary current formulation are readily obtained from (6.10a,b) as follows[2]:

$$\frac{E_{th}}{X_{th}}\sin\theta_{th} - \frac{1}{X_{th}}v_{y1} + B_{14}(v_{y1} - v_{y4}) = 0 \quad (6.16a)$$

$$-\frac{E_{th}}{X_{th}}\cos\theta_{th} + \frac{1}{X_{th}}v_{x1} - B_{14}(v_{x1} - v_{x4}) = 0 \quad (6.16b)$$

---

[2] Note that all currents are taken positive when injected into the system

$$i_{x2} + B_{s24}v_{y2} + B_{24}(v_{y2} - v_{y4}) = 0 \quad (6.16c)$$
$$i_{y2} - B_{s24}v_{x2} - B_{24}(v_{x2} - v_{x4}) = 0 \quad (6.16d)$$
$$i_{x3} + B_{s34}v_{y3} + B_{34}(v_{y3} - v_{y4}) = 0 \quad (6.16e)$$
$$i_{y3} - B_{s34}v_{x3} - B_{34}(v_{x3} - v_{x4}) = 0 \quad (6.16f)$$
$$(B_{s42} + B_{s43})v_{y4} + B_{14}(v_{y4} - v_{y1}) + B_{24}(v_{y4} - v_{y2})$$
$$+ B_{34}(v_{y4} - v_{y3}) = 0 \quad (6.16g)$$
$$-(B_{s42} + B_{s43})v_{x4} - B_{14}(v_{x4} - v_{x1}) - B_{24}(v_{x4} - v_{x2})$$
$$- B_{34}(v_{x4} - v_{x3}) = 0 \quad (6.16h)$$

where i_{x2} and i_{y2} (resp. i_{x3} and i_{y3}) are the components of the current injected by the generator (resp. the load), whose expressions will be detailed in the sequel. The current injected by the Thévenin equivalent at bus 1 is given by:

$$i_{x1} + j\,i_{y1} = \frac{E_{th}(\cos\theta_{th} + j\sin\theta_{th}) - (v_{x1} + jv_{y1})}{j\,X_{th}}$$

The resulting expressions for i_{x1} and i_{y1} have been substituted in (6.16a,b).

6.3.4 Generator modelling

We use I_{fB} in Fig. 3.7 as the base current of the field winding, which yields, according to (3.35) and (3.34):

$$E_f = v_{fd} \quad (6.17)$$
$$E_q = i_{fd} \quad (6.18)$$

Under the above mentioned assumptions, the generator is described by the differential equations:

$$\dot{\delta} = \omega \quad (6.19)$$
$$\dot{\omega} = -\frac{D}{2H}\omega + \frac{\omega_o}{2H}(P_m - P_2) \quad (6.20)$$
$$\dot{E}'_q = \frac{-E'_q + v_{fd} - (X_d - X'_d)i_d}{T'_{do}} \quad (6.21)$$

where P_m is constant (see assumption 8) and v_{fd} has been substituted for E_f. The expression for the active power P_2 produced by the generator is derived as:

$$P_2 = \text{Re}\left\{(v_{x2} + jv_{y2})(i_{x2} - ji_{y2})\right\} = v_{x2}i_{x2} + v_{y2}i_{y2} \quad (6.22)$$

and the current i_d is obtained from (3.18) as:

$$i_d = i_{x2} \sin \delta - i_{y2} \cos \delta \qquad (6.23)$$

At this point, expressions for i_{x2} and i_{y2} are needed for substitution in the above equations (6.22) and (6.23) as well as in the network equations (6.16c,d). They are obtained from (3.20) as follows:

$$
\begin{aligned}
i_{x2} &= \frac{\sin 2\delta}{2} \left(\frac{1}{X_q} - \frac{1}{X_d'} \right) (v_{x2} - E_q' \cos \delta) \\
&\quad - \left(\frac{\cos^2 \delta}{X_q} + \frac{\sin^2 \delta}{X_d'} \right) (v_{y2} - E_q' \sin \delta) \qquad (6.24a)
\end{aligned}
$$

$$
\begin{aligned}
i_{y2} &= \left(\frac{\sin^2 \delta}{X_q} + \frac{\cos^2 \delta}{X_d'} \right) (v_{x2} - E_q' \cos \delta) \\
&\quad + \frac{\sin 2\delta}{2} \left(\frac{1}{X_d'} - \frac{1}{X_q} \right) (v_{y2} - E_q' \sin \delta) \qquad (6.24b)
\end{aligned}
$$

We proceed now with the AVR model. With reference to Fig. 6.7, the state equation takes on the form:

$$
\begin{aligned}
\dot{v}_{fd} &= 0 \quad \text{if } v_{fd} = v_{fd}^{max} \text{ and } G(V_{2o} - V_2 - x_{oxl}) - v_{fd} > 0 \quad (6.25a) \\
&= 0 \quad \text{if } v_{fd} = v_{fd}^{min} \text{ and } G(V_{2o} - V_2 - x_{oxl}) - v_{fd} < 0 \quad (6.25b) \\
&= \frac{-v_{fd} + G(V_{2o} - V_2 - x_{oxl})}{T} \quad \text{otherwise} \quad (6.25c)
\end{aligned}
$$

where the machine terminal voltage V_2 is easily expressed in terms of its components:

$$V_2 = \sqrt{v_{x2}^2 + v_{y2}^2}$$

We derive finally the equations of the OXL model, Fig. 3.12. The OXL input is the field current i_{fd}, which by virtue of (6.18), is equal to E_q in per unit. The latter emf is easily expressed in terms of the state and network variables by using (3.12):

$$E_q = E_q' + (X_d - X_d')i_d$$

and substituting (6.23) for i_d:

$$E_q = E_q' + (X_d - X_d')(i_{x2} \sin \delta - i_{y2} \cos \delta) \qquad (6.26)$$

With reference to Fig. 3.12, the intermediate variable x_2 is given by:

$$
\begin{aligned}
x_2 &= S_1(E_q - I_{fd}^{lim}) \quad \text{if } E_q \geq I_{fd}^{lim} \\
&= S_2(E_q - I_{fd}^{lim}) \quad \text{otherwise}
\end{aligned}
$$

The first OXL state equation is related to the non-windup limited integrator of block 2:

$$\dot{x}_t \;=\; 0 \quad \text{if } x_t = K_2 \text{ and } x_2 \geq 0 \tag{6.27a}$$
$$=\; 0 \quad \text{if } x_t = -K_1 \text{ and } x_2 < 0 \tag{6.27b}$$
$$=\; x_2 \quad \text{otherwise} \tag{6.27c}$$

The intermediate variable x_3 in Fig. 3.12 is given by:

$$x_3 \;=\; E_q - I_{fd}^{lim} \quad \text{if } x_t \geq 0$$
$$=\; -K_r \quad \text{otherwise}$$

and hence the second OXL state equation is related to block 4 as follows:

$$\dot{x}_{oxl} \;=\; 0 \quad \text{if } x_{oxl} = 0 \text{ and } x_3 < 0 \tag{6.28a}$$
$$=\; K_i x_3 \quad \text{otherwise} \tag{6.28b}$$

6.3.5 Load modelling

The load currents i_{x3} and i_{y3} involved in (6.16e,f) are broken down respectively into:

$$i_{x3} \;=\; i_{xE} + i_{xM} + i_{xC} \tag{6.29a}$$
$$i_{y3} \;=\; i_{yE} + i_{yM} + i_{yC} \tag{6.29b}$$

where subscript E refers to the exponential load, M to the induction motor and C to the shunt compensation. Expressions for the various components are derived hereafter.

Exponential load

The exponential load is described by (4.2a,b):

$$P(V_3) \;=\; P_o \left(\frac{V_3}{V_{3o}} \right)^\alpha \tag{6.30a}$$

$$Q(V_3) \;=\; Q_o \left(\frac{V_3}{V_{3o}} \right)^\beta \tag{6.30b}$$

and the complex current injected *into* the network is given by:

$$\bar{I}_E = -\frac{S^\star}{V_3^\star} = -\frac{P(V_3) - jQ(V_3)}{v_{x3} - jv_{y3}}$$

from which one readily obtains:

$$i_{xE} = -\frac{P(V_3)v_{x3} + Q(V_3)v_{y3}}{v_{x3}^2 + v_{y3}^2}$$

$$i_{yE} = \frac{Q(V_3)v_{x3} - P(V_3)v_{y3}}{v_{x3}^2 + v_{y3}^2}$$

where:

$$V_3 = \sqrt{v_{x3}^2 + v_{y3}^2}$$

Motor load

The dynamics of the first-order induction motor model is given by (4.22):

$$\dot{s} = \frac{1}{2H}(T_M - P_M) \tag{6.31}$$

where T_M is constant (see assumption 9) and P_M is the active power absorbed by the motor. The expression of P_M is derived as for the generator:

$$P_M = -\text{Re}\left\{(v_{x3} + jv_{y3})(i_{xM} - ji_{yM})\right\} = -v_{x3}i_{xM} - v_{y3}i_{yM} \tag{6.32}$$

The expressions for i_{xM} and i_{yM} to be substituted in (6.32) as well as in (6.29a,b) are obtained as follows. We define, according to (4.14):

$$R_e + jX_e = \frac{jX_m(\frac{R_r}{s} + jX_r)}{\frac{R_r}{s} + j(X_m + X_r)}$$

The current injected by the motor is given by (4.13), with R_s neglected (see assumption 9) and sign reversed:

$$\bar{I}_M = -\frac{\bar{V}_3}{R_e + j(X_s + X_e)}$$

from which one obtains:

$$i_{xM} = -\frac{v_{x3}R_e + v_{y3}(X_s + X_e)}{R_e^2 + (X_s + X_e)^2}$$

$$i_{yM} = -\frac{v_{y3}R_e - v_{x3}(X_s + X_e)}{R_e^2 + (X_s + X_e)^2}$$

Shunt compensation

Let B_{s3} be the shunt susceptance at bus 3. The complex current injected into the system is:

$$\bar{I}_C = jB_{s3}\bar{V}_3$$

which decomposes into:

$$i_{xC} = -B_{s3}v_{y3}$$
$$i_{yC} = B_{s3}v_{x3}$$

Load tap changer

The LTC is modelled as a discrete device, with the tap changing logic given by (4.33):

$$r_{k+1} = \begin{cases} r_k + \Delta r & \text{if} & V_3 > V_{3o} + d & \text{and} & r_k < r^{max} \\ r_k - \Delta r & \text{if} & V_3 < V_{3o} - d & \text{and} & r_k > r^{min} \\ r_k & \text{otherwise} \end{cases} \qquad (6.33)$$

The tapping delays are assumed fixed (i.e. independent of V_3), but larger for first tap change than for the subsequent ones (sequential LTC model). This is equivalent to (4.32) and (4.34) without the inverse time characteristic, i.e.:

$$\Delta T_o = T_{fo} + T_m$$
$$\Delta T_k = T_f + T_m$$

with $T_{fo} > T_f$.

6.3.6 Summary

By way of summary, Table 6.3 shows how the particular example proposed in this section fits the general model outlined in Section 6.1.

Note that of the two OXL state variables, x_t is considered long-term, because it is the one that controls the activation delay (usually in the order of 20–100 seconds). On the other hand, x_{oxl} is considered a short-term variable, because the gain K_i is usually high, forcing the limitation to take effect (or the OXL to reset) in a short time (usually no more than a few seconds).

Table 6.3 How the example fits the general model of Section 6.1

equations		variables	
Network (6.1)	(6.16a to h)	$\mathbf{y}$:	$v_{x1}, v_{y1}, v_{x2}, \ldots, v_{y4}$
Short-term (6.2)	(6.19), (6.20), (6.21), (6.25a,b or c), (6.28a, b or c), (6.31)	$\mathbf{x}$:	$\delta, \omega, E_q', v_{fd}, x_{oxl}, s$
Long-term continuous (6.3)	(6.27a,b or c)	$\mathbf{z}_c$:	x_t
Long-term discrete (6.4)	(6.33)	$\mathbf{z}_d$:	r

6.4 TIME-SCALE DECOMPOSITION PERSPECTIVE

In the previous section we have outlined a general power system model covering the short and long-term time scales. With modern computer technology, it has become quite feasible to handle the whole set of differential-algebraic, discrete-continuous time equations in digital simulations. However, for the purposes of understanding voltage instability mechanisms, as well as devising faster analysis methods, it is advantageous to exploit the time separation which exists between the short and long-term phenomena.

In Section 5.4 the concept of reduced-order models suitable for each time scale was introduced and discussed in general, mathematical terms. In particular, the *Quasi Steady-State (QSS) approximation* of the slow subsystem was defined. The idea is simple and intuitively appealing: it consists in assuming that the fast subsystem is infinitely fast and can be replaced by its equilibrium equations when dealing with the slow subsystem. Conversely, the fast dynamics can be approximated by considering the slow variables as practically constant during the fast transients. This leads to a significantly simpler analysis of both subsystems and agrees with intuitive reasoning.

Figure 6.9 illustrates the application of this concept. The "coupled" system corresponds to the reference model (6.1 - 6.4). The approximate short-term dynamic model is obtained by assuming that $\mathbf{z}_c$ and $\mathbf{z}_d$ are constant. Note that $\mathbf{z}_d$ is actually constant in the intervals between discrete variable changes (e.g. in between tap changes). The QSS approximation, on the other hand, consists of replacing the short-term dynamic equation (6.2) by the corresponding equilibrium equation:

$$\mathbf{f}(\mathbf{x}, \mathbf{y}, \mathbf{z}_c, \mathbf{z}_d) = 0 \tag{6.34}$$

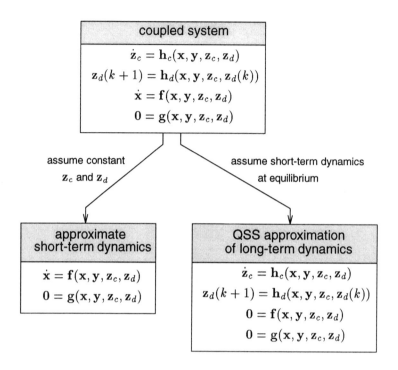

Figure 6.9   Time-scale decomposition and QSS approximation

The applications, as well as the limits of validity of the above time-scale decomposition will be thoroughly discussed in the next chapters, where the concept will be used to analyze instability mechanisms and devise appropriate analysis techniques.

6.5   EQUILIBRIUM EQUATIONS FOR VOLTAGE STABILITY STUDIES

In practice, equilibrium equations are needed in two circumstances:

1. when using the QSS approximation of long-term dynamics (see Fig. 6.9), the short-term dynamics have to be replaced by their equilibrium equations;

2. in studies which involve very slow transitions from one steady-state operating point to another (such as a daily load cycle or a smooth generation reschedul-

ing), the system is modelled with all dynamics, including long-term ones, at equilibrium.

In this section we will concentrate on deriving a practical set of equilibrium equations successively for the short-term and the long-term dynamics. Note that since the response of the network is assumed instantaneous, the latter is always "at equilibrium".

6.5.1 Equilibrium of short-term dynamics

Formally, the equilibrium equations of the short-term dynamics take on the form (6.34). Now, with $\mathbf{f}$ corresponding to the detailed models typically used in dynamic simulation, the equations (6.34) involve a lot of trivial relationships. Therefore, it is desirable to eliminate equations and variables from (6.34) to end up with an equivalent *reduced equilibrium model*, which is easier to handle than the original one.

The problem can be stated as follows. We decompose $\mathbf{f}$ and $\mathbf{x}$ into:

$$\mathbf{x} = \begin{bmatrix} \mathbf{x}_1 \\ \mathbf{x}_2 \end{bmatrix} \qquad \mathbf{f} = \begin{bmatrix} \mathbf{f}_1 \\ \mathbf{f}_2 \end{bmatrix}$$

where $\mathbf{x}_1$ and $\mathbf{f}_1$ have dimension n_1, and $\mathbf{x}_2$ and $\mathbf{f}_2$ dimension n_2. Using this decomposition, we can write the short-term equilibrium equations as:

$$\mathbf{g}(\mathbf{x}_1, \mathbf{x}_2, \mathbf{y}, \mathbf{z}_c, \mathbf{z}_d) \;=\; \mathbf{0}$$
$$\mathbf{f}_1(\mathbf{x}_1, \mathbf{x}_2, \mathbf{y}, \mathbf{z}_c, \mathbf{z}_d) \;=\; \mathbf{0}$$
$$\mathbf{f}_2(\mathbf{x}_1, \mathbf{x}_2, \mathbf{y}, \mathbf{z}_c, \mathbf{z}_d) \;=\; \mathbf{0}$$

Assume that we seek a reduced set of equations with $\mathbf{x}_2$ eliminated. The implicit function theorem (see Section 5.3.1) states that, if the Jacobian of $\mathbf{f}_2$ with respect to $\mathbf{x}_2$ is nonsingular, we can (at least conceptually) obtain $\mathbf{x}_2$ as a function of the other variables. Substituting this function for $\mathbf{x}_2$ in the first two equations, we obtain:

$$\tilde{\mathbf{g}}(\tilde{\mathbf{x}}, \mathbf{y}, \mathbf{z}_c, \mathbf{z}_d) \;=\; \mathbf{0} \qquad\qquad (6.35a)$$
$$\hat{\mathbf{f}}(\tilde{\mathbf{x}}, \mathbf{y}, \mathbf{z}_c, \mathbf{z}_d) \;=\; \mathbf{0} \qquad\qquad (6.35b)$$

where $\tilde{\mathbf{x}} = \mathbf{x}_1$, and $\tilde{\mathbf{g}}$ stands for the network equations expressed in terms of $\tilde{\mathbf{x}}$ (instead of $\mathbf{x}$).

For practice applications, it is desirable to reduce the equilibrium equations to the greatest extent possible without creating unnecessary complexity of the functions $\hat{\mathbf{f}}$.

We present hereafter models for the synchronous generator, the induction motor and the SVC. These models, proposed in [VC93, VJM95], have been thoroughly validated by comparing the resulting quasi steady-state approximation with a full dynamic model [VCM97]. Note that the active-reactive current formulation is used with the equilibrium models.

Synchronous generator. The equilibrium model of the generator and its regulators is reduced to only three equations of the type (6.35b), involving three variables of the type $\tilde{\mathbf{x}}$. This reduced model is detailed in Table 6.4.

The first equation describes saturation. If saturation is neglected ($m = 0$), this equation becomes the trivial relationship $E_q = E_q^s$, which can be eliminated together with the E_q^s variable. This would further reduce the model dimension by one.

The second equation describes excitation control. It depends on the long-term variables associated with limiters, in particular the OXL long-term variable x_t (see Figs. 3.12 and 3.13). If x_t is negative, the variable x_{oxl} is zero at equilibrium and the OXL is inactive. Thus the excitation control equation is the equilibrium equation of the AVR. If x_t is positive, the OXL is active and the equation is the equilibrium condition of the OXL. The same formulation applies to an armature current limiter. Note from Table 6.4 that if neither integral AVRs nor armature current limits are considered, E_q is easily eliminated from all relationships, thus leaving again one less equation of the type (6.35b).

The third equation deals with governor effects. At equilibrium the rotor speeds $\dot{\delta}$ of all the generators are equal. Hence all these components of $\mathbf{x}$ coalesce in a single one, the system angular frequency, which we denote ω_{sys} (see Section 3.2.1).

Finally, the table gives the expressions of currents I_P and I_Q needed in the above reduced model as well as in the network equations (6.13a,b).

Induction motor. The reduced equilibrium model of the induction motor is presented in Table 6.5. It consists of a single equation of the type (6.35b) involving a single variable $\tilde{x}$, namely the slip s.

SVC. The SVC equilibrium model is shown in Table 6.6. The reduced model consists of a voltage controlled, bounded susceptance introduced directly in the expression of the reactive current I_Q. In this way the SVC internal state variables are completely eliminated.

Consider now the network equations (6.13a,b), with the expressions of Tables 6.4 to 6.6 substituted for the currents I_P and I_Q. In all these equations, phase angles appear *through differences only* and hence they are defined up to an additive constant. It is thus required to take one bus as the reference by setting:

$$\theta_r = 0 \qquad\qquad (6.36)$$

Table 6.4 Reduced equilibrium model of the synchronous generator

COMPONENTS OF $\tilde{\mathbf{x}}$	
E_q (emf proportional to field current)	(3.9)
E_q^s (emf behind saturated reactances)	(3.22c)
δ (rotor angle and phase angle of $\bar{E}_q$)	(3.3)

EXPRESSIONS OF CURRENTS TO BE SUBSTITUTED IN (6.13a,b)

$$I_P = \frac{E_q E_q^s}{X_\ell E_q + (X_d - X_\ell)E_q^s} \sin(\delta - \theta) \tag{3.32a}$$

$$+ \frac{E_q V}{2}\left(\frac{1}{X_\ell E_q + (X_q - X_\ell)E_q^s} - \frac{1}{X_\ell E_q + (X_d - X_\ell)E_q^s}\right)\sin 2(\delta - \theta)$$

$$I_Q = \frac{E_q E_q^s}{X_\ell E_q + (X_d - X_\ell)E_q^s} \cos(\delta - \theta) \tag{3.32b}$$

$$- E_q V \left(\frac{\sin^2(\delta - \theta)}{X_\ell E_q + (X_q - X_\ell)E_q^s} + \frac{\cos^2(\delta - \theta)}{X_\ell E_q + (X_d - X_\ell)E_q^s}\right)$$

EQUATIONS $\tilde{\mathbf{f}}(\mathbf{y}, \tilde{\mathbf{x}}, \mathbf{z}_c, \mathbf{z}_d) = 0$

(with I_P and I_Q given by the above expressions)

Saturation: $E_q - (1 + m[(V + X_\ell I_Q)^2 + (X_\ell I_P)^2]^{n/2})E_q^s = 0$ (3.33)

Excitation control: depends on the long-term variable x_t:

    if $x_t \le 0$: one of the following:

for a general AVR:	$E_q - g_{avr}(V_o, V) = 0$	(3.43)
for a linear AVR:	$E_q - G(V_o - V) = 0$	(3.44)
for an integral AVR:	$V - V_o = 0$	

    if $x_t > 0$: one of the following:

for a general OXL:	$E_q - g_{oxl}(V_o, V) = 0$	(3.48)
for an integral OXL:	$E_q - E_q^{lim} = 0$	(3.49)

       for a proportional OXL:

$$E_q - \frac{G}{1 + GK_p}(V_o - V) - \frac{GK_p}{1 + GK_p}E_q^{lim} = 0 \tag{3.50}$$

       for an armature current limit : $\sqrt{I_P^2 + I_Q^2} - I^{max} = 0$ (3.52)

Frequency control: one of the following:

    under governor control: $V I_P - P_o + \gamma\dfrac{\omega_{sys} - \omega_o}{\omega_o} = 0$ (3.36)

    under turbine limit: $V I_P - P^{max} = 0$ (3.37)

Table 6.5   Reduced equilibrium model of the induction motor

COMPONENT OF $\tilde{\mathbf{x}}$	
s (motor slip)	(4.11)

EXPRESSIONS OF CURRENTS TO BE SUBSTITUTED IN (6.13a,b)	
$I_P = -\dfrac{(R_s + R_e)\,V}{(R_s + R_e)^2 + (X_s + X_e)^2}$	(4.21a)
$I_Q = -\dfrac{(X_s + X_e)\,V}{(R_s + R_e)^2 + (X_s + X_e)^2}$	(4.21b)
where $R_e + jX_e = \dfrac{jX_m\!\left(\frac{R_r}{s} + jX_r\right)}{\frac{R_r}{s} + j(X_m + X_r)}$	(4.14)

EQUATION $\tilde{\mathbf{f}}(\mathbf{y}, \tilde{\mathbf{x}}, \mathbf{z}_c, \mathbf{z}_d) = 0$	
Torque balance:	
$\dfrac{\frac{R_r}{s} X_m^2\, V^2}{\left[(R_1 + \frac{R_r}{s})^2 + (X_1 + X_r)^2\right]\left[(R_s^2 + (X_s + X_m)^2)\right]} - T_m(s) = 0$	(4.20)
where $R_1 + jX_1 = \dfrac{jX_m(R_s + jX_s)}{R_s + j(X_s + X_m)}$	(4.12b)
$T_m(s)$ depends upon the type of mechanical load	

Table 6.6   Reduced equilibrium model of the SVC

COMPONENT OF $\tilde{\mathbf{x}}$ AND EQUATION $\tilde{\mathbf{f}}(\mathbf{y}, \tilde{\mathbf{x}}, \mathbf{z}_c, \mathbf{z}_d) = 0$	
none	

EXPRESSIONS OF CURRENTS TO BE SUBSTITUTED IN (6.13a,b)	
$I_{Pi} = 0$	
$I_{Qi} = K(V_{oj} - V_j)V_i \quad$ if $B^{min} < K(V_{oj} - V_j) < B^{max}$	(2.19,2.20)
$\phantom{I_{Qi}} = B^{max}V_i \qquad\quad\;$ if $K(V_{oj} - V_j) \geq B^{max}$	(2.19,2.21)
$\phantom{I_{Qi}} = B^{min}V_i \qquad\quad\;$ if $K(V_{oj} - V_j) \leq B^{min}$	(2.19,2.21)
where j denotes the voltage controlled bus	
$\;\; i$ denotes the bus with variable susceptance	

where r is the number of the reference bus.

To summarize, the short-term equilibrium equations of a system with N buses, g synchronous generators and m induction motors consists of the following equations and variables:

equations		variables	
N	active current (6.13a)	N	bus voltage phase angles
N	reactive current (6.13b)	N	bus voltage magnitudes
1	phase reference (6.36)	1	system frequency ω_{sys}
$3g$	generator (6.35b)	$3g$	generator variables (E_q, E_q^s, δ)
m	motor (6.35b)	m	motor variables (s)

Note that the number of equations of the type (6.35b) depends on the reduction adopted. The above numbers refer to the model shown in Tables 6.4 to 6.6. Alternative formulations are possible with more, or less equations, as pointed out above.

6.5.2 Equilibrium of long-term dynamics

Formally, the long-term equilibrium conditions are obtained by setting:

$$\mathbf{h}_c(\mathbf{x}, \mathbf{y}, \mathbf{z}_c, \mathbf{z}_d) = 0 \qquad (6.37a)$$

$$\mathbf{h}_d(\mathbf{x}, \mathbf{y}, \mathbf{z}_c, \mathbf{z}_d) = \mathbf{z}_d \qquad (6.37b)$$

For simplicity, we group the discrete and continuous variables in a single n_z-dimensional vector:

$$\mathbf{z} = \begin{bmatrix} \mathbf{z}_c \\ \mathbf{z}_d \end{bmatrix}$$

Defining $\mathbf{h}$ as:

$$\mathbf{h} = \begin{bmatrix} \mathbf{h}_c \\ \mathbf{h}_d - \mathbf{z}_d \end{bmatrix}$$

the long-term equilibrium conditions take on the form:

$$\mathbf{f}(\mathbf{x}, \mathbf{y}, \mathbf{z}) = 0 \qquad (6.38a)$$

$$\mathbf{g}(\mathbf{x}, \mathbf{y}, \mathbf{z}) = 0 \qquad (6.38b)$$

$$\mathbf{h}(\mathbf{x}, \mathbf{y}, \mathbf{z}) = 0 \qquad (6.38c)$$

Taking into account the reduction of the short-term equilibrium equations discussed in the previous section, Eqs. (6.38a,b,c) become:

$$\tilde{\mathbf{f}}(\tilde{\mathbf{x}}, \mathbf{y}, \mathbf{z}) = 0 \qquad (6.39a)$$

$$\tilde{\mathbf{g}}(\tilde{\mathbf{x}}, \mathbf{y}, \mathbf{z}) = 0 \qquad (6.39b)$$

$$\tilde{\mathbf{h}}(\tilde{\mathbf{x}}, \mathbf{y}, \mathbf{z}) = 0 \qquad (6.39c)$$

Let us consider more specifically some long-term components or phenomena listed in Table 6.1.

Load Tap changer (LTC). According to (4.33), the equilibrium condition is

$$r_{k+1} = r_k$$

which corresponds to the controlled voltage lying within the LTC dead-band:

$$|V_2 - V_2^o| < d \qquad (6.40)$$

unless the LTC has hit a limit:

$$r_k = r^{max} \quad \text{with} \quad V_2 > V_2^o + d$$

$$\text{or } r_k = r^{min} \quad \text{with} \quad V_2 < V_2^o - d$$

In some computations the inequality (6.40) is conveniently approximated by:

$$V_2 = V_2^o \qquad (6.41)$$

Load self-restoration. Consider the multiplicative generic load model (4.48a,b) (a similar reasoning applies to the additive model (4.51a,b)). Setting the right-hand side of these equations to zero yields the equilibrium conditions:

$$z_P = \left(\frac{V}{V_o}\right)^{\alpha_s - \alpha_t} \qquad (6.42a)$$

$$z_Q = \left(\frac{V}{V_o}\right)^{\beta_s - \beta_t} \qquad (6.42b)$$

which correspond to the load restored to an exponential model with α_s and β_s exponents, as indicated by (4.47a,b), unless the state variables z_P and/or z_Q hit their limits:

$$z_P = z_P^{max} \quad \text{with} \quad \left(\frac{V}{V_o}\right)^{\alpha_s - \alpha_t} > z_P^{max}$$

$$\text{or } z_P = z_P^{min} \quad \text{with} \quad \left(\frac{V}{V_o}\right)^{\alpha_s - \alpha_t} < z_P^{min}$$

with similar relationships for the reactive variable z_Q.

Field and armature current limiters, shunt compensation, etc.. When dealing with long-term equilibrium equations, it may not be obvious to account for discrete-type devices such as OXLs or switched shunt compensation. In some cases, to know the state of these devices, it is required to know how the system has evolved before reaching long-term equilibrium. This holds true especially when several similar devices may interact (e.g. voltage-controlled shunt capacitors or reactors). The way of handling these discrete transitions depends on the use made of the equilibrium equations. Several techniques will be discussed in Chapter 9.

6.5.3   Equilibrium equations reduced to network only

In Section 6.5.1 we have discussed how the equilibrium model of the short-term dynamics can be reduced by eliminating some of the **x** variables. The same technique applies to the elimination of long-term variables. In some cases it is convenient to reduce the equilibrium model up to the point where no $\tilde{\mathbf{f}}$ and no $\tilde{\mathbf{h}}$ equation remains. This yields an equilibrium model reduced to the network equations only, which we will call for short *network-only model*. This takes on the form:

$$\tilde{g}(\mathbf{y}, \omega_{sys}) = 0 \qquad (6.43)$$

These equations are the network relationships (6.13a,b), with appropriate expressions substituted for the active and reactive currents, together with the phase reference relationship (6.36). Consequently, the balance between equations and variables becomes:

equations		variables	
N	active current (6.13a)	N	bus voltage phase angles
N	reactive current (6.13b)	N	bus voltage magnitudes
1	phase reference (6.36)	1	system frequency ω_{sys}

We discuss hereafter how to represent the various power system components at equilibrium in the network-only model.

Synchronous generator. Table 6.7 gives expressions for the injected powers. The one for active power P, accounting for governor effects, is the same as in Table 6.4. For the reactive power Q:
– in the simple case of an unsaturated round-rotor machine, an exact expression for Q is straightforwardly obtained from derivations of Chapter 3. This expression has been proposed for voltage stability calculations is [LAH95];

Table 6.7   Modelling of synchronous generator in network-only models

---

ACTIVE POWER

under governor control: $P = P_o - \gamma \dfrac{\omega_{sys} - \omega_o}{\omega_o}$

under turbine limit: $P = P^{lim}$

REACTIVE POWER

- unsaturated round-rotor machine $(X_d = X_q = X)$:

  under linear AVR control: $Q = -\dfrac{V^2}{X} + \dfrac{1}{X}\sqrt{[G(V_o - V)V]^2 - (XP)^2}$

  under integral OXL control: $Q = -\dfrac{V^2}{X} + \dfrac{1}{X}\sqrt{\left(VE_q^{lim}\right)^2 - (XP)^2}$

  under armature current limit: $Q = \sqrt{(VI^{max})^2 - P^2}$

- saturated machine modelled by equivalent reactances:

  under AVR control: $Q = Q_{bc} + V\left(\dfrac{V_{bc} - V}{X_{mQ}}\right)$

  (subscript bc refers to a base case; see [CG84, BCR84, VC91a])

  under OXL control: $Q = -\dfrac{V^2}{X_{cr}} + \dfrac{1}{X_{cr}}\sqrt{E_{cr}^2 V^2 - (X_{cr}P)^2}$

  (E_{cr} : constant emf; see [BCR84, VC91a])

  under armature current limit: same as above

- saturated machine modelled by curve-fitted polynomials:

  under AVR control: $Q = aV^2 + bV + c$

  under OXL control: $Q = a'V^2 + b'V + c'$

  under armature current limit: same as above

– for the (usual) case of a saturated machine, it is not possible to obtain a closed-form expression for Q and hence network-only models involve some approximation with respect to the model retaining some $\tilde{x}$ variables (see Table 6.4). Consequently, they may not be able to detect the loss of synchronism that may occur under degraded system conditions. A first approach consists in representing the machine under AVR control by a constant voltage behind an equivalent reactance X_{mQ}, and the machine under OXL control by a constant emf behind an equivalent reactance X_{cr}. These two reactances have been introduced in Sections 3.4.1 and 3.4.2, respectively;

– a second approach consists in fitting polynomials on VQ curves of the type shown in Fig. 3.17. A quadratic form as shown in Table 6.7 is sufficient. Note that for the machine under OXL control, the a', b', c' coefficients vary with active power P, as can be seen from Fig. 3.17.

Induction motor. Variations of P and Q with V for the induction motor have been illustrated in Figs. 4.6, 4.8 and 4.12 in Chapter 4. Here too, a polynomial model can be fitted to these curves. Again, such a model may not be able to reproduce the motor stalling at low voltage.

SVC. The SVC model shown in Table 6.6 is already of the network-only type.

HVDC. The simplified representation of an HVDC link by special loads applied to the AC system (as outlined in Section 4.7) is well suited to network-only models. An alternative formulation has been proposed in [CAD92].

LTC. The approximate equilibrium condition (6.41) can be implicitly taken into account by modelling the LTC-controlled load as a constant power load in the network equations. Indeed the restoration of the secondary voltage means the restoration of the voltage dependent load power, as explained in Section 4.4.3.

This very common technique applies to LTCs feeding distribution systems. It does not apply to LTCs between transmission and sub-transmission (see Fig. 2.22), if the sub-transmission system is modelled explicitly. If it is not, constant power loads can again be used to represent both the sub-transmission and distribution systems as seen from the transmission level (see also Section 2.8).

The constant power assumption stems from (6.41), which neglects the effects of the tap deadband and limits. In practice, a large number of LTC deadbands may have a non negligible (usually stabilizing) effect. The tap limits should be checked and, if reached, the load should be represented by its short-term equilibrium characteristic behind the limited transformer, which contributes in general to stabilizing the system. Finally, the constant power load should be located either behind the leakage reactance of the explicitly modelled transformer (bus 2 in Fig. 4.16) or at the high-voltage bus (bus 1 in the same figure), with the load power including the transformer losses, according to (4.38a,b). The second technique is computationally simpler and is convenient for (sub-)transmission utilities gathering real-time data on the primary side of the transformers.

Some publications dealing with *short-term* voltage stability analysis, use constant power loads in the system model in an attempt to approximate LTC effects. While this is correct for equilibrium calculations, let us emphasize that this leads very often to non-physical results concerning the dynamic response of the system. Indeed, the assumption of constant power loads makes the *long-term* load restoration by LTCs faster (in fact, infinitely faster) than the short-term dynamics of generators, SVCs, etc. As a consequence, time simulation or eigenvalue analysis may reveal oscillatory instabilities or singularities that do not exist in the real system. Reference [VCV96] discusses the limits of validity of the constant power load model and provides examples of the above artificial instabilities.

Load self-restoration. The equilibrium conditions (6.42a,b) yield the long-term equilibrium behaviour (4.47a,b). Hence in the network-only equilibrium model, the short-term exponential load model (4.46a,b) should be replaced by the long-term model (4.47a,b). This holds true provided that no limit on z_P or z_Q are met, as indicated in the previous section.

6.5.4   On the use of load flow equations

The load flow (or power flow) equations describe the steady-state operation of a power system as seen from the network. Load flow programs are used to determine system operating points (including initial conditions of dynamic simulations) as well as to evaluate the impact of contingencies on branch currents and bus voltages. The load flow program is certainly the most widely used power system analysis software. For these reasons, load flow equations have been widely used by industry in voltage stability analyses based on steady-state equations. In this section we first recall their formulation and then discuss their limit of validity.

The N system buses are classified into:

- N_{PQ} buses of the type "PQ", where the injected active and reactive powers are specified. This applies to buses without generation or with reactive power limited generators. Equations (6.12a,b) with constant P and Q are used for these buses;

- N_{PV} buses of the type "PV", where the injected active power and the bus voltage magnitude are specified. Equation (6.12a) with constant P is used, together with:

$$V_i = V_i^o \tag{6.44}$$

- one slack-bus: serving as a phase reference and injecting the power needed to satisfy the active power balance of the system. This power is supplied by a generator connected to this bus. Hence, assuming that the r-th bus is the slack-bus, we have:

$$V_r = V_r^o \tag{6.45}$$
$$\theta_r = 0 \tag{6.46}$$

This leads to $2N_{PQ} + 2N_{PV} + 2 = 2N$ equations. Note that by substituting the known values for the voltages at the slack and PV buses as known parameters, (6.44 - 6.46) may be eliminated from the equations, thereby reducing their number from $2N$ to $2N_{PQ} + N_{PV}$.

Clearly, the load flow equations belong to the family of network-only models. However, with respect to the network-only model discussed in the previous section, they involve the following approximations:

■ *Loads are treated as constant power.* Refer to the previous section for a discussion of load modelling in equilibrium equations;

■ *generators have constant voltage or constant reactive power*, which corresponds to the simple VQ diagram of Fig. 6.10. A truly constant voltage is adequate for integral-type AVRs (which exist on some machines) but for proportional AVRs with low gain and/or saturated exciters, the voltage droop should be taken into account. Under field - and even more under armature - current limit, the reactive power varies with the voltage. Figure 6.10 should be compared to the accurate VQ diagram of Fig. 3.17. As indicated in the previous section, a voltage-dependent reactive power injection is more adequate. More important, the reactive limit used in the load flow calculation must be updated with the active power, as shown by the capability curves of Fig. 3.19 or the VQ curves of Fig. 3.17;

■ *generators other than the slack-bus have constant active power.* Any change in generation level is compensated by the slack-bus, instead of being shared by a certain number of generators through governor or load frequency control. In the network-only model of Table 6.7, this is equivalent to making the permanent speed droop R infinite for all but the slack generators, in which case the ω_{sys} variable only appears in the slack-bus active power equation. The latter can be removed and solved separately for ω_{sys}. This decreases the number of equations from $2N + 1$ (general model of Section 6.5.3) to $2N$ (load flow approximation). In load flow calculations, the active power productions can be adjusted "externally" to share the effort over several generators [LSI85]. However, sensitivity analyses based on the load flow Jacobian do not take this into account.

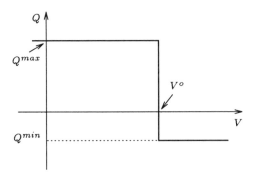

Figure 6.10   Simplified generator VQ diagram for load flow calculations

Reference [SP90] further discusses the approximations involved in stability conditions derived from the Jacobian of the load flow equations (as proposed originally in [VSI75]) instead of the exact equilibrium equations.

6.6   DETAILED EXAMPLE (CONTINUED) : EQUILIBRIUM FORMULATION

We continue now the detailed example of Section 6.3 considering the equilibrium modelling of the simple system previously described in that section.

6.6.1   Network equations

The network equations in active and reactive current formulation are readily obtained from (6.13a,b) as follows:

$$-\frac{E_{th}}{X_{th}}\sin(\theta_1 - \theta_{th}) + B_{14}V_4\sin(\theta_1 - \theta_4) \;=\; 0 \quad (6.47a)$$

$$\frac{E_{th}}{X_{th}}\cos(\theta_1 - \theta_{th}) - \frac{V_1}{X_{th}} + B_{14}V_1 - B_{14}V_4\cos(\theta_1 - \theta_4) \;=\; 0 \quad (6.47b)$$

$$I_{P2} + B_{24}V_4\sin(\theta_2 - \theta_4) \;=\; 0 \quad (6.47c)$$

$$I_{Q2} + (B_{s24} + B_{24})V_2 - B_{24}V_4\cos(\theta_2 - \theta_4) \;=\; 0 \quad (6.47d)$$

$$I_{P3} + B_{34}V_4\sin(\theta_3 - \theta_4) \;=\; 0 \quad (6.47e)$$

$$I_{Q3} + (B_{s34} + B_{34})V_3 - B_{34}V_4 \cos(\theta_3 - \theta_4) = 0 \quad (6.47\text{f})$$

$$B_{14}V_1 \sin(\theta_4 - \theta_1) + B_{24}V_2 \sin(\theta_4 - \theta_2) + B_{34}V_3 \sin(\theta_4 - \theta_3) = 0 \quad (6.47\text{g})$$

$$(B_{14} + B_{s42} + B_{24} + B_{s43} + B_{34})V_4 - B_{14}V_1 \cos(\theta_4 - \theta_1)$$
$$- B_{24}V_2 \cos(\theta_4 - \theta_2) - B_{34}V_3 \cos(\theta_4 - \theta_3) = 0 \quad (6.47\text{h})$$

where I_{P2} and I_{Q2} (resp. I_{P3} and I_{Q3}) are the components of the current injected by the generator (resp. the load), whose expressions will be detailed in the sequel. The current $I_{P1} + j\, I_{Q1}$ of the Thévenin equivalent at bus 1 has been directly substituted in (6.47a,b). Note that all currents are taken positive when injected into the system.

Since frequency is kept constant by the remote system connected to bus 1, the phase angle of the Thévenin emf is constant. It is thus natural to use this as a reference angle, by setting[3]:

$$\theta_{th} = 0 \qquad (6.48)$$

6.6.2 Generator equilibrium equations

The generator equilibrium equations are taken from Table 6.4. They are easily shown to be equivalent to the generator, AVR and OXL equations (6.20, 6.21, 6.25c, 6.28b) with time derivatives set to zero.

In the absence of saturation (assumption 6 of Section 6.3.2), the E_q and E_q^s variables coincide and the saturation relationship is eliminated.

The active and reactive current expressions of Table 6.4 become:

$$I_{P2} = \frac{E_q}{X_d}\sin(\delta - \theta_2) + \frac{V_2}{2}\left(\frac{1}{X_q} - \frac{1}{X_d}\right)\sin 2(\delta - \theta_2) \qquad (6.49\text{a})$$

$$I_{Q2} = \frac{E_q}{X_d}\cos(\delta - \theta_2) - V_2\left(\frac{\sin^2(\delta - \theta_2)}{X_q} + \frac{\cos^2(\delta - \theta_2)}{X_d}\right) \qquad (6.49\text{b})$$

The excitation control equation depends on the OXL status at long-term equilibrium:

- if $x_t < 0$, the OXL is inactive ($x_{oxl} = 0$) and the equilibrium equation is that of the AVR:

$$E_q - G(V_{2o} - V_2) = 0 \qquad (6.50)$$

---

[3] in fact, the Thévenin equivalent introduces a 5-th bus in the system. We do not write network equations at this bus, but rather take into account the equivalent through the current injected at bus 1. Four phase angles out of the five are free to vary.

- if $x_t \geq 0$, the OXL is active. At equilibrium, the input of the integrator block 4 in Fig. 3.12 is zero and hence:

$$E_q - I_{fd}^{lim} = 0 \qquad (6.51)$$

Since the generator does not participate in frequency control (assumption 8 of Section 6.3.2), its active power is constant (and equal to mechanical power) at equilibrium and the following equation holds:

$$I_{P2}V_2 - P_m = 0 \qquad (6.52)$$

with I_{P2} given by (6.49a).

The expressions (6.49a,b) of the currents must be substituted in the network equations (6.47c,d). They involve two internal state variables of the type $\tilde{x}$, namely E_q and δ, balanced by two equations of the type (6.35b), namely (6.52) and one of the two equations (6.50) or (6.51).

Note finally that for this simplified machine model, E_q can be replaced directly in (6.49a,b), so that the sole remaining generator variable is δ.

6.6.3   Load equilibrium equations

For the load, it is important to distinguish between short and long-term equilibrium. In the former the LTC ratio r is considered constant while in the latter it is assumed to have reached a value that restores the voltage V_3 to the setpoint value V_{3o}.

Short-term equilibrium

The load currents I_{P3} and I_{Q3} involved in (6.47e,f) are broken down respectively into:

$$
\begin{aligned}
I_{P3} &= I_{PE} + I_{PM} & (6.53a) \\
I_{Q3} &= I_{QE} + I_{QM} + I_{QC} & (6.53b)
\end{aligned}
$$

where subscript E refers to the exponential load, M to the induction motor and C to the shunt compensation. Expressions for the various components are derived hereafter.

The exponential load currents are readily obtained from (6.30a):

$$I_{PE} = P_o \frac{V_3^{\alpha-1}}{V_{3o}^{\alpha}}$$

$$I_{QE} = Q_o \frac{V_3^{\beta-1}}{V_{3o}^{\beta}}$$

and the shunt capacitor current is given by:

$$I_{QC} = B_{s3} V_3$$

The motor currents are obtained from Table 6.5, neglecting R_s (see assumption 9 of Section 6.3.2):

$$I_{PM} = -\frac{R_e V_3}{R_e^2 + (X_s + X_e)^2}$$

$$I_{QM} = -\frac{(X_s + X_e) V_3}{R_e^2 + (X_s + X_e)^2} \qquad (6.54)$$

with:

$$R_e + jX_e = \frac{jX_m(\frac{R_r}{s} + jX_r)}{\frac{R_r}{s} + j(X_m + X_r)}$$

The above expressions involve one variable of the type $\tilde{x}$, the motor slip s, which is balanced by the motor equilibrium equation, taken from the same table:

$$\frac{\frac{R_r}{s} X_m^2 V_3^2}{[(\frac{R_r}{s})^2 + (X_1 + X_r)^2] [(X_s + X_m)^2]} - T_M = 0 \qquad (6.55)$$

with

$$X_1 = \frac{X_m X_s}{X_m + X_s}$$

Long-term equilibrium

Neglecting the deadband and the limits, the LTC equilibrium condition is:

$$V_3 = V_{3o} \qquad (6.56)$$

This is equivalent to replacing the load at bus 3 by the constant power $P(V_{3o})$ (resp. $Q(V_{3o})$) it consumes under voltage V_{3o}, i.e.

$$I_{P3} = -\frac{P(V_{3o})}{V_{3o}} \qquad (6.57a)$$

$$I_{Q3} = -\frac{Q(V_{3o})}{V_{3o}} \qquad (6.57b)$$

The additional equation (6.56) is balanced by an additional unknown : the LTC ratio r (which appears in the network equations through B_{34}, B_{s34} and B_{s43}: see Table 6.2).

A different approach consists in shifting the constant power load at bus 4, using (4.38a,b). In this case, the reactive load at bus 4 includes the reactive losses in the X_{34} reactance. Moreover, r disappears from the equations, together with (6.56).

6.7 NUMBER-CRUNCHING PROBLEM

Here are data for the simple system detailed in Section 6.3 and 6.6.

System reactances (in pu on 100-MVA base):

$$X_{th} = 0.01 \quad X_{14} = 0.0277 \quad X_{24} = 0.016 \quad X_{34} = 0.004$$

Generator step-up transformer ratio : $n = 1.04$
Machine parameters (reactances in pu on machine 500-MVA rated power):

$$X_d = 2.1 \quad X_q = 2.1 \quad X'_d = 0.4$$
$$T'_{d0} = 8 \text{ s} \quad \omega_o = 2\pi 50 \text{ rad/s} \quad H = 3.5 \text{ s} \quad D = 4 \text{ pu}$$

AVR parameters (on the exciter pu system):

$$G = 50 \quad T = 0.1 \text{ s} \quad v_{fd}^{min} = 0. \text{ pu} \quad v_{fd}^{max} = 5. \text{ pu}$$

OXL parameters (pu system detailed in the text):

$$I_{fd}^{lim} = 2.825 \text{ pu} \quad S_1 = 1 \quad S_2 = 2$$
$$K_1 = 20 \quad K_2 = 0.1 \quad K_r = 1 \quad K_i = 0.1$$

LTC data:

$$r^{min} = 0.8 \quad r^{max} = 1.1 \quad \Delta r = 0.01$$
$$T_{fo} + T_m = 20 \text{ s} \quad T_f + T_m = 10 \text{ s} \quad V_{3o} = 1 \text{ pu} \quad d = 0.01 \text{ pu}$$

Motor data (reactances and resistance in pu on motor 800-MVA rated power):

$$X_s = 0.1 \quad X_r = 0.18 \quad X_m = 3.2 \text{ pu} \quad R_r = 0.018 \text{ pu} \quad H = 0.5 \text{ s}$$

Three cases are considered, corresponding to the load characteristics and operating point data detailed in Table 6.8.

Table 6.8 Load characteristics and operating point data

	Case 1	Case 2	Case 3
load characteristics			
α coefficient of exponential load	1.5	1.5	2.0
β coefficient of exponential load	2.5	2.5	2.0
Q_o/P_o ratio of exponential load	0.5	0.5	0.5
proportion of total load active power			
taken initially by motor	0.	0.	0.4
B_{s3} susceptance (pu on 100-MVA base)	6.0	6.0	6.822 $\star$
load flow data			
V_1 (pu)	1.08	1.10	1.08
V_2 (pu)	1.01	1.00	1.01
P_2 (MW)	300	450	200
P_3 (MW)	1500	1500	1500
Q_3 (Mvar)	150	150	150
$\star$ computed from the initial motor state to match the value of Q_3			

Using any convenient numerical time integration tool, determine the system response to the tripping of one circuit between buses 1 and 4 (reactance X_{14} doubled), in each of the three cases shown in Table 6.8.

The corresponding outputs are shown in Section 8.5, to illustrate long-term instability mechanisms. Refer to this section regarding the outputs to consider, the simulation time interval, etc.

7

LOADABILITY, SENSITIVITY AND BIFURCATION ANALYSIS

"Festina lente"[1]
Caesar Augustus

One key point in the analysis of a power system is the identification of the operating points where significant changes occur in voltage stability. As discussed in Chapter 5, these points are *bifurcation points* of the nonlinear system model. As we will show, these points correspond also to the *loadability limits* of the power system.

In this chapter we will discuss global problems, such as the identification of critical operating conditions, as well as local properties of an operating point (such as sensitivities), which are investigated with linearized models. Eigenvalue techniques are a special tool in this investigation.

When discussing bifurcations in this chapter, we will concentrate on saddle-node bifurcations, which are the most commonly encountered in cases of voltage instability and collapse.

In bifurcation analysis we assume smooth, slow parameter variations. The effect of large, sudden disturbances on the stability properties of the power system will be discussed in the next chapter.

---

[1] Hurry up slowly

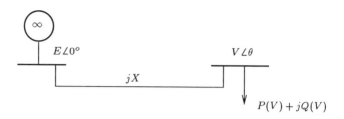

Figure 7.1 Two-bus system

7.1 LOADABILITY LIMITS

7.1.1 The effect of load characteristic

In Section 2.2 we have derived the maximum power that can be delivered at the load end of a two-bus system, while the concept of loadability limit was introduced in Section 2.5.2. We will now investigate further the loadability limits. We will start by analyzing the two-bus system first introduced in Fig. 2.2 using exponential and polynomial load characteristics. A one-line diagram of the system is shown in Fig. 7.1.

Exponential load

Consider first the exponential load model defined in (4.2a,b):

$$P \;=\; z\,P_o\left(\frac{V}{V_o}\right)^{\alpha} \tag{7.1a}$$

$$Q \;=\; z\,Q_o\left(\frac{V}{V_o}\right)^{\beta} \tag{7.1b}$$

In Section 2.5 we have seen how the load characteristic interacts with the network characteristic to produce the operating point of the system for various changes in demand. If we incorporate the exponential load (7.1a,b) into the equations (2.10a,b) of the two-bus system of Fig. 7.1, we obtain:

$$-\frac{EV}{X}\sin\theta \;=\; z\,P_o\left(\frac{V}{V_o}\right)^{\alpha} \tag{7.2a}$$

$$-\frac{V^2}{X}+\frac{EV}{X}\cos\theta \;=\; z\,Q_o\left(\frac{V}{V_o}\right)^{\beta} \tag{7.2b}$$

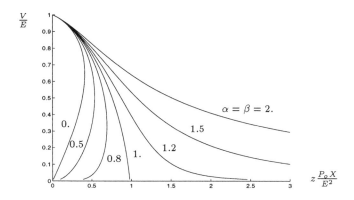

Figure 7.2 Voltage vs. load demand for various exponential load models

These equations have a trivial solution $V = 0$ (short circuit) which is of little interest here. Assuming $V \neq 0$ we may divide by V to get current instead of power equations:

$$-\frac{E}{X}\sin\theta = z\frac{P_o}{(V_o)^\alpha}V^{\alpha-1} \qquad (7.3a)$$

$$-\frac{V}{X} + \frac{E}{X}\cos\theta = z\frac{Q_o}{(V_o)^\beta}V^{\beta-1} \qquad (7.3b)$$

Let us investigate now how the number of solutions (V, θ) changes with the loading parameter z. For a given value of z, the system of equations (7.3a,b) cannot be solved analytically, except for some special values of α and β (for instance the integer values $\alpha = \beta = 0, 1$, or 2). In the general case, one has to resort to numerical computation.

A numerically computed plot of V as a function of z is shown in Fig. 7.2. In this figure, we chose $\alpha = \beta$ for simplicity and a constant power factor corresponding to $Q_o/P_o = 0.2$. Two radically different behaviours are observed in the solution curves:

- for $\alpha = \beta \geq 1$, there is a single solution V for every value of the demand z

- for $\alpha = \beta < 1$, there is a threshold value z^{max} such that for $z < z^{max}$ there are two solutions, for $z = z^{max}$ there is a single solution, and for $z > z^{max}$ there is none. As can be seen, z^{max} decreases with α $(= \beta)$. We will refer to the case $z = z^{max}$ as the *loadability limit* of the system, i.e. the operating condition corresponding to the largest load demand, for which a solution exists.

Note that in the first case ($\alpha = \beta \geq 1$) there is no loadability limit. Further mathematical discussion of this example may be found in [LSP92, LSP98], where it is shown that the above results are valid for all passive networks with loads having $P_0 > 0$. In the more general case, where $\alpha \neq \beta$, the following is observed:

■   for $\alpha > 1$: there is a value $\beta^\star$ such that for $\beta > \beta^\star$ there is a solution for every z; the larger α, the smaller $\beta^\star$

■   for $\alpha < 1$: whatever the value of β, there is a threshold z^{max} such that for $z > z^{max}$ there is no solution.

Polynomial load

Let us examine now the loadability limit of the two-bus system of Fig. 7.1 using the ZIP load model defined in Section 4.1.1. Substituting the polynomial load characteristics (4.4a,b) in equations (2.10a,b) we obtain:

$$\frac{EV}{X} \sin\theta + z P_o \left[a_P \left(\frac{V}{V_o}\right)^2 + b_P \frac{V}{V_o} + c_P \right] = 0 \qquad (7.4a)$$

$$\frac{V^2}{X} - \frac{EV}{X} \cos\theta + z Q_o \left[a_Q \left(\frac{V}{V_o}\right)^2 + b_Q \frac{V}{V_o} + c_Q \right] = 0 \qquad (7.4b)$$

As in the exponential load case, these equations are not easy to solve analytically. However, we can determine the operating point graphically as the intersection of network and load characteristics. This is shown in Fig. 7.3, where we have assumed for simplicity $a_P = a_Q, b_P = b_Q, c_P = c_Q$. With this assumption the load power factor remains constant and the network characteristic is easily drawn for the given power factor. For instance, the network characteristic drawn with a solid line in Fig. 7.3 corresponds to $Q/P = 0.2$. In the same figure a few load characteristics are drawn with dotted lines. It is assumed that $V_o = E$, $P_o = E^2/X$, in order to have both characteristics on the same scale.

As can be seen in Fig. 7.3, due to the constant power terms, a loadability limit z^{max} always exists for nonzero c_P or c_Q. The corresponding operating point is marked as point A. Note that the loadability limit A may be reached on the lower part of the network characteristic.

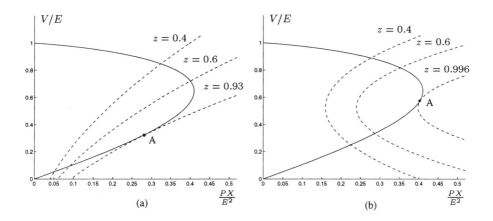

Figure 7.3 Loadability limits for ZIP load characteristic: (a) $a = 0.4$, $b = 0.5$, $c = 0.1$, (b) $a = 2.2$, $b = -2.3$, $c = 1.1$

7.1.2 Properties of loadability limits

We will investigate now a general property of loadability limits. We will concentrate on loadability limits brought about by an increase in load demand, but the techniques basically apply to other types of system stresses related to voltage stability, such as generation reschedulings which tend to decrease the production in a load area and increase a remote generation.

For the purpose of this analysis we assume that the power system is described (in steady state) by a set of n algebraic equations in n algebraic variables which we denote by the vector $\mathbf{u}$:

$$\varphi(\mathbf{u}, \mathbf{p}) = 0 \qquad (7.5)$$

where φ is a vector of *smooth* functions, and $\mathbf{p}$ is an n_p-dimensional vector of parameters.

Equation (7.5) is general, so that φ and $\mathbf{u}$ can correspond to the steady state of the power system under various conditions. For instance, φ can be any of the following:

- the set of network equations $\mathbf{g}$ with both short-term and long-term state variables considered constant ($\mathbf{x} = \mathbf{x}_o, \mathbf{z} = \mathbf{z}_o$), in which case $\mathbf{u} = \mathbf{y}$

■   the unreduced set of equations $[\mathbf{f}^T \quad \mathbf{g}^T]^T$ corresponding to short-term equilibrium
    with the long-term variables $\mathbf{z}$ considered constant, in which case $\mathbf{u}$ is made up
    of both $\mathbf{x}$ and $\mathbf{y}$

■   the reduced set of short-term equilibrium equations with $\tilde{\mathbf{f}}$ replacing $\mathbf{f}$, in which
    case $\mathbf{u}$ is made up of $\tilde{\mathbf{x}}$ and $\mathbf{y}$

■   the unreduced set of long-term equilibrium equations $[\mathbf{h}^T \quad \mathbf{f}^T \quad \mathbf{g}^T]^T$, where
    $\mathbf{u}^T = [\mathbf{z}^T \quad \mathbf{x}^T \quad \mathbf{y}^T]$

■   the network-only equations derived in Chapter 6, where again $\mathbf{u} = \mathbf{y}$.

In the last two cases some assumptions concerning the discrete variables $\mathbf{z}$ may be
necessary, so that the functions φ remain smooth. This problem will be addressed in
the last section of this Chapter.

Obviously, the results to be obtained depend on the set of equations used. For instance, by taking (7.5) to be the set of short-term equilibrium conditions, a short-term
loadability limit is determined. Similarly, when (7.5) corresponds to the long-term
equilibrium equations, the long-term loadability limit is obtained.

When discussing loadability, we will assume that the independent parameters $\mathbf{p}$ correspond to load demand variables. By *loadability limit* we have defined the point where
the demand reaches a maximum value, after which there is no solution of (7.5). If the
load is considered as constant power the loadability limit corresponds to the maximum
deliverable power to a bus, or a set of buses.

In a general power system there is an infinite number of ways to reach a loadability
limit, each "combination" of $\mathbf{p}$ variables yielding one particular limit. Translated
into mathematical terms, a loadability limit corresponds to the maximum of a scalar
function ζ of $\mathbf{p}$, over all solutions of equations (7.5). We thus consider the optimization
problem:

$$\max_{\mathbf{p},\mathbf{u}} \quad \zeta(\mathbf{p}) \tag{7.6}$$

$$\text{subject to} \quad \varphi(\mathbf{u},\mathbf{p}) = 0$$

The solution of the optimization problem satisfies the optimality conditions given
hereafter and known as the Kuhn-Tucker conditions [KT51]. First we define the
Lagrangian:

$$\mathcal{L} = \zeta(\mathbf{p}) + \mathbf{w}^T \varphi(\mathbf{u},\mathbf{p}) = \zeta(\mathbf{p}) + \sum_i w_i \varphi_i(\mathbf{u},\mathbf{p})$$

where $\mathbf{w}$ is a vector of Lagrange multipliers. Then we set to zero the derivatives of $\mathcal{L}$ with respect to $\mathbf{w}, \mathbf{p}$ and $\mathbf{u}$, thus obtaining the following necessary, first-order optimality conditions:

$$\nabla_{\mathbf{w}}\mathcal{L} = 0 \quad \Leftrightarrow \quad \varphi(\mathbf{u}, \mathbf{p}) = 0 \tag{7.7a}$$

$$\nabla_{\mathbf{p}}\mathcal{L} = 0 \quad \Leftrightarrow \quad \nabla_{\mathbf{p}}\zeta + \varphi_{\mathbf{p}}^T\mathbf{w} = 0 \tag{7.7b}$$

$$\nabla_{\mathbf{u}}\mathcal{L} = 0 \quad \Leftrightarrow \quad \varphi_{\mathbf{u}}^T\mathbf{w} = 0 \tag{7.7c}$$

where $\varphi_{\mathbf{u}}$ and $\varphi_{\mathbf{p}}$ are the Jacobians of φ with respect to $\mathbf{u}$ and $\mathbf{p}$, and $\nabla_{\mathbf{x}}\zeta$ is the gradient of the scalar function ζ with respect to the vector $\mathbf{x}$, i.e. the vector of partial derivatives such that:

$$[\nabla_{\mathbf{x}}\zeta]_i = \frac{\partial\zeta}{\partial x_i}$$

The first optimality condition (7.7a) yields the original set of constraints (7.5). Regarding the second condition (7.7b), note that ζ can be chosen to be such that $\nabla_{\mathbf{p}}\zeta \neq 0$. A typical such function is, for instance, the weighted sum of the components of $\mathbf{p}$. By assuming $\nabla_{\mathbf{p}}\zeta \neq 0$ it follows from (7.7b) that $\mathbf{w}$ cannot be zero at the solution. This, together with the third equation (7.7c) indicates that there is a linear combination between the columns of $\varphi_{\mathbf{u}}^T$ (i.e. the rows of $\varphi_{\mathbf{u}}$) and hence the following important result is obtained:

at a loadability limit, the Jacobian $\varphi_{\mathbf{u}}$ of the steady-state equations (7.5) is singular.

Thus a necessary loadability condition is:

$$\det \varphi_{\mathbf{u}} = 0 \tag{7.8}$$

This result can be translated in terms of eigenvalues and eigenvectors. Recall that the eigenvalues λ of a square matrix $\mathbf{A}$ are the solution of

$$\det(\mathbf{A} - \lambda\mathbf{I}) = 0$$

where $\mathbf{I}$ is the unity matrix. The right eigenvector $\mathbf{v}$ of an eigenvalue λ is such that:

$$\mathbf{A}\mathbf{v} = \lambda\mathbf{v}$$

while its left eigenvector satisfies:

$$\mathbf{A}^T\mathbf{w} = \lambda\mathbf{w}$$

Hence the above result can be restated as follows:

at a loadability limit, the Jacobian φ_u of equations (7.5) has a zero eigenvalue and the corresponding left eigenvector $\mathbf{w}$ is the vector of Lagrange multipliers of the optimization problem (7.6).

The singularity condition of various Jacobian matrices of the general form φ_u, monitored through the determinant, the smallest singular value, or the eigenvalue closest to the origin, has been used extensively in the voltage stability and collapse literature [VSI75, TMI83, KPB86, TT88, AJ88, SCS88, IWG90, GMK92, LSA92, IWG93, BBM96].

In this section we have assumed that the power system is represented in the steady state by the smooth set of equations (7.5). The results stated above are therefore valid for loadability limits that are not enforced by discrete events, such as generator limitation. The effect of discontinuities on the loadability limits will be discussed in the last section of this chapter.

7.1.3 Illustration on a two-bus system

Constant power load

Consider again the two-bus system of Fig. 7.1, with a constant power load at steady state. The set of steady-state equations (7.5) are obtained from (2.10a,b) as:

$$P + \frac{EV}{X} \sin \theta = 0 \tag{7.9a}$$

$$Q + \frac{V^2}{X} - \frac{EV}{X} \cos \theta = 0 \tag{7.9b}$$

This system may represent the steady-state conditions of either a constant excitation, round rotor generator with its synchronous reactance added to X, or a machine with an integral AVR. The load P, Q can be considered constant due to LTC action (deadband and ratio limits neglected). Thus φ in this case is a long-term equilibrium condition. Incidentally, this coincides with the power flow equations with one PQ and one slack bus.

In this example, the vector $\mathbf{u}$ consists of θ and V, the parameter vector is $\mathbf{p} = [P\ Q]^T$, and the Jacobian φ_u is given by:

$$\varphi_u = \begin{bmatrix} \dfrac{EV}{X} \cos \theta & \dfrac{E}{X} \sin \theta \\[3mm] \dfrac{EV}{X} \sin \theta & \dfrac{2V}{X} - \dfrac{E}{X} \cos \theta \end{bmatrix} \tag{7.10}$$

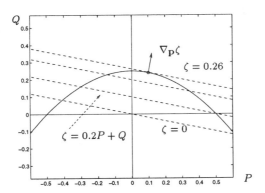

Figure 7.4   PQ parabola as a locus of optimal points

The singularity condition (7.8) is

$$\frac{2EV^2}{X^2}\cos\theta - \frac{E^2V}{X^2}\cos^2\theta - \frac{E^2V}{X^2}\sin^2\theta = 0$$

which can be simplified into:

$$E - 2V\cos\theta = 0 \qquad (7.11)$$

Using (7.9b) to eliminate θ and replacing V^2 by its expression (2.13), the above singularity condition becomes:

$$\pm\sqrt{\frac{E^4}{4} - X^2P^2 - XE^2Q} = 0$$

or:

$$\left(\frac{E^2}{2X}\right)^2 - P^2 - \frac{E^2}{X}Q = 0 \qquad (7.12)$$

which is equation (2.12) of the parabola that was drawn in Fig. 2.5 and is shown again here in Fig. 7.4. We can thus redefine the loadability limit parabola as the set of points where the Jacobian φ_u is singular.

To illustrate the optimization approach in determining loadability limits we consider the problem of maximizing the scalar function:

$$\zeta = aP + Q$$

subject to (7.9a,b). The optimization process is shown in Fig. 7.4 for $a = 0.2$. The dashed lines correspond to various values of ζ. Note that the gradient $\nabla_\mathbf{p}\zeta = [a \; 1]^T$ is perpendicular to the equal ζ lines. By changing the value of a different loadability limits are obtained.

Voltage sensitive load

In the general case, where the power consumed by the load depends on voltage, an appropriate load model has to be substituted for P and Q in (7.9a,b), as was done for the exponential load in (7.2a,b) and for the ZIP load in (7.4a,b). The vector $\mathbf{u}$ remains the same (θ and V), but the parameter vector $\mathbf{p}$ is now the scalar z, where z is the load demand. The singularity condition is different and corresponds to the maximum value of z, for which the appropriate equations have a solution. As pointed out in Section 7.1.1, for certain load models there are no loadability limits.

Consider, for instance, the steady-state equations for the ZIP load (7.4a,b). For $V_o = 1$ the Jacobian of this system is:

$$
\varphi_\mathbf{u} =
\begin{bmatrix}
\dfrac{EV}{X}\cos\theta & zP_o(b_P + 2a_PV) + \dfrac{E}{X}\sin\theta \\[2mm]
\dfrac{EV}{X}\sin\theta & zQ_o(b_Q + 2a_QV) + \dfrac{2V}{X} - \dfrac{E}{X}\cos\theta
\end{bmatrix}
$$

The corresponding singularity condition of $\varphi_\mathbf{u}$ is:

$$
E + zP_o(b_P + 2a_PV)\sin\theta - [zQ_ob_Q + 2(1 + zQ_oa_Q)V]\cos\theta \qquad (7.13)
$$

and is obviously different from (7.11) for nonzero a, or b coefficients.

For a given combination of P_o, Q_o, the singularity condition (7.13) and the original equations (7.4a,b) determine a loadability limit z^{max} and the corresponding values of V and θ. Using these values the load consumption P and Q for z^{max} can be calculated. Plotting these values in the (P, Q) plane (as shown in Fig. 7.5a with a dashed line), one obtains a loadability curve lying inside the parabola (7.12), which is plotted again in Fig. 7.5a with a solid line. This has to be expected, since no matter how P and Q change with voltage, they cannot increase beyond the parabola limit.

The relative position of the two limits in Fig. 7.5a may convey the wrong impression that the limits for voltage sensitive loads are more restrictive than those for constant power loads. To clarify things, one has to resort to the PV curves, like those of Fig. 7.3, from which it is obvious that the loading parameter z is smaller at the maximum power point (nose of the PV curve) than at the loadability limit. When z increases beyond its value at the nose point, the consumed power decreases.

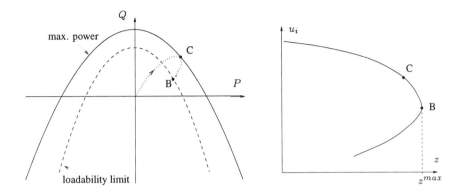

Figure 7.5a Loadability in PQ space   **Figure 7.5b** Parameter space

To show this effect in the load space we can draw a trajectory of the consumed P and Q as z increases. This is done in Fig. 7.5a with a dotted line. The trajectory reaches the parabola at point C and turns back before hitting the loadability limit at point B, at which $z = z^{max}$. This picture can be simplified by drawing the loadability limit in the independent parameter space instead of the load space. Since in our case the parameter z is scalar, the limit is a point z^{max} on the z axis. Plotting any dependent variable u_i against z one obtains Fig. 7.5b. The resulting curve folds at the loadability limit B for $z = z^{max}$ and not at point C where the consumed power is maximized.

7.2 SENSITIVITY ANALYSIS

7.2.1 Derivation of sensitivities

Consider the steady-state condition of a power system operating at some point characterized by a given $\mathbf{p}$ and $\mathbf{u}$ satisfying (7.5). Let η be some quantity of interest that can be expressed as a function of $\mathbf{u}$ and possibly $\mathbf{p}$.

If some change in the parameter $\mathbf{p}$ takes place, the system will generally operate at some other point still satisfying (7.5). As a result, η will also change. For small changes in $\mathbf{p}$ we are interested in determining the sensitivity of η to each p_i:

$$S_{\eta p_i} = \lim_{\Delta p_i \to 0} \frac{\Delta \eta}{\Delta p_i} \tag{7.14}$$

Differentiating $\eta(\mathbf{u}, \mathbf{p})$ according to the chain rule yields:

$$d\eta = d\mathbf{p}^T \nabla_{\mathbf{p}} \eta + d\mathbf{u}^T \nabla_{\mathbf{u}} \eta \tag{7.15}$$

Note that $\nabla_{\mathbf{p}} \eta = 0$ if η does not depend explicitly on $\mathbf{p}$. On the other hand, differentiating (7.5) gives:

$$\varphi_{\mathbf{u}} d\mathbf{u} + \varphi_{\mathbf{p}} d\mathbf{p} = 0$$

or, assuming that $\varphi_{\mathbf{u}}$ is nonsingular:

$$d\mathbf{u} = -\varphi_{\mathbf{u}}^{-1} \varphi_{\mathbf{p}} d\mathbf{p} \tag{7.16}$$

Introducing (7.16) into (7.15) yields:

$$d\eta = d\mathbf{p}^T [\nabla_{\mathbf{p}} \eta - \varphi_{\mathbf{p}}^T (\varphi_{\mathbf{u}}^T)^{-1} \nabla_{\mathbf{u}} \eta]$$

Hence the sensitivity (7.14) is the i-th component of the vector of sensitivities:

$$S_{\eta \mathbf{p}} = \nabla_{\mathbf{p}} \eta - \varphi_{\mathbf{p}}^T (\varphi_{\mathbf{u}}^T)^{-1} \nabla_{\mathbf{u}} \eta \tag{7.17}$$

Note that the sensitivities depend on the inverse of $\varphi_{\mathbf{u}}$.

Equation (7.17) can be expressed in terms of the eigenvalues and eigenvectors of $\varphi_{\mathbf{u}}$. Using the definitions of Section 5.1.3 this matrix can be written as:

$$\varphi_{\mathbf{u}} = \mathbf{V} \mathbf{\Lambda} \mathbf{W} \tag{7.18}$$

where $\mathbf{V}$ is a matrix, whose columns are the right eigenvectors of $\varphi_{\mathbf{u}}$, $\mathbf{\Lambda}$ is a diagonal matrix containing its eigenvalues (which are assumed distinct), and $\mathbf{W}$ is a matrix whose rows are its left eigenvectors. Assuming that eigenvectors are normalized so that:

$$\mathbf{w}_i^T \mathbf{v}_i = 1 \qquad i = 1, \ldots, n$$

and using property (5.17) it is easily seen that:

$$\mathbf{V} = \mathbf{W}^{-1}$$

Thus, inverting (7.18) we get:

$$\varphi_{\mathbf{u}}^{-1} = \mathbf{V} \mathbf{\Lambda}^{-1} \mathbf{W} = \sum_{i=1}^{n} \frac{\mathbf{v}_i \mathbf{w}_i^T}{\lambda_i} \tag{7.19}$$

Substituting the transpose of (7.19) in the sensitivity formula (7.17) we get:

$$S_{\eta \mathbf{p}} = \nabla_{\mathbf{p}} \eta - \varphi_{\mathbf{p}}^T \left[\sum_{i=1}^{n} \frac{\mathbf{w}_i \mathbf{v}_i^T}{\lambda_i} \right] \nabla_{\mathbf{u}} \eta \tag{7.20}$$

Since according to (7.8) at a loadability limit the Jacobian $\varphi_{\mathbf{u}}$ is singular, and thus has a zero eigenvalue, we can conclude that in the general case:

as a loadability limit is approached, all sensitivities tend to infinity.

Note that there are isolated exceptions to this rule. One such exception is the peculiar case of a radial system having a block diagonal φ_u with a single singular block and all components of $\nabla_u \eta$ relative to this block being zero (see discussion to [YSM92]).

The above property has been exploited in several sensitivity-type indices proposed in the literature [TMI83, CGS84, BCR84, FJC85, FOC90, BP92, BBM96].

7.2.2 Illustration on the two-bus system

Constant power load revisited

To illustrate the behaviour of sensitivities, we come back to the two-bus system of Fig. 7.1 and we consider $S_{Q_g Q}$, the sensitivity of the reactive power generation Q_g to the reactive load Q. Simply stated, $S_{Q_g Q}$ indicates how many var's the generator has to produce if the load consumes 1 var more, with all other parameters (including the active power load) remaining constant. $S_{Q_g Q}$ can be interpreted as the reciprocal of the efficiency of transmitting an increment of reactive power from the generator to the load.

The steady-state equations for the two-bus system with constant power load are (7.9a,b) with $\mathbf{u} = [\theta \ V]^T$. The sensitivity can be derived from the general formula (7.17) with:

$$\eta = Q_g = \frac{E^2}{X} - \frac{EV}{X} \cos \theta$$

The Jacobian matrix φ_u is the one given in (7.10). Considering the active load P constant, the only remaining parameter is Q and hence:

$$\nabla_p \eta = 0 \qquad \nabla_u \eta = \begin{bmatrix} \dfrac{EV}{X} \sin \theta \\ -\dfrac{E}{X} \cos \theta \end{bmatrix} \qquad \varphi_p = \begin{bmatrix} 0 \\ 1 \end{bmatrix}$$

Alternatively, the sensitivity can be derived from the expression (2.16) directly as:

$$S_{Q_g Q} = \frac{\partial Q_g}{\partial Q} = \pm \frac{E^2}{2X \sqrt{\left(\frac{E^2}{2X}\right)^2 - \frac{QE^2}{X} - P^2}}$$

which clearly tends to infinity as one approaches any point of the parabola (7.12).

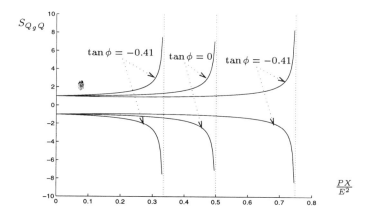

Figure 7.6   Sensitivity S_{Q_gQ} vs active load power P

As the operating point changes we can plot the sensitivity as a function of the loading. For instance, Fig. 7.6 shows S_{Q_gQ} as a function of the loading level P. The sensitivity depends also on the reactive loading, therefore it is plotted for various power factors. The sensitivity tends to $+\infty$ as the power limit is approached from the upper part of the PV curve and to $-\infty$ as it is approached from the lower part.

Voltage sensitive load

Note that when the load power depends upon voltage, the infinite sensitivities occur at the loadability limit, where the Jacobian φ_u becomes singular, and not at the maximum power point. Referring to Figs. 7.5a and 7.5b, the sensitivities become infinite at point B and not at point C, where the Jacobian φ_u is nonsingular.

7.3   BIFURCATION ANALYSIS

7.3.1   Modelling considerations

The reader may already have noticed the similarity between the necessary loadability condition (7.8) and the necessary Saddle-Node Bifurcation (SNB) condition (5.38) derived in Chapter 5. We will show in this section how these two problems can be related by properly selecting a set of equations.

Bifurcation analysis has been applied to voltage stability problems in many papers. Let us quote nonexhaustively [AFI82, KPB86, DC89, SP90, VSZ91, CAD92, AL92, PSL95a]. In this section we consider SNB analysis in a time-scale decomposition perspective following the approach introduced in [YSM92, VCV96]. To this purpose we derive SNB conditions for the complete, detailed model of a power system, as well as for the dynamic subsystems stemming from the time-scale decomposition defined in Section 6.4.

Since the necessary SNB condition of a system is the singularity of the state Jacobian (5.38), we will have to deal in this section with several Jacobian matrices, corresponding to different time scales. We give in Fig. 7.7 an overview of the linearized equations of the coupled power system model, as well as of the short-term and long-term approximate subsystems. The unreduced Jacobians (corresponding to the D-A system) and the reduced ones (or state Jacobians) introduced in Fig. 7.7 will be defined as we proceed. Note that there is a one-to-one correspondence between this figure and Fig. 6.9.

We will start by discussing the bifurcation conditions of the short-term and long-term subsystems (resulting from the quasi-steady-state approximation), following which we will review the dynamics of the coupled two-time-scale system and the limits of validity of the time-scale decomposition.

7.3.2   SNB of short-term dynamics

As discussed in Chapter 6 the short-term model consists of the following set of differential and algebraic equations, to which we have added an n_p-dimensional vector of parameters $\mathbf{p}$:

$$\dot{\mathbf{x}} = \mathbf{f}(\mathbf{x}, \mathbf{y}, \mathbf{z}, \mathbf{p}) \tag{7.21a}$$
$$0 = \mathbf{g}(\mathbf{x}, \mathbf{y}, \mathbf{z}, \mathbf{p}) \tag{7.21b}$$

where (7.21b) are the network equations, and $\mathbf{z}$ are the long-term variables, which are considered as slowly varying parameters. Thus both $\mathbf{p}$ and $\mathbf{z}$ can contribute to bifurcations.

According to the discussion of Section 7.1, the singularity of $\mathbf{g_y}$ is a necessary condition for a network loadability limit *with state variables fixed*. In the case of generators, a fixed value of $\mathbf{x}$ corresponds to a model made up of voltage sources (emf's) in series

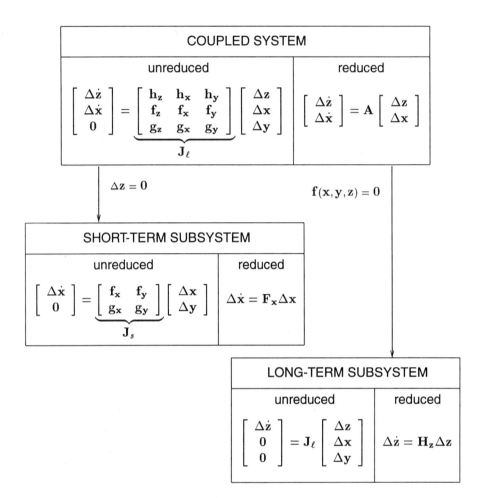

Figure 7.7 Jacobian matrices

with impedances. The same is true for induction motors. Thus equations (7.21b) refer to a passive network and as pointed out in Section 7.1.1, if all loads are assumed to be more sensitive to voltage than constant current ($\alpha, \beta > 1$), there is no loadability limit and g_y is never singular [LSP98]. This facilitates our analysis because, in the event of a singularity of (7.21b), we would have to resort to more detailed modelling of the network to capture the instability mechanism involved.

Assuming that g_y is nonsingular, we can use (7.21b) to eliminate y thus obtaining the following set of ODEs:

$$\dot{x} = F(x, z, p) \tag{7.22}$$

The equilibrium conditions of (7.22) are:

$$f(x^*, y^*, z, p) = 0 \tag{7.23a}$$
$$g(x^*, y^*, z, p) = 0 \tag{7.23b}$$

At such an equilibrium we can define the unreduced Jacobian matrix of the short-term subsystem:

$$J_s = \begin{bmatrix} f_x & f_y \\ g_x & g_y \end{bmatrix} \tag{7.24}$$

As discussed in Section 5.2.2, the necessary SNB condition of the system (7.22) is the singularity of the state Jacobian:

$$\det F_x(x^*, z, p) = 0 \tag{7.25}$$

When g_y is nonsingular, F_x is related to the submatrices of the unreduced Jacobian J_s as was shown in deriving equation (5.46):

$$F_x = f_x - f_y g_y^{-1} g_x \tag{7.26}$$

Now using Schur's formula (5.48) the determinant of the unreduced Jacobian J_s is related to that of the state matrix F_x through:

$$\det J_s = \det g_y \ \det[f_x - f_y g_y^{-1} g_x] = \det g_y \ \det F_x \tag{7.27}$$

Thus, for a nonsingular g_y the following singularity conditions are equivalent:

$$\det J_s = 0 \ \Leftrightarrow \ \det F_x = 0$$

or, in other words:

> the necessary condition for a saddle-node bifurcation of the short-term dynamics is the singularity of the unreduced Jacobian J_s.

Let us investigate now the relation of the SNB condition with that for a loadability limit. Using the notation of Section 7.1.2 we consider φ to be the set of short-term equilibrium conditions (7.23a,b) and u to correspond to x and y. The Jacobian of the optimization problem (7.6) is accordingly $\varphi_u = J_s$ and thus (for g_y nonsingular):

$$\det F_x = 0 \ \Leftrightarrow \ \det \varphi_u = 0$$

In other words:

the necessary condition for a saddle-node bifurcation of the short-term dynamics coincides with that for a loadability limit with the short-term dynamics at equilibrium.

7.3.3 Examples

Stability and SNB of induction motors

One major component affecting voltage stability in the short-term time scale is the induction motor load. Some of the cases discussed in Chapter 4, involving three-phase and single-phase induction motors under constant terminal voltage, can serve as examples of short-term dynamics stability and bifurcation analysis.

It was seen, for instance, in Fig. 4.4 that a three-phase induction motor with a constant mechanical torque has two equilibrium points (for a torque $T_0 < T_{max}$) of which the one closer to synchronous speed is stable, and the other one unstable. For $T_0 = T_{max}$ there is only one equilibrium (saddle-node), and for $T_0 > T_{max}$ there is no equilibrium.

When the mechanical torque varies with the square of motor speed, the motor has either one or three equilibrium points (Fig. 4.7). In the case of a single equilibrium point it was found that it is always stable, whereas in the case of three equilibrium points the middle one is unstable and the other two are stable.

The single-phase motor with constant mechanical torque behaves similarly to the 3-phase one having two equilibria (for torque smaller than T_{max}), of which the one with lower slip is stable. Contrary to the 3-phase motor, however, the single-phase motor with quadratic mechanical torque has always only one equilibrium point which is stable (Fig. 4.13).

More formally the stability of motor equilibria can be determined by obtaining the linearized version of the motor acceleration equation:

$$2H \Delta \dot{s} = \left(\frac{\partial T_m}{\partial s} - \frac{\partial T_e}{\partial s} \right) \Delta s \tag{7.28}$$

The corresponding state Jacobian is:

$$F_x = \frac{1}{2H} \left(\frac{\partial T_m}{\partial s} - \frac{\partial T_e}{\partial s} \right) \tag{7.29}$$

Since F_x must be negative for stability, when the slope of the electrical torque is larger than that of the mechanical torque, the equilibrium is stable. This confirms the intuitive reasoning used in Chapter 4.

Keeping the supply voltage constant and considering a slow variation of the torque parameter (T_0, or T_2), we can find the points where a stable and an unstable equilibrium coalesce and disappear. These are SNB points, for which the mechanical and electrical torque characteristics are tangent:

$$\frac{\partial T_m}{\partial s} = \frac{\partial T_e}{\partial s}$$

and therefore the state Jacobian is zero:

$$F_x = 0$$

Such a saddle-node bifurcation is, for instance, the point corresponding to T_{max} in Fig. 4.4. Also, points A and B of Fig. 4.7 are both SNB points. Note that depending on the direction of T_2 variation, in one of them a pair of equilibrium points (one stable and one unstable) are generated instead of disappearing. Finally, point C in Fig. 4.13 is a SNB point for the single-phase motor with a composite mechanical torque.

Also, the terminal voltage V of the motor can be used as a slowly varying parameter to produce SNB conditions. So, if we consider the torque parameter fixed, a SNB is encountered at the stalling point A of a three-phase motor with constant mechanical torque (Fig. 4.6). In the slip-torque diagram of Fig. 7.8 the process of lowering V is shown as a succession of electrical torque characteristics with decreasing T_{max}. At the SNB point A, T_{max} becomes equal to T_0 and stability is lost.

Motor fed through a line

Consider the system of Fig. 7.9, which consists of an infinite bus with a constant voltage E feeding a motor load through a line with reactance X. This configuration is more typical of a power system since the motor voltage changes with the loading conditions.

For simplicity we neglect the motor stator resistance R_s and assume constant mechanical torque, in which case the real power P drawn by the motor is independent of voltage. Thus we can draw the network QV characteristic for a given P and couple it with the motor QV characteristic, like the one drawn in Fig. 4.6. The coupling of the two characteristics is shown in Fig. 7.10. The motor parameters are those of the first row of Table 4.3, with stator resistance neglected. The network parameters are $E = 1, X = 0.4, B = 0.5$, all in per unit on the motor rating.

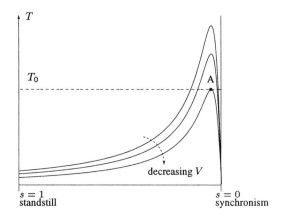

Figure 7.8 Slip-torque curves as V varies

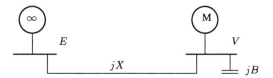

Figure 7.9 Motor fed through a line

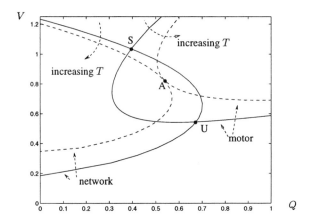

Figure 7.10 Network and motor QV curves

For a constant mechanical torque of 0.5 pu there are two equilibrium points, given as the intersections of the network and motor characteristics drawn in Fig. 7.10 with solid lines. Point S is stable, and point U is unstable. As the torque increases, the network characteristic shrinks (due to the increased active power demand), while the motor characteristic moves to the right (refer also to Fig. 4.6). Thus eventually the two equilibrium points approach each other and coalesce at the SNB point. The corresponding QV characteristics are plotted with dashed lines in Fig. 7.10 and are tangent to each other at point A, which is the stalling point of the motor.

Note that the bifurcation point is on the upper part of the network QV curve, which means that the induction motor is more restrictive than a constant Q load connected to the same bus. The latter would produce a bifurcation at the nose of the QV curve. Note also that the stalling point of the motor lies on its own QV characteristic before the minimum voltage point, which is a bifurcation point under constant terminal voltage. Thus the interaction of the network and the motor is detrimental for both components and should not be ignored.

Other examples of SNB of short-term dynamics have been given in [TT87, SO90, CAD92, RLS92, Pal93].

7.3.4   SNB of long-term dynamics

Let us analyze now the SNB condition for the long-term dynamics. In this section we consider that the short-term dynamics are at equilibrium, assuming a good separation of time scales and using the QSS approximation. For the sake of simplicity we will cover here only the case of continuous long-term dynamics. Under these assumption the long-term dynamics are described by the following Differential-Algebraic (D-A) system:

$$\dot{z} = h(x, y, z, p) \qquad (7.30a)$$
$$0 = f(x, y, z, p) \qquad (7.30b)$$
$$0 = g(x, y, z, p) \qquad (7.30c)$$

Consider the unreduced Jacobian of this system:

$$J_\ell = \left[\begin{array}{ccc} h_z & h_x & h_y \\ f_z & \left(\begin{array}{cc} f_x & f_y \\ g_x & g_y \end{array} \right) \end{array} \right] \qquad (7.31)$$

The submatrix in parentheses is the unreduced Jacobian J_s of the short-term subsystem (7.24). As discussed above, the singularity of J_s is a necessary condition for a SNB of short-term dynamics. Note that in the terminology of D-A systems of Section 5.3.2, the singularity of $\mathbf{J}_s$ at an equilibrium point results in a singularity induced bifurcation.

Assuming now that $\mathbf{J}_s$ is nonsingular, the implicit function theorem can be applied and thus the reduced long-term state equations take on the form:

$$\dot{\mathbf{z}} = \mathbf{H}(\mathbf{z}, \mathbf{p}) \tag{7.32}$$

The equilibrium condition of long-term dynamics is:

$$\mathbf{H}(\mathbf{z}, \mathbf{p}) = \mathbf{0} \tag{7.33}$$

which is equivalent to:

$$\mathbf{h}(\mathbf{x}, \mathbf{y}, \mathbf{z}, \mathbf{p}) = \mathbf{0} \tag{7.34}$$

in conjunction with (7.30b,c).

The state matrix of the long-term dynamics can be computed from the components of the unreduced long-term Jacobian $\mathbf{J}_\ell$ as:

$$\mathbf{H_z} = \mathbf{h_z} - [\mathbf{h_x} \quad \mathbf{h_y}]\mathbf{J}_s^{-1} \begin{bmatrix} \mathbf{f_z} \\ \mathbf{g_z} \end{bmatrix} \tag{7.35}$$

Using a similar analysis to that of Section 7.3.2, we can show that, when $\mathbf{J}_s$ is nonsingular:

$$\det \mathbf{H_z} = 0 \quad \Leftrightarrow \quad \det \mathbf{J}_\ell = 0$$

Thus:

> the necessary condition for a saddle-node bifurcation of the long-term dynamics is the singularity of the unreduced long-term Jacobian $\mathbf{J}_\ell$.

The unreduced long-term Jacobian is also the Jacobian of the algebraic equations (7.34, 7.30b,c) forming the long-term equilibrium constraints. Thus, in this case :

$$\varphi_\mathbf{u} = \mathbf{J}_\ell$$

and we can conclude that:

> the long-term loadability limits are met in general at a saddle-node bifurcation of long-term dynamics.

Alternative Jacobians

As pointed out in Chapter 6, the long-term equilibrium equation model (7.34, 7.30b,c) can be rewritten in the form of a network-only model (6.43), from which both short-term and long-term state variables have been eliminated. Assuming that one can eliminate system frequency, the network variables **y** are the only remaining ones. We will denote this set of equations as:

$$G(y) = 0 \tag{7.36}$$

The elimination of **x** and **z** requires that:

$$\det \begin{bmatrix} h_z & h_x \\ f_z & f_x \end{bmatrix} \neq 0 \tag{7.37}$$

Under this assumption, the Jacobian of (7.36) is given by:

$$G_y = g_y - \begin{bmatrix} g_z & g_x \end{bmatrix} \begin{bmatrix} h_z & h_x \\ f_z & f_x \end{bmatrix}^{-1} \begin{bmatrix} h_y \\ f_y \end{bmatrix} \tag{7.38}$$

It is easily shown using once again Schur's formula that:

$$\det J_\ell = \det \begin{bmatrix} h_z & h_x \\ f_z & f_x \end{bmatrix} \cdot \det G_y$$

Thus the Jacobian G_y becomes singular at a long-term loadability limit, just as the unreduced one J_ℓ and H_z.

In [GMK92] the set of equilibrium conditions (7.36) is written in the form of power flow equations:

$$P(\theta, V) = 0 \tag{7.39a}$$
$$Q(\theta, V) = 0 \tag{7.39b}$$

From (7.39a,b) we can write:

$$G_y = \begin{bmatrix} P_\theta & P_V \\ Q_\theta & Q_V \end{bmatrix} \tag{7.40}$$

Using this formulation we can proceed one step further, in order to calculate the effect that a small change in voltage ΔV will produce on the reactive powers ΔQ. Assuming that P_θ is nonsingular, this particular sensitivity matrix is given by:

$$J_{QV} = Q_V - Q_\theta P_\theta^{-1} P_V \tag{7.41}$$

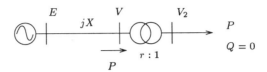

Figure 7.11   Two-bus LTC system

This matrix is helpful for network calculations [GMK92]. By following a similar approach to that followed for the other Jacobians, it is easily shown that:

$$\det \mathbf{G_y} = \det \mathbf{P}_\theta \ \det \mathbf{J}_{QV}$$

Thus, with the assumptions mentioned in this section, the reduced matrix $\mathbf{J}_{QV}$ becomes singular together with $\mathbf{G_y}$, $\mathbf{J}_\ell$ and $\mathbf{H_z}$.

7.3.5   Examples

Generator–LTC

Consider the simple system of Fig. 7.11. A load is fed by a generator through a line and an LTC transformer. For the sake of simplicity, we assume a single reactance X incorporating leakage and line reactance. We assume also a purely active (unity power factor) load. The LTC tries to keep the low voltage V_2 at the V_2^o setpoint value.

We assume that the load characteristic is exponential. In accordance with Chapter 4, we call this the *transient* or *short-term* load characteristic:

$$P = P_o \left(\frac{V_2}{V_2^o} \right)^{\alpha_t} \tag{7.42}$$

where we assume $\alpha_t > 1$ and take the LTC setpoint V_2^o as the voltage reference of the exponential load. The transformer being ideal, P is also the power entering the transformer (see Fig. 7.11). Since r is the ideal transformer ratio the secondary voltage is:

$$V_2 = V/r \tag{7.43}$$

Hence the load power can be expressed as a function of V:

$$P = P_o \left(\frac{V}{r V_2^o} \right)^{\alpha_t} \tag{7.44}$$

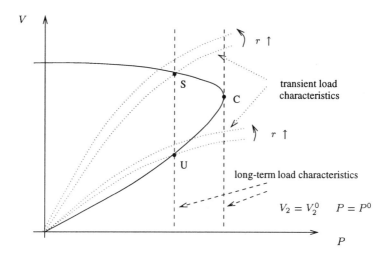

Figure 7.12 PV curve of two-bus system

To each value of r corresponds one transient load characteristics as seen from the network. The latter are shown with dotted lines in Fig. 7.12.

In this system the long-term dynamics are due to the LTC. The long-term equilibrium is such that $V_2 = V_2^o$ (ignoring the deadband), which means $P = P_o$. In other words, for a given load demand, the long-term load characteristics is constant power, as shown by the vertical dashed line in Fig. 7.12.

Assuming that the short-term dynamics are stable, the short-term equilibrium characteristic of the network and generator is the well-know PV curve shown with solid line in Fig. 7.12. As there are no long-term generator dynamics (we do not consider excitation limitation in this example), this is also the long-term equilibrium characteristic of the network. Point C corresponds to the maximum power that can be delivered to the load. For a given load demand P_o, less than this maximum power, we thus have two long-term equilibria, denoted by S and U in Fig 7.12.

Assuming a continuous LTC model, like the one of (4.35) the long-term state equation of the system is:

$$T_c \dot{r} = V_2 - V_2^o$$

The long-term state Jacobian is:

$$H_z = \frac{1}{T_c} \frac{\partial V_2}{\partial r}$$

Thus, the stability condition for an LTC is that the increased tap ratio r results in decreased secondary voltage:

$$\frac{\partial V_2}{\partial r} < 0 \tag{7.45}$$

The stability of equilibrium points can be checked intuitively, without resorting to the above condition, by slightly disturbing the LTC ratio. Figure 7.12 shows for instance the effect of a small increase in r (the same conclusions would be drawn considering a decrease in r):

- at point S, the increase in r yields a lower load power and hence a lower secondary voltage. Therefore the LTC will react by decreasing r and the operating point will come back to S. The operating point S is thus stable;

- at point U, the increase in r yields a higher load power and hence a higher secondary voltage. Therefore, the LTC will react by further increasing r, making the operating point further depart from U. The operating point U is thus unstable.

If the demand P_o increases gradually, the two equilibria converge to each other. At point C they coalesce and disappear. Point C is thus a *saddle-node bifurcation of the long-term dynamics*. The necessary condition for a SNB of the LTC is:

$$\frac{\partial V_2}{\partial r} = 0 \tag{7.46}$$

It is easy to show that when this condition holds, the long-term load characteristic is tangent to the network characteristic.

Restorative load

In the previous example, the long-term load characteristic was constant power due to LTC action. Therefore, the saddle-node bifurcation and maximum power points coincided. This is no longer true if the load restores to a long-term characteristic other than constant power.

To illustrate this, we consider the same system as in Fig. 7.11 but we replace the transformer and passive load with a self-restoring generic load of the multiplicative

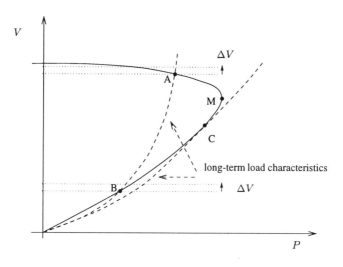

Figure 7.13 Load restoration with $\alpha_s < 1$

type described in Section 4.6.2. The steady-state (or long-term) load characteristic is:

$$P_s = P_o \left(\frac{V}{V_o}\right)^{\alpha_s} \tag{7.47}$$

where we take $0 < \alpha_s < 1$. (In the previous example α_s was equal to zero).

The corresponding PV curves are shown in Fig. 7.13. The dashed curves are two long-term characteristics. If the demand P_o increases the two equilibria A and B coalesce at point C, which is a saddle-node bifurcation. Thus when reaching point C the system becomes unstable. Note that the long-term loadability limit is point C, and not the maximum power point M.

The differential equation of the generic multiplicative load model is given by (4.48a) and can be rewritten as:

$$P_o T \dot{z} = P_s - P$$

where P_s is given by (7.47) and P is the power consumed by the load. The state Jacobian of the system is:

$$H_z = \frac{1}{TP_o} \left(\frac{\partial P_s}{\partial z} - \frac{\partial P}{\partial z}\right) \tag{7.48}$$

The first term in the parenthesis is due to the fact that a change in z will create a change in V and thus through (7.47) a change in P_s.

To investigate the stability of points A and B of Fig. 7.13 we consider the consumed power P as a function of V obtained from the *network* PV characteristic. We can thus apply the chain rule to (7.48) to obtain:

$$H_z = \frac{1}{TP_o} \left(\frac{\partial P_s}{\partial V} - \frac{\partial P}{\partial V} \right) \frac{\partial V}{\partial z}$$

Assuming any transient load model with $\alpha_t > 1$, it is clear from Fig. 7.13 that as the load parameter z increases, the voltage V drops. Thus $\partial V/\partial z < 0$ and the stability condition is:

$$\frac{\partial P_s}{\partial V} > \frac{\partial P}{\partial V}$$

i.e. the slope of the long-term load characteristic $\partial P_s/\partial V$ must be larger than the slope of the network characteristic $\partial P/\partial V$. If we assume a positive ΔV at point A of Fig. 7.13, we obtain $\Delta P_s > \Delta P$ and thus point A is stable. At point B the opposite holds true, i.e. $\Delta P_s < \Delta P$ and thus this point is unstable.

At the SNB point C the two slopes are equal:

$$\frac{\partial P_s}{\partial V} = \frac{\partial P}{\partial V}$$

and thus the load and network characteristics are tangent to each other.

Further discussion of SNB of long-term dynamics can be found in [Pal92, GMK92, Hil93, XM94].

7.3.6   Interaction between time scales

SNB of the coupled system

The full model of the two time-scale power system (considering for simplicity only continuous long-term dynamics) is made up of the following set of D-A equations:

$$\dot{z} = \mathbf{h}(\mathbf{x}, \mathbf{y}, \mathbf{z}) \qquad (7.49a)$$
$$\dot{\mathbf{x}} = \mathbf{f}(\mathbf{x}, \mathbf{y}, \mathbf{z}) \qquad (7.49b)$$
$$0 = \mathbf{g}(\mathbf{x}, \mathbf{y}, \mathbf{z}) \qquad (7.49c)$$

The analysis of this system is certainly more demanding computationally than the decoupled one consisting of the short-term and long-term subsystems. Therefore, the use of the coupled system is recommended only when the short-term dynamics start becoming slow, so that there is considerable interaction between the time scales.

Assuming as usual that the algebraic Jacobian $\mathbf{g_y}$ is nonsingular, the state Jacobian of the coupled system is given by:

$$\mathbf{A} = \begin{bmatrix} \mathbf{h_z} & \mathbf{h_x} \\ \mathbf{f_z} & \mathbf{f_x} \end{bmatrix} - \begin{bmatrix} \mathbf{h_y} \\ \mathbf{f_y} \end{bmatrix} \mathbf{g_y}^{-1} \begin{bmatrix} \mathbf{g_z} & \mathbf{g_x} \end{bmatrix} \qquad (7.50)$$

Using once more Schur's formula, we can establish an equivalence of the singularity conditions of the coupled system state matrix $\mathbf{A}$ and the long-term Jacobian $\mathbf{J}_\ell$, which are linked by the relation:

$$\det \mathbf{J}_\ell = \det \mathbf{g_y} \det \mathbf{A} \qquad (7.51)$$

Thus, for nonsingular $\mathbf{g_y}$:

$$\det \mathbf{A} = 0 \quad \Leftrightarrow \quad \det \mathbf{J}_\ell = 0$$

Using the equivalence obtained earlier between the singularity of $\mathbf{J}_\ell$ and $\mathbf{H_z}$ (when $\mathbf{J}_s$ is nonsingular) we arrive at the following important conclusion:

> the necessary condition for a saddle-node bifurcation of the coupled system dynamics coincides with that for the long-term subsystem.

Thus the bifurcation points predicted by the long-term dynamics subsystem are also in general bifurcation points of the coupled system and no approximation error is involved in predicting a SNB from the long-term subsystem alone.

Interaction of short- and long-term dynamics

As long-term dynamics evolve (especially after an instability of long-term dynamics) they may induce a SNB of the short-term subsystem. In this way, a slow long-term instability can produce a sudden collapse occurring at the SNB point of short-term dynamics.

As we have seen in the oscillator example of Section 5.4.4, near a SNB of the fast subsystem dynamics, the trajectory departs from the slow manifold (which in our case is the short-term equilibrium manifold). This can be approximated as a discrete transition occurring at the SNB point (Fig. 5.15). Since, however, an eigenvalue of short-term dynamics approaches the origin, the short-term dynamics become *slow* before the SNB. Therefore, an accurate analysis requires that we go back to the coupled power system equations (7.49a,b,c). The approximation of a discrete transition becomes better as the long-term dynamics become infinitely slow.

Note that it may not be always necessary to resort to the complete coupled model when the short-term Jacobian $\mathbf{J}_s$ becomes singular. Indeed, the singularity may be local, in

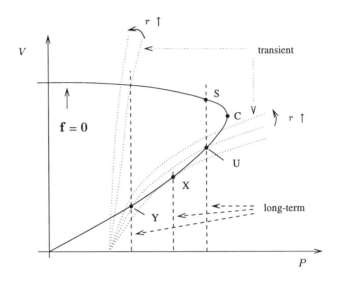

Figure 7.14   Effect of constant power load

which case only a subset of the short-term equations f need to be included. In the case of induction motors this procedure is analyzed in [VM98].

7.3.7   Examples

LTC with ZIP load

We revert back to the system of Fig. 7.11, where we assume now that the short-term load equilibrium characteristics has some constant power part. Although this load model is questionable at low voltage, it serves here as an illustration of the problems encountered by long-term models, when short-term stability is lost.

The long-term load characteristic is constant power due to LTC action. The corresponding curves are shown in Fig. 7.14. Consider the long-term equilibria S and U in Fig. 7.14. As we have seen in Section 7.3.5, equilibrium point S is stable, U is unstable, and point C is a saddle-node bifurcation of long-term dynamics, separating stable and unstable long-term equilibrium points.

If we apply the intuitive stability check of Section 7.3.5 (see Fig. 7.12) to the long-term equilibria below X (e.g. Y), they appear to be stable. As we will see, however, this conclusion is erroneous.

Let us first investigate the particular nature of the long-term equilibrium X. At this point, the *short-term* load characteristic is tangent to the network PV curve. This means that both $\mathbf{F_x}$ and $\mathbf{J}_s$ are singular. If the long-term state variable r is decreased slightly, the short-term load characteristics does no longer intersect the network PV curve. Thus, point X is an SNB of short-term dynamics brought about by the change in the long-term variable r. At this point two short-term equilibria (one with a positive, and the other with a negative real eigenvalue) coalesce and disappear. Assuming that the short-term equilibria on the upper part of the PV curve are stable up to point X, it follows that those below X are unstable. Accordingly, the long-term stability check of Section 7.3.5, which assumed implicitly that the short-term equilibrium is stable, cannot be applied below X.

Since the assumption of a stable short-term dynamics equilibrium stops being valid at point X, the time-scale decomposition breaks down at this point. The differential-algebraic system corresponding to the QSS approximation of the long-term dynamics undergoes a singularity induced bifurcation (see Section 5.3.2). This is revealed by an infinite H_z (in the general case an infinite eigenvalue of $\mathbf{H_z}$). When X is approached from the direction of U, H_z tends to $+\infty$. When X is approached from the direction of Y, H_z tends to $-\infty$.

Using the terminology of [HH91], point X is a *non-causal* point separating the constraint manifold (which in this case is the PV curve of Fig. 7.14) into two causal regions, one above and one below X.

For a detailed analysis near point X one should resort to the coupled modelling including short-term dynamics, as illustrated in the following example.

Generator-LTC-infinite bus

In this example we illustrate the effect of the interaction between short-term and long-term dynamics. Consider the power system of Fig. 7.15 consisting of a constant-excitation generator located near a purely resistive load fed through an LTC. The load exceeds the local generator capacity and is also fed from an infinite bus through a tie-line. The short-term dynamics are those of the generator, the long-term dynamics are those of the LTC, which is considered continuous.

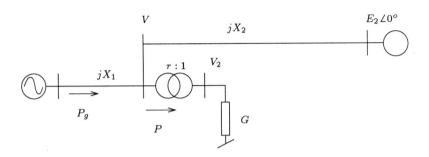

Figure 7.15   Generator-LTC-infinite bus

The network PV curve showing the primary LTC voltage V as a function of load power P (see Fig. 7.15) is shown in Fig. 7.16 with a solid line. This curve is plotted considering generator equilibrium conditions. In the same figure the load PV curves, as seen from the primary side of the LTC transformer, corresponding to two values of the tap ratio r are plotted with dashed lines.

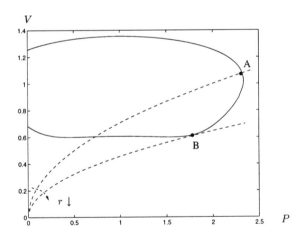

Figure 7.16   Network and load PV curves

Let us assume that the tap ratio r is decreasing slowly trying to regulate the secondary voltage V_2, thus bringing the operating point towards B, which is a SNB point of short-term dynamics. Further decrease in r causes the short-term equilibrium to disappear.

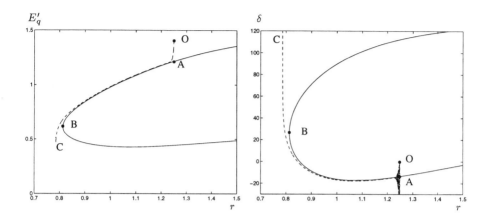

Figure 7.17 Generator equilibrium manifold

To illustrate the effect of the time-scale decomposition, we plot in Fig 7.17 with a solid line the short-term equilibrium manifold of the generator state variables E'_q and δ, as functions of the long-term state variable r. Using the QSS approximation, as the tap ratio r decreases, the system moves along the solid lines from point A to point B where the short-term equilibrium is lost. By simulating the full dynamics of the coupled system, and assuming initial conditions corresponding to point O lying outside the short-term equilibrium manifold, we obtain the trajectory marked with dotted lines in Fig 7.17. Clearly the QSS approximation is valid until close to the SNB point B. The final outcome of the loss of short-term equilibrium is in this case the loss of generator synchronism. A similar case is analyzed in [VSP96], where it is also shown that before losing synchronism, the generator undergoes a phase of slow demagnetization.

Note that the rate of tap movement determines the proximity of the coupled system response to points A and B of Fig 7.17. If the tap-changer response is made slower (better time-scale separation) the response of the coupled system will come closer to both these points. If, on the other hand, the tap movement becomes faster, the interaction between time scales increases and the trajectory will depart further away from points A and B.

7.4 EIGENVECTOR AND SINGULAR VECTOR PROPERTIES

We have up to now established the simultaneous occurrence of singularity in the reduced and unreduced Jacobian matrices. One important property of the unreduced Jacobian is that it is very sparse. This makes it an attractive candidate for various practical computations in a large-scale system. A typical example is the computation of eigenvalues and eigenvectors [Mar86, WS90] or singular values and vectors [LSA92].

7.4.1 On the use of eigenvectors

As discussed in Chapter 5, the eigenvalues and eigenvectors of the state Jacobian determine the response of the system close to an equilibrium. We have also pointed out that at a SNB a real eigenvalue becomes zero as the equilibrium point disappears. The corresponding eigenvectors contain valuable information on the nature of the bifurcation, the response of the system and the effectiveness of control measures [GMK92, Dob92, MXA94, RVC96]. The right eigenvector shows the direction in state space along which the states will evolve due to the SNB. On the other hand, the left eigenvector shows which states have a prominent effect on the zero eigenvalue, i.e. which states are more effective in order to control the bifurcation.

This applies equally to short-term and long-term dynamics. For instance:

- in the *short-term* the right eigenvector corresponding to a zero eigenvalue indicates which generator angles, or motor slips will increase due to the bifurcation, and consequently which machines are most likely to lose synchronism, or which motors are prone to stalling [VCV96, VM98];

- in the *long-term*, the left eigenvector indicates which tap-changers should be blocked, or reversed, to contain a voltage instability [RVC96].

We will now show that, at the point of singularity, the left and right eigenvectors corresponding to the zero eigenvalue of the unreduced, sparse Jacobian are directly related to those of the reduced state matrix. For later reference, we consider here only the long-term model, the derivation for the short-term one being identical.

Assume a long-term equilibrium point where the long-term Jacobian $\mathbf{J}_\ell$ is singular and the short-term one $\mathbf{J}_s$ is nonsingular. The equation defining the right eigenvector of

$\mathbf{J}_\ell$ corresponding to the zero eigenvalue is:

$$
\begin{bmatrix}
\mathbf{h_z} & \mathbf{h_x} & \mathbf{h_y} \\
\mathbf{f_z} & & \\
\mathbf{g_z} & & \mathbf{J}_s
\end{bmatrix}
\begin{bmatrix}
\mathbf{v}_z \\
\mathbf{v}_x \\
\mathbf{v}_y
\end{bmatrix}
=
\begin{bmatrix}
0 \\
0 \\
0
\end{bmatrix}
\tag{7.52}
$$

where $\mathbf{v}_z$, $\mathbf{v}_x$ and $\mathbf{v}_y$ are partitions of the right eigenvector corresponding to long-term state variables $\mathbf{z}$, short-term state variables $\mathbf{x}$, and network variables $\mathbf{y}$ respectively. The elimination of $\mathbf{v}_x$ and $\mathbf{v}_y$ from equation (7.52) gives:

$$
\left(\mathbf{h_z} - [\mathbf{h_x} \ \ \mathbf{h_y}] \mathbf{J}_s^{-1}
\begin{bmatrix}
\mathbf{f_z} \\
\mathbf{g_z}
\end{bmatrix}
\right) \mathbf{v}_z = \mathbf{H_z} \mathbf{v}_z = 0
\tag{7.53}
$$

Thus, the elements of the right eigenvector of $\mathbf{J}_\ell$ corresponding to long-term state variables $\mathbf{v}_z$ form the right eigenvector of $\mathbf{H_z}$ relative to the zero eigenvalue. The same is true for the left eigenvector as well:

$$
\begin{bmatrix} \mathbf{w}_z^T & \mathbf{w}_x^T & \mathbf{w}_y^T \end{bmatrix}
\begin{bmatrix}
\mathbf{h_z} & \mathbf{h_x} & \mathbf{h_y} \\
\mathbf{f_z} & & \\
\mathbf{g_z} & & \mathbf{J}_s
\end{bmatrix}
= \begin{bmatrix} 0 & 0 & 0 \end{bmatrix}
\tag{7.54}
$$

from which after elimination:

$$
\mathbf{w}_z^T \left(\mathbf{h_z} - [\mathbf{h_x} \ \ \mathbf{h_y}] \mathbf{J}_s^{-1}
\begin{bmatrix}
\mathbf{f_z} \\
\mathbf{g_z}
\end{bmatrix}
\right) = \mathbf{w}_z^T \mathbf{H_z} = 0
\tag{7.55}
$$

Using this property, the calculation of the eigenvectors corresponding to a zero eigenvalue of $\mathbf{H_z}$ can be performed indirectly and more efficiently through the eigenvectors of the sparse Jacobian matrix $\mathbf{J}_\ell$.

Applying the same reasoning, we can conclude that for a zero eigenvalue, the eigenvector elements of $\mathbf{J}_\ell$ corresponding to the network variables ($\mathbf{v}_y$ and $\mathbf{w}_y$) form the respective eigenvectors of matrix $\mathbf{G_y}$ introduced in (7.36). Moreover, the eigenvector elements of $\mathbf{J}_\ell$ corresponding to voltage magnitudes form the eigenvectors of $\mathbf{J}_{QV}$ defined in 7.41.

Similarly, it can be shown that the eigenvectors of $\mathbf{J}_{QV}$ coincide with the elements corresponding to voltage of the eigenvectors of $\mathbf{G_y}$. This correspondence applies only for a zero eigenvalue.

Beside computational efficiency, the eigenvectors of the unreduced Jacobians carry much information regarding the *algebraic* variables, e.g. those relative to load buses. A practical application is the identification of weak areas in a power system and the determination of the best location for adding reactive support, shedding load, or taking other countermeasures. These aspects will be discussed in Sections 7.5.3 and 8.6.2.

7.4.2   On singular value decomposition

The Singular Value Decomposition (SVD) is an approach to measure proximity of matrices to singularity[2]. In voltage stability calculations, SVD has been proposed as an alternative to eigenanalysis for investigating Jacobian singularity at loadability limits [TT88, LSA92, LAH93, IWG93, BBM96].

Similarly to eigenvalue decomposition (7.18) a square matrix can be expressed in terms of singular values and vectors as:

$$\mathbf{A} = \mathbf{W}_\sigma \mathbf{\Sigma} \mathbf{V}_\sigma^T \tag{7.56}$$

where $\mathbf{V}_\sigma$ and $\mathbf{W}_\sigma$ are matrices whose columns are the *right* and *left singular vectors* respectively, while $\mathbf{\Sigma}$ is a diagonal matrix whose entries are the *singular values* σ_i $(i = 1, \ldots n)$ of $\mathbf{A}$. The singular vectors are such that:

$$\mathbf{A}\mathbf{v}_{\sigma i} = \sigma_i \mathbf{w}_{\sigma i} \tag{7.57a}$$
$$\mathbf{w}_{\sigma i}^T \mathbf{A} = \sigma_i \mathbf{v}_{\sigma i}^T \tag{7.57b}$$

It is easily shown from the above equations that the squares of the singular values are the eigenvalues of $\mathbf{A}\mathbf{A}^T$ (as well as of $\mathbf{A}^T\mathbf{A}$) and that right singular vectors are the eigenvectors of $\mathbf{A}^T\mathbf{A}$, while the left singular vectors are eigenvectors of $\mathbf{A}\mathbf{A}^T$. Thus:

■   singular values are always real nonnegative numbers;

■   the inverse of $\mathbf{V}_\sigma$ is equal to its transpose ($\mathbf{V}_\sigma^{-1} = \mathbf{V}_\sigma^T$). The same holds for $\mathbf{W}_\sigma$;

■   when $\mathbf{A}$ is symmetric its eigenvalues and singular values are equal.

For a singular matrix the smallest singular value is zero. Thus from (7.57a,b):

$$\mathbf{A}\mathbf{v}_{\sigma i} = 0$$
$$\mathbf{w}_{\sigma i}^T \mathbf{A} = 0$$

Comparing with the corresponding relations for eigenvalues, it follows that (for a simple zero eigenvalue) the eigenvectors coincide with the corresponding singular vectors and both carry essentially the same information.

The singular vectors corresponding to a zero singular value can be computed from the unreduced Jacobian, as shown for the eigenvectors.

---

[2] SVD applies to rectangular, as well as to square matrices. Here we consider only the latter case

7.5 LOADABILITY OR BIFURCATION SURFACE

In this section we derive further properties of SNB points in the space of independent parameters. Early investigations of this type can be traced back to [JG81]. The material of this section is largely based on [Dob92, DL93, ADH94, GDA97].

We will rely on the general equilibrium model (7.5) discussed in Section 7.1.2:

$$\varphi(\mathbf{u}, \mathbf{p}) = 0 \tag{7.58}$$

This can be the set of either short-term, or long-term equilibrium equations, that can be reduced, or unreduced. Most practical applications deal with the long-term equilibrium equations.

7.5.1 Parameter space

From the previous discussion it has become clear that loadability limits are in general SNB points of an appropriate power system model. These points satisfy the necessary condition of a singular Jacobian:

$$\det \varphi_{\mathbf{u}} = 0 \tag{7.59}$$

which is a smooth scalar equation. From a mathematical point of view, the natural space to picture the loadability limits is the space of independent parameters $\mathbf{p}$. As seen in the two-bus example (Fig. 7.5b), any curve showing a dependent variable as a function of $\mathbf{p}$ will fold with respect to this space at the loadability limit, and the corresponding sensitivity will become infinite and change sign.

In the n_p-dimensional parameter space the points satisfying (7.59) belong to a manifold of dimension $n_p - 1$, which we will call *loadability* or *bifurcation surface*.

From the point of view of secure system operation, it may be of interest to locate the loadability surface with respect to the current, or an anticipated, operating point. In normal operating conditions this will provide a security margin, whereas during emergency conditions (resulting from significant outages) this will give a general direction for the corrective actions necessary to restore an operating point.

It is common to characterize an operating point in terms of MW's or Mvar's. Hence a particular space of interest is that of load active and reactive powers, which we will call *power space* for short. As we have seen, for loads recovering to the exponential model (7.1a,b), the independent parameters are P^o, Q^o or z. Now, when the loads recover

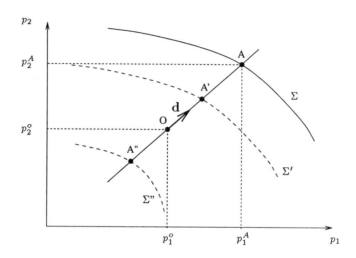

Figure 7.18   Loadability limits in the parameter space

to constant power (due for instance to LTC action), load powers can be considered as independent parameters and thus, the power space becomes a parameter space. An example of bifurcation surface in the power space is the loadability parabola shown in Fig. 7.4 for a single bus with a PQ load.

Figure 7.18 shows a two-dimensional parameter space. The two parameters are denoted as p_1 and p_2, with p_1^o and p_2^o corresponding to an initial operating point O. The loadability or bifurcation surface is denoted by Σ. A point on Σ can be thought of as corresponding to a loadability limit in a given direction of parameter change. In Fig. 7.18 for instance, point A lying on Σ corresponds to the direction of parameter change defined by the vector **d**. Along this direction, one has:

$$\mathbf{p} = \mathbf{p}^o + \mu\mathbf{d} \tag{7.60}$$

and at the SNB point A:

$$\mathbf{p}^\star = \mathbf{p}^o + \mu^\star\mathbf{d} \tag{7.61}$$

where $\mu^\star$ is the maximum value of μ such that (7.58) has a solution (see Section 7.1.2). For a different direction another point on Σ is encountered. We will call $\mu^\star$ the *loadability margin* of point $\mathbf{p}^o$ along direction **d**.

If the considered parameter space is the power space, a *power margin* to instability is obtained. For instance, assuming that p_1 (resp. p_2) corresponds to load active (resp. reactive) power, the active power margin is given by $p_1^\star - p_1^o = \mu^\star d_1$ and the reactive one by $p_2^\star - p_2^o = \mu^\star d_2$, where d_1, d_2 are the components of **d**.

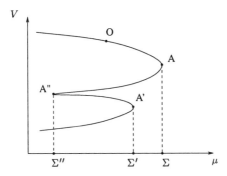

Figure 7.19 Multiple local optima

Note that along the chosen direction there may exist multiple intersection points with the bifurcation surface, which correspond to different local maxima or minima of μ. For instance, points A', A" shown in Fig. 7.18 belong to other branches Σ', Σ'' of the bifurcation surface. This is better illustrated in Fig. 7.19, which shows a typical case of a two-load system. Points A and A' are local maxima, whereas point A" is a local minimum of μ. There are two solutions of (7.58) below Σ'', four solutions between Σ'' and Σ', two solutions again between Σ' and Σ and no solutions outside Σ. More exotic diagrams of this type, including Hopf bifurcation points as well, are shown in [CAD92].

In the remaining of this book we will refer to Σ as the part of the bifurcation surface, outside which there is no solution of (7.58). Very often, the other branches do not correspond to viable operating points.

7.5.2 Normal vector and closest bifurcation points

We investigate now a geometrical property of the bifurcation surface Σ. Starting from a point $\mathbf{p}$ on Σ, assume a small, arbitrary change $\mathbf{dp}$ in parameters *such that* $\mathbf{p} + \mathbf{dp}$ *still belongs to this surface*. Hence, we have :

$$\varphi(\mathbf{u} + \mathbf{du}, \mathbf{p} + \mathbf{dp}) = 0 \qquad (7.62)$$

together with the singularity condition (7.59). Expanding (7.62) while taking into account (7.58) yields:

$$\varphi_{\mathbf{u}}\mathbf{du} + \varphi_{\mathbf{p}}\mathbf{dp} = 0$$

Premultiplying by $\mathbf{w}^T$, where $\mathbf{w}$ is the left eigenvector corresponding to the zero eigenvalue of $\varphi_{\mathbf{u}}$, we get:

$$\mathbf{w}^T\varphi_{\mathbf{u}}\mathbf{du} + \mathbf{w}^T\varphi_{\mathbf{p}}\mathbf{dp} = \mathbf{w}^T\varphi_{\mathbf{p}}\mathbf{dp} = \mathbf{n}^T\mathbf{dp} = 0 \tag{7.63}$$

where:

$$\mathbf{n} = \varphi_{\mathbf{p}}^T\mathbf{w} \tag{7.64}$$

Equation (7.63) expresses that $\mathbf{n}$ is orthogonal to any small $\mathbf{dp}$ lying on Σ. It is thus the normal vector of Σ at the point considered. Equivalently, (7.63) is the equation of the hyperplane $\mathcal{H}$ tangent to the hypersurface Σ (see Fig. 7.21)[3].

Equation (7.64) sheds a new light on the rôle played by the left eigenvector of a zero eigenvalue of $\varphi_{\mathbf{u}}$. Beside being a vector of Lagrange multipliers for the optimization problem (7.6), the left eigenvector is related to the normal vector $\mathbf{n}$, from which information on the bifurcation surface can be obtained. Another use of this eigenvector will be shown in the next section.

The normal vector provides a way to determine the "shortest distance" to the bifurcation surface from a point in the parameter space. Consider for instance point S in Fig. 7.20: the direction $\mathbf{d}_S$ used to reach this point is collinear with the normal vector $\mathbf{n}_S$ at this point. Therefore S is the point of Σ closest to O in the sense of the Euclidean distance

$$\|\mathbf{p} - \mathbf{p}^o\|_2 = \sum_{i=1}^{n_p}(p_i - p_i^o)^2$$

The same figure suggests a procedure to obtain point S. Starting from an initially guessed direction $\mathbf{d}_A$, the bifurcation point A is determined, along with the corresponding normal vector $\mathbf{n}_A$. This vector is used as the next direction to reach Σ from O, which yields point B. The procedure is repeated until the direction of parameter increase coincides with the resulting normal vector, within some tolerance. This and alternative algorithms have been discussed in [DL93, ADH94].

The convergence and the result of the above procedure strongly depend on the shape of the Σ surface, which should be smooth and convex. There is however no proof yet of such a convexity for a general power system model. Moreover, in practice, different directions of parameter changes yield different generators under reactive limit at the bifurcation point (see Section 7.6) and hence Σ is only piecewise smooth. Finally,

---

[3]Comparing (7.64) to (7.7b) it is clear that the normal vector $\mathbf{n}$ is collinear with the gradient of the function ζ for which the point in question is a local optimum:

$$\mathbf{n} = -\nabla_{\mathbf{p}}\zeta$$

This was also shown in Fig. 7.4

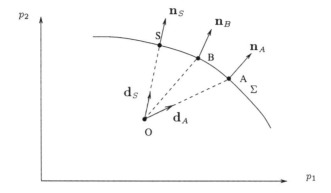

Figure 7.20   Finding the closest bifurcation point

when applying the above procedure in the power space, it may be necessary to correct successive directions in order to avoid convergence towards unrealistic power increase patterns.

7.5.3   Sensitivity of margins to parameters

We have introduced the power margin as a measure of system robustness at a given operating point. We show hereafter how sensitivities of this margin to various parameters can be obtained from the left eigenvector corresponding to the zero eigenvalue of the Jacobian $\varphi_{\mathbf{u}}$. Note that the derivation applies to any margin but the power margin is probably the one of greatest practical interest.

We consider an initial operating point $\mathbf{p}^o$ and a direction $\mathbf{d}$ in parameter space as in (7.60). Along this direction we have a margin $\mu^\star$ corresponding to a parameter value $\mathbf{p}^\star$ given by (7.61). At this point the following equations are satisfied:

$$\varphi(\mathbf{u}^\star, \mathbf{p}^\star) = \varphi(\mathbf{u}^\star, \mathbf{p}^o + \mu^\star \mathbf{d}) = \mathbf{0} \qquad (7.65)$$

together with the singularity condition (7.59).

Assume now that we change the parameters $\mathbf{p}^o$ defining the initial operating point by the small amount $\mathbf{dp}^o$. As a result, the margin changes from $\mu^\star$ to $\mu^\star + d\mu^\star$, as sketched in Fig. 7.21. At the new SNB point, we have:

$$\varphi(\mathbf{u}^\star + \mathbf{du}^\star, \mathbf{p}^\star + \mathbf{dp}^\star) = \varphi(\mathbf{u}^\star + \mathbf{du}^\star, \mathbf{p}^o + \mathbf{dp}^o + d\mu^\star \mathbf{d}) = \mathbf{0} \qquad (7.66)$$

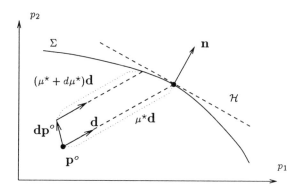

Figure 7.21   Effect of small parameter changes on the power margin

together with the singularity condition (7.59). Expanding (7.66) while taking (7.65) into account yields:

$$\varphi_u du^\star + \varphi_p(dp^o + d\mu^\star d) = 0$$

Premultiplying by $\mathbf{w}^T$ and assuming $\mathbf{w}^T \varphi_p \mathbf{d} \neq 0$ we obtain :

$$d\mu^\star = -\frac{\mathbf{w}^T \varphi_p d\mathbf{p}^o}{\mathbf{w}^T \varphi_p \mathbf{d}} \tag{7.67}$$

Thus the sensitivities of the margin with respect to the initial parameters $\mathbf{p}^o$ is given by:

$$S_{\mu^\star \, \mathbf{p}^o} = -\frac{\varphi_p^T \mathbf{w}}{\mathbf{w}^T \varphi_p \mathbf{d}} \tag{7.68}$$

We can give a geometrical interpretation to (7.67) by substituting in it the expression (7.64) for the normal vector. Thus we can write:

$$d\mu^\star = -\frac{\mathbf{n}^T d\mathbf{p}^o}{\mathbf{n}^T \mathbf{d}} \tag{7.69}$$

Assuming that $\mathbf{n}$ has unit length the nominator of (7.69) is the projection of the parameter change $\mathbf{p}^o$ on the direction of the normal vector. The denominator of (7.69) is a measure of the angle formed between the vectors $\mathbf{n}$ and $\mathbf{d}$.

In certain cases it may be desirable to calculate the effect on the load margin of some other parameters, not involved in the loadability margin calculation. In this case we have to partition the parameter and direction vector as follows:

$$\mathbf{p} = \begin{bmatrix} \mathbf{p}_1 \\ \mathbf{p}_2 \end{bmatrix} \qquad \mathbf{p}^o = \begin{bmatrix} \mathbf{p}_1^o \\ \mathbf{p}_2^o \end{bmatrix} \qquad \mathbf{d} = \begin{bmatrix} \mathbf{d}_1 \\ 0 \end{bmatrix} \tag{7.70}$$

where $\mathbf{p}_1$ is the vector of parameters involved in margin calculation and $\mathbf{p}_2$ is the vector of the other parameters, which are kept constant while computing the loadability margin. Introducing the partition (7.70) into (7.67) one gets:

$$d\mu^\star = -\frac{\mathbf{w}^T \varphi_{\mathbf{p}_1} d\mathbf{p}_1^o + \mathbf{w}^T \varphi_{\mathbf{p}_2} d\mathbf{p}_2^o}{\mathbf{w}^T \varphi_{\mathbf{p}_1} \mathbf{d}_1} \qquad (7.71)$$

We thus conclude that the sensitivity of $\mu^\star$ to the parameters $\mathbf{p}_1^o$ is the same that was given by (7.68):

$$S_{\mu^\star \, \mathbf{p}_1^o} = -\frac{\varphi_{\mathbf{p}_1}^T \mathbf{w}}{\mathbf{w}^T \varphi_{\mathbf{p}_1} \mathbf{d}_1} \qquad (7.72)$$

while the sensitivity of $\mu^\star$ with respect to $\mathbf{p}_2^o$ is:

$$S_{\mu^\star \, \mathbf{p}_2^o} = -\frac{\varphi_{\mathbf{p}_2}^T \mathbf{w}}{\mathbf{w}^T \varphi_{\mathbf{p}_1} \mathbf{d}_1} \qquad (7.73)$$

From applications viewpoint, the largest sensitivities $S_{\mu^\star \, \mathbf{p}_1^o}$ indicate the load or generator powers which have the strongest influence on the power margin, or equivalently, the most efficient ones to manipulate in order to increase the power margin. Similarly, (7.73) may be used to identify other parameters with the greatest effect on the power margin. A typical example is the sensitivity to branch impedances, which allows one to identify the critical branches in a given instability scenario, or the best location for compensation.

Using linear extrapolation, the above sensitivities allow one to determine the change in margin caused by a given change in parameter or conversely, to find the amount of parameter variation required to increase the margin by a given amount. This information is obtained without re-computing the margin with the modified parameters. The limit of validity of this extrapolation is that of the linearization. Second-order sensitivities have been proposed in [GDA97]. Note finally that $\varphi_{\mathbf{p}_1}$ and $\varphi_{\mathbf{p}_2}$ are often very simple matrices.

7.6 LOADABILITY LIMITS IN THE PRESENCE OF DISCONTINUITIES

7.6.1 Inequality constraint formulation

In Section 7.1.2 we have seen that a loadability limit corresponds to a singularity of the steady-state Jacobian matrix $\varphi_{\mathbf{u}}$, and hence to a SNB condition, provided that functions φ_i are smooth. In practical power system models, however, there are various discontinuities present, which we have modelled by the discrete state variables z_d.

Some of these discrete phenomena can be reasonably approximated by continuous models : see for instance the LTC in Section 4.4.2. Continuous models are acceptable approximations of discrete events, provided that the step changes of the actual discrete system are small enough. When the discrete steps become relatively large, the existence of a stable equilibrium is not a sufficient condition for the stability of the system : one has to take into account the region of attraction problem, as discussed in Section 5.1.5. More examples of the problems caused by large disturbances will be given in the next chapter.

Other discontinuities, however, cannot be circumvented in the same way. A typical example is the generator limitation, which we will detail in the next section. One way of dealing with these discontinuities in loadability analysis is by considering that the smooth steady-state equations (7.5) are subject to *inequality constraints*:

$$\psi_i(\mathbf{u}, \mathbf{p}) \geq 0 \quad i = 1, \ldots, m \tag{7.74}$$

where all ψ_i's are smooth functions. We will write the above inequalities in compact vector form as:

$$\psi(\mathbf{u}, \mathbf{p}) \geq 0 \tag{7.75}$$

In the presence of inequality constraints, we can reformulate our optimization problem, Equation (7.6), as follows:

$$\max_{\mathbf{p}, \mathbf{u}} \quad \zeta(\mathbf{p}) \tag{7.76a}$$

$$\text{subject to} \quad \varphi(\mathbf{u}, \mathbf{p}) = 0 \tag{7.76b}$$

$$\psi(\mathbf{u}, \mathbf{p}) \geq 0 \tag{7.76c}$$

The constraints (7.76c) that are satisfied as equalities are called *active* constraints [GMW81, Rao78] and those that are satisfied as strict inequalities are called *inactive*. We define the Lagrangian:

$$\mathcal{L} = \zeta(\mathbf{p}) + \mathbf{w}_\varphi^T \varphi(\mathbf{u}, \mathbf{p}) + \mathbf{w}_\psi^T \psi(\mathbf{u}, \mathbf{p}) \tag{7.77}$$

where $\mathbf{w}_\varphi$, $\mathbf{w}_\psi$ are the vectors of Lagrange multipliers corresponding to equality and inequality constraints respectively. The Kuhn-Tucker [KT51] necessary optimality conditions for this problem are:

$$\varphi(\mathbf{u}, \mathbf{p}) \quad = \quad 0 \tag{7.78a}$$

$$\psi(\mathbf{u}, \mathbf{p}) \geq 0 \qquad (7.78\text{b})$$

$$\nabla_\mathbf{p}\zeta + \varphi_\mathbf{p}^T\mathbf{w}_\varphi + \psi_\mathbf{p}^T\mathbf{w}_\psi = 0 \qquad (7.78\text{c})$$

$$\varphi_\mathbf{u}^T\mathbf{w}_\varphi + \psi_\mathbf{u}^T\mathbf{w}_\psi = 0 \qquad (7.78\text{d})$$

$$w_{\psi i}\psi_i(\mathbf{u}, \mathbf{p}) = 0 \quad i = 1, \ldots, m \qquad (7.78\text{e})$$

$$w_{\psi i} \geq 0 \quad i = 1, \ldots, m \qquad (7.78\text{f})$$

which should be compared to the conditions (7.7a,b,c) of the equality-constrained problem. The additional optimality conditions brought by the inequality constraints are easily interpreted as follows. Condition (7.78b) expresses that the optimum must satisfy these inequalities. Condition (7.78e) means that either $\psi_i(\mathbf{u}, \mathbf{p}) = 0$ (when the inequality is active) or $w_{\psi i} = 0$ (when it is inactive). In other words, the Lagrange multipliers of inactive inequality constraints are zero, while for active constraints they just play the same role as for equality constraints. Finally, (7.78f) requires that Lagrange multipliers of active inequality constraints should be positive in order to achieve an optimum. If the Lagrange multiplier of an active inequality constraint turns out to be negative, there is a direction along which ζ increases while $\psi_i(\mathbf{u}, \mathbf{p}) > 0$.

Regarding our loadability problem, the most interesting result is given by (7.78d). Comparing the latter to (7.7c) we see that when one or more inequality constraints are active, the argument we used to prove the singularity of $\varphi_\mathbf{u}$ does no longer apply (it would apply if $\mathbf{w}_\psi = 0$, i.e. if there was no active inequality constraints). We thus conclude that in the presence of inequality constraints, the Jacobian $\varphi_\mathbf{u}$ *can be nonsingular* at the loadability limit. This is an important exception to the rule we derived in Section 7.1, which is valid only in the case of smooth equality constraints.

7.6.2 Application to limited generators

We will now apply this analysis to derive the loadability of a system incorporating the effect of generator limiters. This effect was studied in [DL91, VC91b]. Recall the generator equations of Table 6.4. As shown in this table, the excitation system of generator i is in one of the following two states:

- under AVR control: $\psi_i^{avr}(\mathbf{u}, \mathbf{p}) = 0$

- under OXL control: $\psi_i^{oxl}(\mathbf{u}, \mathbf{p}) = 0$

where both ψ_i^{avr} and ψ_i^{oxl} are smooth functions. The case of a generator on armature current limit is not considered here, but can be analyzed in a similar manner.

We now formulate the following optimization problem:

$$\max_{\mathbf{p},\mathbf{u},\mathbf{E}_q} \quad \zeta(\mathbf{p}) \tag{7.79a}$$

$$\text{subject to} \quad \tilde{\varphi}(\mathbf{u},\mathbf{p}) = \mathbf{0} \tag{7.79b}$$

$$\psi_i^{avr}(\mathbf{u},\mathbf{p}) \cdot \psi_i^{oxl}(\mathbf{u},\mathbf{p}) = 0 \quad i = 1,\ldots,n_g \tag{7.79c}$$

$$\psi_i^{avr}(\mathbf{u},\mathbf{p}) \geq 0 \quad i = 1,\ldots,n_g \tag{7.79d}$$

$$\psi_i^{oxl}(\mathbf{u},\mathbf{p}) \geq 0 \quad i = 1,\ldots,n_g \tag{7.79e}$$

where n_g is the number of generators. In this formulation we have written the n_g equations φ corresponding to generator excitation explicitly and we denote the remaining ones as $\tilde{\varphi}$. Denoting as n the total number of equations φ and variables $\mathbf{u}$, $\tilde{\varphi}$ has dimension $n - n_g$. Equality constraints (7.79c) are introduced to force at least one of the inequality constraints (7.79d,e) of each generator to be active.

If at an optimum point *only one* inequality constraint per generator is active, conditions (7.79b,c,d,e) become:

$$\tilde{\varphi}(\mathbf{u},\mathbf{p}) = \mathbf{0}$$
$$\psi_i^{avr}(\mathbf{u},\mathbf{p}) = 0 \qquad \text{for the machines under AVR control}$$
$$\psi_i^{oxl}(\mathbf{u},\mathbf{p}) = 0 \qquad \text{for the machines under OXL control}$$

which makes up a set of n steady-state equations. The derivation of Section 7.1.2 applies to this set of equations and its Jacobian is therefore singular. Such a loadability limit corresponds to an SNB condition.

On the other hand, if at least one generator has both inequalities (7.79d,e) active at the loadability limit, the total number of equations exceeds n and the derivation of Section 7.1.2 does not apply. The Jacobian of the algebraic equations in this case is rectangular and the loadability limit *does not* correspond to an SNB.

Figure 3.21 of Section 3.6 has shown the effect of generator field limitation for a single generator. A similar case is shown in Fig. 7.22a, where we plot the PV curves corresponding to the generator on AVR control, and under field current limitation. We assume a load restoring to constant power in the long term. The machine gets limited at point A. The loadability limit is point B, where the Jacobian of equilibrium equations with the generator limited is singular. This is clearly an SNB of long-term dynamics.

Let us now consider the case where the generator has both inequalities active at the solution of the optimization problem (7.79a). This corresponds to a loadability limit that cannot be related to a singular Jacobian and thus it does not correspond to an SNB of power system dynamics. This type of loadability limit is shown graphically

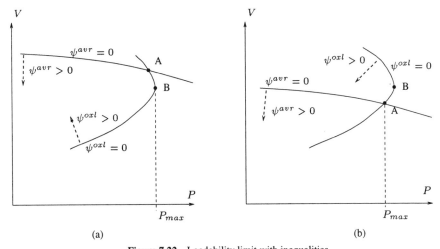

Figure 7.22 Loadability limit with inequalities

in Fig. 7.22b. Clearly the loadability limit is the intersection point A of the two PV curves. Even though this point does not correspond to an SNB, the stability of the system *does change*, because on the PV curve corresponding to the limited generator, the intersection point A lies below the SNB of long-term dynamics, which is point B.

So, even though we can find loadability limits that are not saddle-node bifurcations, the stability properties do change at these loadability limits abruptly.

7.6.3 Effect on sensitivities

Let us now investigate the effect of discontinuities on sensitivity functions. In Fig. 7.6 we showed how the sensitivity (7.17) of any variable with respect to a parameter becomes infinite at a loadability limit corresponding to an SNB. When taking into account the discontinuities caused by generators switching from AVR to OXL control, the sensitivities undergo discrete steps at the breaking points where the two characteristics (before and after the limitation) intersect.

In the case of breaking points similar to point A shown in Fig. 7.22a, which are not loadability limits, the discrete step does not correspond to a change of sensitivity sign. The shape of a typical sensitivity function in this case is shown in Fig. 7.23a. The sensitivity becomes infinite and changes sign at the loadability limit, which is the SNB point B of Fig. 7.22a.

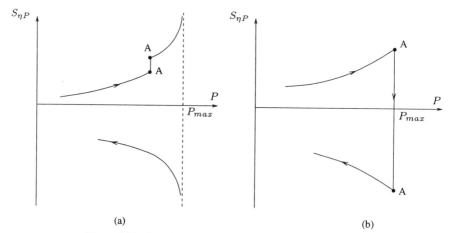

Figure 7.23   Sensitivity functions in the presence of discontinuities

Consider now a breaking point corresponding to a loadability limit, such as point A in Fig. 7.22b. The sensitivity now will change sign when switching from the one curve to the other at point A, although it will not become infinite at this type of loadability limit. The shape of a sensitivity curve for this case is shown in Fig. 7.23b.

We thus conclude that both the stability of the system and the sign of sensitivities change abruptly at a loadability limit, even when this limit is not a saddle-node bifurcation. This property is very useful for practical applications.

7.7   PROBLEMS

7.1. Obtain the loadability limits in the (P, Q) plane for the ZIP loads of Fig. 7.3 and compare with Fig. 7.5a.

7.2. Try to find power system counterexamples for:
(i) $\mathbf{J}$ and $\varphi_{\mathbf{u}}$ becoming singular for the the same conditions, so that (7.27) is not applicable. Hint: read the discussion on [YSM92] and use a radial system.
(ii) $\mathbf{F_x}$ becoming singular without having a SNB. Hint: see Fig. 5.11
(iii) $\varphi_{\mathbf{u}}$ becoming singular without encountering a loadability limit.

7.3. Discuss the region of attraction of the stable equilibrium point of an induction motor under constant mechanical torque (Fig. 4.4), and under quadratic mechanical torque (Fig. 4.7). What are the consequences in terms of motor starting?

7.4. Draw a bifurcation diagram for an induction motor with constant torque. You can choose to use constant T_0 and variable terminal voltage V, or constant V and variable T_0. Repeat for quadratic mechanical torque.

7.5. Write down the coupled set of equations for the system of Fig. 7.15, i.e. network equations $g = 0$, short-term differential equations, long-term difference or differential equations. Simulate the system with initial conditions: $\delta = 0$, $E'_q = 1.4$, rotor speed equal to synchronous, and $r = 1.25$.

7.6. Write down the QSS set of equations for the system of Fig. 7.15, i.e. network equations $g = 0$, short-term equilibrium equations, and long-term difference or differential equations. Simulate the system with initial condition $r = 1.25$.

7.7. The parameters of the system of Fig. 7.15 are (in per unit on a common base):

$$E_f = 2.5 \quad X_d = X_q = 1.4 \quad X'_d = 0.2 \quad P_m = 1.0$$
$$X_1 = 0.1 \quad E_2 = 1.0 \quad G = 3.125 \quad V_2^o = 1.0$$

The transformer leakage reactance is neglected for simplicity. Produce the short-term equilibrium network PV curve of Fig. 7.16 and simulate the system for different initial conditions using the generator differential equations (3.4,3.15) and the continuous LTC model (4.35).

7.8. In a two-dimensional parameter space derive the expression (7.69) for the differential margin geometrically.
Hint: In Fig. 7.21 draw a line parallel to $\mathbf{n}$ from $\mathbf{p}^o$ and project $d\mathbf{p}^o$ on it. Then write an expression involving this projection, the margin differential $d\mu^*$ and $\cos\theta$, where θ is the angle formed by vectors $\mathbf{d}$ and $\mathbf{n}$.

7.9. Find a counterexample, for which $\mathbf{n}^T\mathbf{d} = 0$, so that (7.69) is not valid. What is the shape of Σ in this case?

8

INSTABILITY MECHANISMS AND COUNTERMEASURES

"Abyssus abyssum invocat" [1]
David XLI,8

The objective of this chapter is to provide a classification of the basic loss of stability mechanisms relevant to voltage phenomena and to point out appropriate countermeasures. In the previous chapter we have seen that a gradual parameter variation may result in a loss of stability through a saddle-node bifurcation. However, in many voltage incidents experienced so far, instability occurred following large disturbances, such as short-circuits, line outages, or generator trippings. The emphasis in this chapter is on instability mechanisms induced by abrupt, large variations in the structure and/or the parameters of the system.

We start from an overview of countermeasures against voltage instability. Following this, we proceed to classify the various voltage instability mechanisms. Our presentation follows the time-scale decomposition used in Chapters 6 and 7. First, we give detailed examples of short-term instability, followed by a discussion of relevant corrective actions. We then proceed to present long-term instability mechanisms using the detailed test system modelled in Chapter 6. Finally, we investigate several aspects of corrective actions against long-term instability.

8.1 TYPES OF COUNTERMEASURES

The practical importance of voltage stability analysis is that it helps in designing and selecting countermeasures so as to avoid voltage collapse and enhance stability. A

---

[1] Instability invokes collapse (free translation)

Table 8.1   Countermeasures against voltage instability

design stage	action to be taken

1. power system planning	transmission system reinforcement
	series compensation (Section 2.6.1)
	shunt compensation (Section 2.6.2)
	    SVC (Section 2.6.3)
	construction of generating stations
	    near load centers
	    with line-drop compensation (Section 3.2.2)
	    with low power factor (Section 3.6)
	    with step-up transformer control (Section 2.8)

2. system protection design	reactive compensation switching
	    capacitors
	    reactors
	HVDC modulation (Section 4.7)
	LTC emergency control
	load shedding

3. operational planning	commitment of out-of-merit units voltage security assessment (Chapter 9)
4. real-time	maintain voltage profile maintain reactive reserves generation rescheduling starting-up of gas turbines on-line voltage security assessment (Chapter 9)

detailed description of available countermeasures against voltage collapse is found in [CTF94b].

Countermeasures can be taken at various design stages ranging from power system planning to real-time. Between these two we identify the system protection design, and the operational planning stages. In Table 8.1 we outline a number of countermeasures, relevant to each design stage, that can contribute to enhance voltage stability. For these actions already discussed in this book we also give the corresponding section.

At the power system planning stage transmission reinforcement is the most obvious countermeasure. However, environmental considerations leave little room for new transmission lines near heavily populated load centers, so that other solutions must be sought.

System protection against voltage collapse consists of automatic control actions based on local, or area-wide measurements, that aim at avoiding or containing voltage instability. The system protection scheme has to be designed complementary to, and in coordination with equipment protection, such as that of generators, or transmission lines.

In designing system protection against voltage instability, load reduction is always the ultimate countermeasure. Load reduction can be implemented either directly as load-shedding, or indirectly through a deliberate distribution-side voltage drop brought about by controlling the bulk power delivery LTCs. Such emergency control can be achieved by LTC blocking, by bringing back the tap to a predetermined position, or by reducing LTC setpoint.

As we will see in the appropriate section, apart from reducing load, the above actions remove also one driving force of long-term instability.

During normal operation, real-time control involves typically generator AVR setpoint modification, reactive device switching, generator rescheduling, etc. These actions are usually performed manually by system operators. Alternatively, in some countries some of them are implemented automatically through the secondary voltage control (see Section 3.2.3). In emergency situations, some of the corrective actions listed under item 2 of Table 8.1 can be taken by the system operator.

In addition to the above controls, there is a growing need for voltage security assessment. In Chapter 9 we will describe methods that can be used both in operational planning and real time to assess voltage security.

In the remaining sections of this chapter we analyze voltage instability mechanisms and we focus on the countermeasures that can become part of a system protection scheme against voltage collapse (item 2 of Table 8.1).

8.2 CLASSIFICATION OF INSTABILITY MECHANISMS

Voltage phenomena evolve in different time scales and, as already mentioned, we use the time-scale decomposition of Section 6.4 for the classification of basic voltage instability mechanisms.

8.2.1   Short-term voltage instability

Short-term voltage instability is also known as *transient voltage collapse* [Tay94]. In the short-term period immediately following a disturbance the slow, long-term variables z do not respond yet and can be considered constant, resulting in the approximate short-term subsystem shown in Fig. 6.9. The three major instability mechanisms relating to this subsystem are the following:

ST1: loss of post-disturbance equilibrium of short-term dynamics;

ST2: lack of attraction towards the stable post-disturbance equilibrium of short-term dynamics;

ST3: oscillatory instability of the post-disturbance equilibrium.

As discussed in Chapter 1, the short-term time scale is the time frame of transient *angle* stability. In this respect, the loss of synchronism of one or more generators following too slow a fault clearing is a typical ST2 mechanism [PM94].

In the case of short-term *voltage* stability, the driving force of instability is the tendency of dynamic loads to restore consumed power in the time frame of a second. A typical load component of this type is the induction motor, analyzed in detailed in Section 4.3. Another example is an HVDC link with fast power control, described in Section 4.7.

A typical case of ST1 voltage instability is the stalling of induction motors after a disturbance increasing the total transmission impedance. Due to the increased impedance, the motor mechanical and electrical torque curves may not intersect after the disturbance, leaving the system without post-disturbance equilibrium. As a result the motor(s) stall and the network voltage collapses. This mechanism will be illustrated in Section 8.3.1.

An ST2 voltage instability is, for instance, the stalling of induction motors after a short-circuit. For heavily loaded motors and/or slowly cleared fault conditions, motors cannot reaccelerate after the fault is cleared. The motor mechanical and electrical torque curves intersect in this case, but at fault clearing the motor slip exceeds the unstable equilibrium value. This type of instability will be illustrated in Section 8.3.2.

Oscillatory voltage instability (ST3 type) is less common[2], but cases of generator-motor oscillations have been reported in practice [dMF96]. We will give an example of this type of instability in Section 8.3.3.

---

[2] Some cases of oscillatory voltage instability found in the literature are due to the use of constant power load models in the short-term time scale. Refer to Section 6.5.3

8.2.2   Long-term voltage instability

Let us assume that the system has survived the short-term period following the initial disturbance. From there on the system is driven by the long-term dynamics captured by the z variables. In this section we assume also that the short-term dynamics respond in a stable manner to the changes of z. Hence we can use the QSS approximation of long-term dynamics, as indicated in Fig 6.9. The possibility of subsequent loss of short-term stability will be considered in the next section.

Similarly to short-term dynamics, long-term dynamics may become unstable in the following ways:

LT1: through a loss of equilibrium of the long-term dynamics;

LT2: through a lack of attraction towards the stable long-term equilibrium;

LT3: through slowly growing voltage oscillations.

The above mechanisms lead to what is known as *long-term voltage instability*. LT1 is the most typical instability mechanism, with the loads trying either to recover their pre-disturbance powers through LTC actions or to reach their long-term characteristics through self-restoration. This important instability scenario will be discussed in detail in Section 8.5.1.

A typical example of LT2 instability would be an LT1 scenario followed by a delayed corrective action (e.g. shunt compensation switching or load shedding) which restores a stable equilibrium but not soon enough for the system to be attracted by the stable post-control equilibrium. This will be further explained in Section 8.6.3.

To the authors' knowledge, the third instability mechanism, LT3, has not been observed in a real power systems. It is mentioned here for completeness, but will not be given any further consideration in this chapter. The existence of long-term voltage oscillations is due to cascaded load restoration, as in the case of incorrectly tuned two-level LTCs (Section 4.4.4). Even when correctly tuned, cascaded LTCs may exhibit oscillations for large enough disturbances, but most likely not up to the point of becoming oscillatory unstable. A scenario of unstable voltage oscillations for a single load bus system with a restorative load behind an LTC is documented in [VVC95].

8.2.3   Short-term instability induced by long-term dynamics

We now consider the case where the evolution of long-term variables, usually after a long-term instability, leads to a short-term instability. We can distinguish again three types of instability:

S-LT1: loss of short-term equilibrium caused by long-term dynamics;

S-LT2: loss of attraction to the stable short-term equilibrium due to shrinking region of attraction caused by long-term dynamics;

S-LT3: oscillatory instability of the short-term dynamics caused by long-term dynamics.

A typical case of S-LT1 instability is when the system degradation caused by long-term instability LT1 or LT2 results in either loss of synchronism of field current limited generators, or induction motors stalling.

In such cases, the long-term instability is the cause, the short-term instability being the ultimate result. When an S-LT1 instability is encountered, the slow degradation due to long-term instability leads to a sudden transition in the form of a collapse. If the long-term dynamics are slow enough, this can be seen as a bifurcation of the short-term dynamics with the long-term variables considered as parameters. Detailed examples of this type of instability will be given in Section 8.5.2.

In practice, an S-LT2 instability will be encountered before reaching the actual saddle-node of short-term dynamics, due to the shrinking region of attraction of the stable equilibrium point as it is approached by the unstable one. Thus, a random parameter variation, or a small discrete transition will result in a lack of attraction.

It should be noted that LT1, or LT2 instability may occur without triggering S-LT1 or S-LT2 instabilities. These are less severe cases of long-term instability usually ending in a steady state where both taps and load self-restoration have reached their limits [HTL90]. Such cases are sometimes referred to as cases of "partial voltage collapse" [Tay94].

There are some cases reported in the voltage stability literature showing evidence of the instability mechanism S-LT3. This type of instability could occur, for instance, in systems having both voltage and electromechanical oscillation problems. A possible scenario is the field limitation by a takeover OXL (see Section 3.3.1) that makes the PSS of the limited generator inactive. The resulting loss of stabilizing action can

Figure 8.1 Motor fed through 4 transmission line circuits

produce an oscillatory instability of short-term dynamics, thus an S-LT3 instability. Such a case is documented in [CTF95].

8.3 EXAMPLES OF SHORT-TERM VOLTAGE INSTABILITY

Consider the system of Fig. 8.1. Bus 1 is an infinite bus feeding a three-phase motor load through 4 parallel transmission lines and a step-down transformer. The motor can be thought as the aggregate of an industrial load, or a largely commercial load dominated by air conditioning. The motor mechanical torque is considered constant for simplicity. The motor is represented with a more detailed model than that described in Sections 4.3.2 and 6.3. The model used here includes the rotor electrical transients, but as we will see, the dominant dynamics remain those of rotor motion, captured by (4.22).

8.3.1 Example 1: ST1 instability

The effect of tripping 2, or 3 circuits at time $t = 1$ s is shown in Fig. 8.2. If two circuits are tripped, a new stable equilibrium with increased slip and reduced voltage is obtained. If three circuits are tripped, the equilibrium disappears and the motor stalls. The slips keeps on increasing until the motor stops completely ($s = 1$). The result is unacceptably low voltage at the motor bus and a large current on the remaining line.

The ST1 mechanism (loss of short-term equilibrium) is illustrated in Fig. 8.3, where the slip-torque curves with all 4 circuits, two circuits, and one circuit in service are shown. The dashed line corresponds to the constant mechanical torque of the motor. Obviously, there is no short-term equilibrium with 3 circuits tripped.

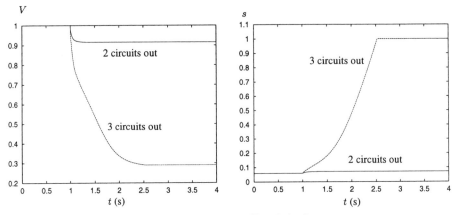

Figure 8.2   Tripping of 2, or 3 circuits

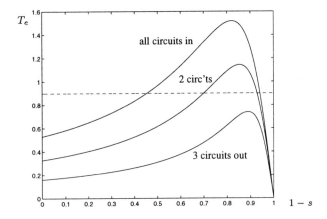

Figure 8.3   Slip-torque curves

8.3.2   Example 2: ST2 instability

Consider again the system of Fig. 8.1. A three-phase short-circuit is imposed (at time $t= 1$ s) at the receiving end of one of the transmission lines, next to bus 2. As seen in Fig. 8.4, during the fault the voltage V drops to zero and the motor decelerates rapidly. The fault is cleared by opening the faulted circuit, which remains open.

We know from the previous example that there is a stable equilibrium with a single circuit open. This is shown as point S in the slip-torque diagram of Fig. 8.5. If the fault is cleared rapidly the system is attracted by this equilibrium. If, on the other hand, the fault remains on for long enough, the motor will decelerate beyond the unstable

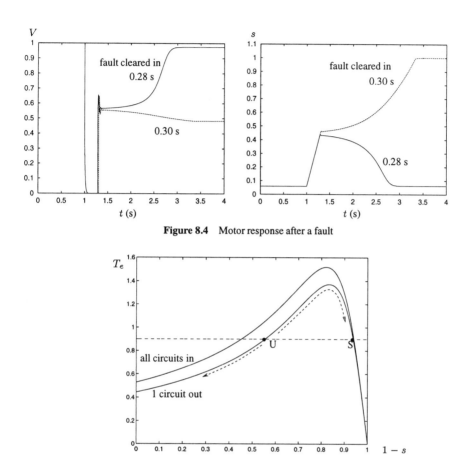

Figure 8.4 Motor response after a fault

Figure 8.5 Slip-torque curve with post-disturbance equilibrium points

equilibrium point U shown in Fig. 8.5. Beyond this point the mechanical torque is larger than the electrical one and the motor is unable to reaccelerate even after the fault is cleared. In mathematical terms point U is the boundary of the region of attraction of the post-disturbance stable equilibrium S. Motor stalling after a slowly cleared fault is, therefore, a typical case of ST-2 voltage instability.

The motor responses for fault clearing times of 0.28 s and 0.30 s are shown in Fig. 8.4. The critical clearing time lies obviously in between the two. Note that the final voltage with the motor stalled is higher than that of the ST1 case shown in Fig. 8.2, because there are now more circuits in service.

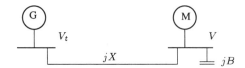

Figure 8.6   Isolated generator - induction motor system

8.3.3   Example 3: ST3 instability

To illustrate a case of short-term oscillatory voltage instability we consider the system of Fig. 8.6 consisting of a synchronous generator feeding an isolated three-phase induction motor load. The mechanical torque of the motor is considered constant. The generator has a first-order excitation system with a proportional automatic voltage regulator (see Fig. 6.7). We assume that the frequency transients have no impact on the response of the system and are thus neglected,

The system is represented by the following set of differential equations:

$$T'_{do}\dot{E}'_q = E_f - E'_q - (X_d - X'_d)i_d \tag{8.1}$$

$$T\dot{E}_f = G(V_o - V_t) - E_f \tag{8.2}$$

$$2H\dot{s} = T_m - T_e \tag{8.3}$$

on the corresponding state variables E'_q, E_f, s. The algebraic variables i_d, V_t, T_e entering the above state equations can be calculated as functions of the network and state variables, as explained in Chapter 6.

By computing the equilibrium points of this system for various mechanical torques T_m we can obtain the bifurcation diagram drawn with a solid line in Fig. 8.7a. The SNB point is at the nose of this diagram and is marked as point C. The actual stability limit, however, is point A, which is a Hopf bifurcation (see Section 5.2.3). The effect of adding capacitive compensation at the motor bus is shown by the dashed curve. Note that although the SNB of the compensated system (C') occurs at higher load torque, the Hopf Bifurcation (HB) point (A') moves slightly to the left actually reducing the stability margin.

The variation of the three eigenvalues of the state matrix $\mathbf{F_x}$ as the equilibrium moves from point O to point C on the uncompensated $(B = 0)$ torque-voltage curve of Fig. 8.7a is shown in Fig. 8.7b. At the HB point A the complex eigenvalue pair crosses the imaginary axis[3], whereas at the SNB point C one eigenvalue becomes zero. Note

---

[3]For point A, as well as for the initial operating point O, the third, real eigenvalue is very far to the left and is not shown in Fig. 8.7b

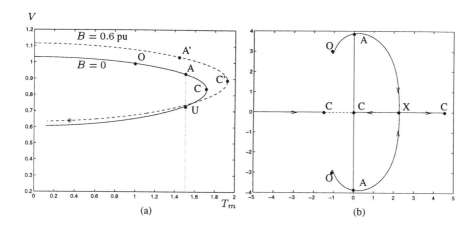

Figure 8.7 (a) Generator-motor bifurcation diagram; (b) Eigenvalue locus

that, in this case, the system is unstable both before and after the SNB, due to the other positive eigenvalue.

Coming back to large disturbances, assume that the post-disturbance condition corresponds to an unstable operating point immediately after the HB point A. The time response of the motor bus voltage V and normalized speed $(1 - s)$ is shown in Fig. 8.8. This is a case of an ST3 instability. The unstable voltage oscillations lead eventually to the stalling of the motor when the oscillation amplitude grows beyond the other unstable equilibrium of the system (point U in Fig. 8.7a).

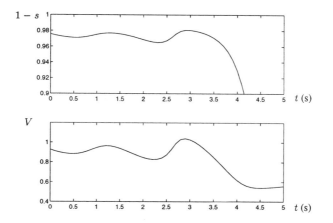

Figure 8.8 Simulation of motor speed and voltage after the Hopf bifurcation

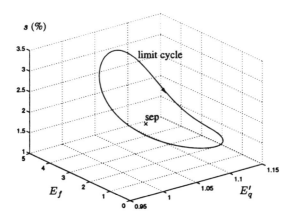

Figure 8.9 Limit cycle surrounding the stable equilibrium before Hopf bifurcation

From Fig. 8.8 it is clear that there is no attracting limit cycle after the Hopf bifurcation. This signifies that we have a case of subcritical HB and, as discussed in Section 5.2.3, the bifurcation is brought about by the shrinking of an unstable limit cycle that exists *before* the bifurcation. This unstable limit cycle is drawn in the three-dimensional state space of Fig. 8.9 together with the stable equilibrium point (shown as "sep") for comparison.

As a general remark, the oscillatory behaviour of the generator-motor system is due to the interaction of two load restoration processes acting in the same (short-term) time scale. More specifically, the load restoration due to the slip adjustment (load admittance increase) races against the voltage regulation due to the AVR (load restored indirectly through restoring voltage).

A short-term oscillatory instability, such as the one in this example, is usually due to incorrect settings of control equipment, such as generator AVRs. The guide rule for their proper tuning is to avoid the interaction with fast load recovery mechanisms, such as induction motors. For instance by making the AVR faster one can wipe out the oscillatory instability (see problem 8.3). The objective should be to restore the motor bus voltage before the motor has time to decelerate considerably after a voltage drop. In practical situations voltage oscillations may arise when a generator, whose AVR was normally tuned for interconnected operation is obliged to operate in isolation after some disturbance leading to islanding, or during the recovery phase after a black-out.

8.4 COUNTERMEASURES TO SHORT-TERM INSTABILITY

When a power system is prone to short-term voltage instability (large induction motor load and/or HVDC fed from a weak AC system), fast reactive support near the load center is essential. The reactive support should be able to restore a stable equilibrium point under the worst contingency considered, and it should be fast enough to act before the motors decelerate beyond the post-control unstable equilibrium point.

The most common fast reactive support devices are generators, synchronous condensers, and SVCs. In practice, to get the automatic support of these devices they should be operating prior to the disturbance with sufficient reactive power margins. Note that the assistance provided by generators and synchronous condensers in excess of their rated field (or armature) current capability is time-limited and it will only transform a short-term voltage problem into a long-term one.

8.4.1 Fast capacitor switching

Mechanically switched and thyristor controlled capacitors can be operated through undervoltage relays and can be fast enough to stabilize a short-term unstable system. Reference [IWG96] gives examples of switched capacitors in the range of 0.15-0.75 seconds. Some logic must be incorporated into the controller in order to avoid capacitor switching during system transients not related to voltage problems.

In systems where shunt reactors are normally in operation, fast reactor tripping [BTS96] can be used instead of capacitor switching to achieve the same result.

8.4.2 Static Var Compensators

These devices constitute probably the best countermeasure to avoid short-term voltage instability. A detailed study of how SVCs can prevent induction motor instability was reported in [HEZ89]. As explained in Section 2.6.3, SVCs can maintain the regulated voltage very close to its setpoint and their response is practically instantaneous. Thus deceleration of induction motors during a disturbance is avoided. Since the regulating range of an SVC is limited, other slower acting reactive compensation devices should be switched on manually by the operator in a way to maximize the availability of SVCs during emergencies.

8.4.3 HVDC modulation

As discussed in Section 4.7, HVDC links are equipped with fast control devices, which can in principle be modulated in order to enhance short-term voltage stability. The most obvious countermeasure in cases of short-term voltage instability is the reduction of the reactive power absorbed by the HVDC which is achieved by lowering its active power or direct current setpoint [PSH92]. In cases where active power is exported from a weak area via the HVDC, the active power reduction helps also to reduce the AC power transfer into the weak area.

Although the actual HVDC controls are fast, the corresponding changes in active power are acceptable only when they are restricted within the regulating range of primary frequency controllers in both sides of the HVDC link, so that active power balance is maintained without undesirable frequency excursions.

8.4.4 Fast fault clearing

In the case of instabilities caused by transient faults, such as described in Section 8.3.2, a possible remedy is faster fault clearing. Small induction motors, such as the ones on residential air condition units will usually stall when subjected to low voltages due to a fault, even for a very fast clearing time [WSD92]. Larger, industrial motors, as well as small motors away from the fault location can ride through faults without stalling, when the fault is cleared fast enough. Thus a coordination of protective devices is essential for systems prone to ST2 instabilities.

Compared to EHV, HV subtransmission systems have usually slower fault clearing times for cost reasons. Therefore, when such systems are feeding motor dominated load, load shedding appears as the cost-effective solution [DDF97].

8.4.5 Fast load shedding

Fast, automatic undervoltage load shedding is the ultimate countermeasure to short-term voltage instability. Several fast undervoltage load shedding schemes have been designed similarly to existing, widely used underfrequency ones. These schemes can be either decentralized, based on local voltage measurements [Tay92, TCN94], or they can be coordinated by the EMS using the SCADA system [NM92].

Fast load shedding requires response times as fast as 1.5 second. This is fast enough to prevent the stalling of large induction motors, but in areas with a large amount of air conditioning load stalling of some motors in the meantime cannot be ruled out. Load shedding systems involving centralized control through the SCADA system tend to be slower, and are thus less suitable to counteract short-term voltage instability.

The security of an automatic load shedding system against misoperation can be enhanced by requiring manual arming of the load shedding scheme by the operator based upon indications of a possible voltage instability.

A similar countermeasure is the selective load shedding, accomplished by tripping induction motor load. In this way only the load prone to short-term instability is shed and the rest of the system is saved. As mentioned in Section 4.3.3, thermal motor protection cannot be relied upon to disconnect stalled motors, and thus undervoltage relays is the alternative. Industrial motors are usually equipped with undervoltage protection, while [WSD92] suggests that similar devices could be installed in single-phase air conditioners as well.

8.5 CASE STUDIES OF LONG-TERM VOLTAGE INSTABILITY

In this section we consider examples of long-term instability using the simple system described in Section 6.3 (Fig. 6.6). The disturbance simulated is a circuit tripping between buses 1 and 4 at time $t = 1$ s. This disturbance is studied under various conditions of local generation and load composition, which constitute Cases 1, 2, and 3 considered in this section. The system responses are obtained by numerical integration of the model derived in Section 6.3. System and operating point data were given in Section 6.7.

8.5.1 Case 1: LT1 instability

In Case 1 the load is represented with an exponential model and the local generator is operating at 2/3 of rated turbine power. The evolution of the system after the circuit tripping is illustrated in Fig. 8.10, which shows the voltage at bus 4, the transmission-side bus of the LTC transformer feeding the load.

The initial fast transient caused by the disturbance dies out soon after, showing that the short-term dynamics are stable. Thus a short-term equilibrium is established, with V_4 settling down close to 0.96 pu. After this point the mechanism driving the system response is the LTC, which tries to restore the load-side voltage by lowering the tap

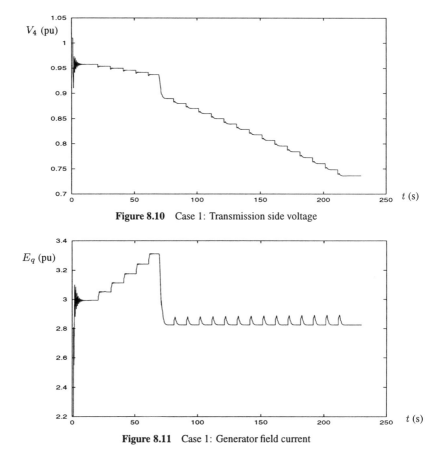

Figure 8.10   Case 1: Transmission side voltage

Figure 8.11   Case 1: Generator field current

ratio r. The operation of the LTC starts after an initial time delay of 20 s, and continues at 10 s intervals. This results in a further reduction of the transmission-side voltage V_4. At about $t = 70$ s, the generator OXL is activated and the loss of local reactive support has a considerable impact on system voltage, which drops below 0.9 pu. Subsequent tap changes cause the voltage V_4 to drop finally below 0.75 pu. At this point the LTC has reached its limit and the voltage decline stops, since no other dynamic mechanism is involved.

The effect of the overexcitation limiter is evident in Fig. 8.11 showing the response of generator emf E_q, which is proportional to field current. As seen in the figure, after the disturbance the generator field current jumps to 3 pu, which exceeds the rotor current limit E_q^{lim}, set to 2.825 pu. This initiates the inverse time mechanism of the OXL. LTC operation after the disturbance imposes an even heavier reactive demand

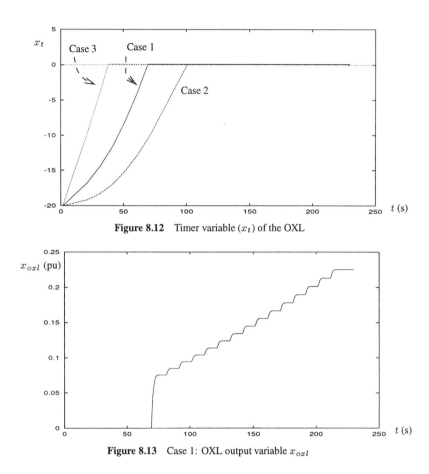

Figure 8.12 Timer variable (x_t) of the OXL

Figure 8.13 Case 1: OXL output variable x_{oxl}

on the generator. This aggravates further the rotor overload, until eventually the OXL is activated bringing the field current back to its rated value. Note that the OXL is of the integral type, so that E_q is forced to E_q^{lim}. Subsequent tap changes result in a transient field current rise, which is quickly sensed and corrected by the OXL.

The OXL operation is illustrated by the responses shown in Figs. 8.12 and 8.13 showing the state variables x_t and x_{oxl} respectively (see Fig. 3.12). The curve marked as "Case 1" in Fig. 8.12 shows that x_t starts increasing from its lower limit (-20) at time $t = 1$ s, i.e. right after the generator becomes overexcited. As the stress on the generator increases, so does the slope $\dot{x}_t$, until at about $t = 70$ s, x_t becomes positive and the OXL is activated.

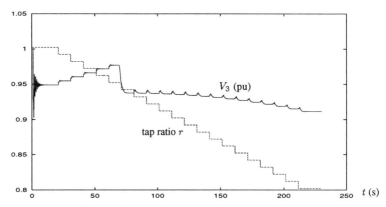

Figure 8.14   Case 1: Load voltage and tap ratio

The limitation of the field current is imposed by subtracting the signal x_{oxl}, whose response is shown in Fig. 8.13, from the AVR input (see Fig. 6.7). Note that the response of x_{oxl} is fast and that its value increases in order to keep E_q constant as the network voltage deteriorates. This response is typical of an integral OXL.

Up to this point we have seen that the simulated disturbance results in generator field limitation and produces unacceptably low voltage at the transmission system, but we have not yet justified the existence of a long-term instability. To do this, we have to observe the response of the load-side voltage V_3 and the LTC ratio r, shown in Fig. 8.14. As seen in this figure, the response of the load voltage exhibits two radically different patterns. Before the generator limitation each tap movement produces the intended effect of rising the secondary voltage. After the limitation of the generator field, the tap changes have initially almost no effect on V_3, and subsequently they even produce the reverse effect by lowering it. Several aspects of this unstable LTC operation have been investigated in [VL88, MIC87, OYS91, VL92].

A deeper understanding of the long-term instability mechanism, can be obtained from PV curves, such as those given in Fig. 8.15, In these curves we show voltage V_4 as a function of the power P transmitted to the load through the LTC (see Fig. 6.8). All curves are drawn considering the short-term dynamics at equilibrium. Three network PV curves (pre-disturbance, post-disturbance before and after OXL action) are shown. These curves have been obtained by varying r continuously, solving the short-term equilibrium equations of Section 6.6, and recording the values of V_4 and P. We also show three short-term load characteristics corresponding to the indicated values of the tap ratio r.

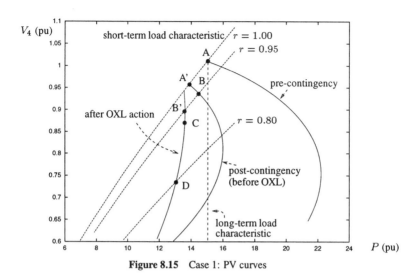

Figure 8.15 Case 1: PV curves

Before the disturbance the system operates at point A. The disturbance causes the network characteristic to shrink. Consequently, the system operates at the short-term equilibrium point A' corresponding to the current value of the tap. Subsequent LTC operation brings the system to point B. At this point the generator field gets limited by the OXL and the system jumps to point B'. From there on, the LTC keeps on decreasing the tap until it finally reaches its limit at point D.

Note that during this transition the system has crossed point C which is the nose of the network PV curve with the generator limited. We call this point the *critical point* on the PV curve. The corresponding voltage is called *critical voltage*. After this point both voltage and load restoration by the LTC fail.

The nature of instability is revealed by observing that the long-term load characteristic, which is the vertical line passing through the pre-disturbance operating point A (see Fig. 4.17) does not intersect the network PV curve with the generator field limited. This is clearly a case of LT1 instability, for which the long-term equilibrium equations corresponding to the final system configuration, as described in Section 6.6, have no solution.

The final outcome of LT1 instability in this case was the pseudo-stabilization at low voltage due to LTC limitation. While this condition allows some time for operator intervention, it should not be mistaken for a stable steady state. Other load recovery mechanisms, such as distribution regulating transformers, thermostatic loads, etc. may become active driving the voltage decline further towards a collapse. Thus, it is more reasonable to consider the final operating condition as unstable, since any attempt to restore load will drive the system to further degradation. Note that it is also possible for the generator to be tripped by undervoltage protection, which would initiate a black-out.

It is of some interest to discuss the qualitative behaviour of the system at this tap-limited unstable situation. Consider a demand increase corresponding to a new load device being connected to the system. The device switched on will of course receive some power. The resulting voltage drop, however, will cause a significant decrease of the power drawn by all other connected devices, so that the total power consumed will decrease.

Remark

In systems having a limited range of LTC action (e.g. a single level between transmission and distribution), or a relatively low critical voltage, LTCs could hit their limits before reaching the critical point. In such systems, it is essential to take into account load self-restoration effects which will continue to degrade system conditions.

On the other hand, in systems having two or more levels of LTCs in cascade, or having a relatively high critical voltage, the critical point is usually crossed before LTCs reach their limits. In this case, load self-restoration effects are often "hidden" behind LTC effects. They become significant if LTCs are blocked.

8.5.2   Case 2: S-LT1 generator instability

In the previous section we saw a case of long-term instability, which ended up with the LTC reaching its limit and the system settling down at unacceptably low voltages. We give now two more cases, which both result in the loss of short-term equilibrium after the long-term instability. In Case 2 the short-term instability is associated with the local generator at bus 2, while in Case 3, which we will present in the following section, the short-term instability is related to the equivalent motor at bus 3.

In Case 2, the system initial conditions are modified by increasing the local generator active production to full rated turbine power. This reduces the initial power flow on line 1-4, and consequently the maximum power that can be transferred to the load is slightly increased, as can be seen by comparing the PV curves shown in Fig. 8.16 with those of Fig. 8.15.

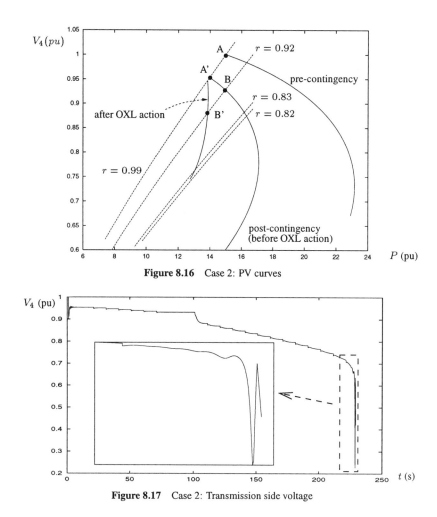

Figure 8.16 Case 2: PV curves

Figure 8.17 Case 2: Transmission side voltage

As in the previous case, the operating point follows the path AA' (circuit tripping), A'B (LTC operation), BB' (generator limitation). Again there is no long-term equilibrium with the generator limited. However, the radical difference with respect to the previous case is that the network characteristic with the generator under field current limitation does not intersect the *short-term* load characteristic for $r = 0.82$. Thus, as the LTC reduces the tap ratio from 0.83 to 0.82, the system loses short-term equilibrium, which constitutes an S-LT1 instability.

The sequence of events is shown in Fig. 8.17, where the simulated response of the transmission-side voltage V_4 is plotted as a function of time. This time the generator

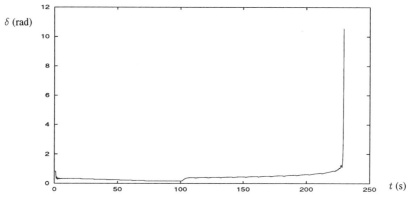

Figure 8.18   Case 2: Generator rotor angle

field current gets limited at about $t = 100$ s, as seen also in Fig. 8.12. As the LTC keeps reducing the tap ratio, the generator eventually loses synchronism at about $t = 220$ s. The loss of synchronism is evident in Fig. 8.18 showing the response of generator rotor angle. The subsequent pole-slips result in the voltage oscillations observed after the collapse, and shown in the magnified section of Fig. 8.17. The out-of-step relay will eventually trip the generator. The final outcome of such a case is likely to be a blackout of the load area, caused by the tripping of the remaining line due to overload.

Note that while in the previous case the limited range of the LTC tap ratio contained somehow the long-term instability, allowing time for operator actions, in the S-LT1 case all corrective measures have to be taken before the instability of short-term dynamics initiates a collapse.

8.5.3   Case 3: S-LT1 motor instability

In this case 40% of the load at bus 3 is made up of an equivalent induction motor. The capacitor is adjusted accordingly, so that at the initial operating point the power factor of the composite load remains unchanged. The local active generation is reduced below 50% of rated turbine power.

The corresponding PV curves are shown in Fig. 8.19 and the response of V_4 in Fig. 8.20. The circuit tripping causes now a more severe generator overexcitation problem due to the increased reactive consumption of the induction motor at lower voltage (see Fig. 4.6). The increased overload forces the OXL to act faster. As seen in Fig. 8.12 the long-term OXL variable x_t becomes positive in this case at about $t = 40$ s. The

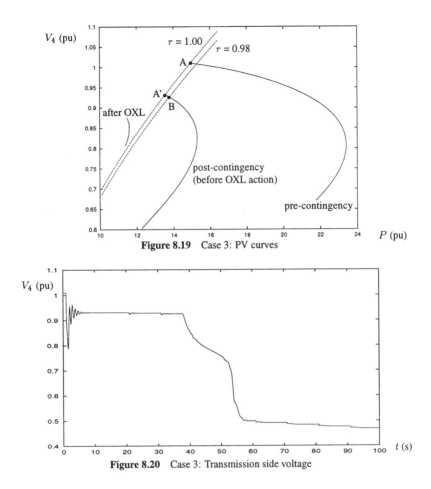

Figure 8.19   Case 3: PV curves

Figure 8.20   Case 3: Transmission side voltage

LTC has time to take only two steps before this occurs. Figure 8.19 shows that after generator limitation there is no intersection between the short-term load characteristic and the network PV curve. Thus, short-term equilibrium is lost and we have again a case of S-LT1 instability.

This time, the loss of short-term equilibrium takes on the form of motor stalling. This is evident in the response of motor slip, shown in Fig. 8.21. As voltage declines after the reduction of generator field current caused by the OXL, the motor decelerates until it finally stalls at about $t = 55$ s. After stalling the motor comes to a complete stop ($s = 1$) and remains locked at standstill. The high reactive current drawn by the stalled motor results in a very low voltage that will probably lead to a black-out caused by generator undervoltage, or line overcurrent tripping. Note that the generator does not

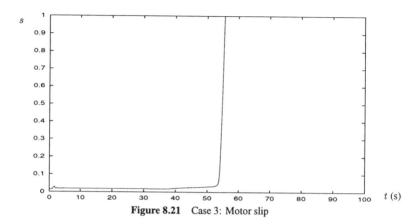

Figure 8.21 Case 3: Motor slip

lose synchronism in this case. It is possible, however, for an S-LT1 instability to result in both motor stalling and generator loss of synchronism.

8.6 CORRECTIVE ACTIONS AGAINST LONG-TERM INSTABILITY

8.6.1 Countermeasure objectives

Before discussing the corrective actions that can be taken in a case of long-term voltage instability, we will distinguish three types of objectives to be achieved by these countermeasures:

1. restore a long-term equilibrium (fast enough so that it is attracting);

2. avoid an S-LT instability of short-term dynamics;

3. stop system degradation.

The first objective is an obvious requirement for the system to return to normal operation. The second objective is simply to maintain the system in operation, while the third and more ambitious objective [AAH97] stems from the concern that the long-term voltage degradation could result in a black-out, due to a cascade of events, such as:

a. overcurrent tripping of lines;

b. undervoltage tripping of field-limited generators;

c. line tripping due to impedance (zone 3) relays.

The above objectives can be achieved using a variety of control actions, such as reactive compensation switching, generation rescheduling, or load reduction. As discussed in Section 8.1, load can be reduced either by direct shedding, or indirectly through emergency LTC control.

As already mentioned, LTC emergency controls can be implemented in several ways. LTC blocking can be used for slowing down the system degradation by stopping the load restoration process. However, as pointed out at the end of Section 8.5.1, other load restoration mechanisms may be present downstream of the bulk power delivery transformer. LTC setpoint reduction is preferable from the viewpoint of customer voltage quality, since although the voltage remains low, it is on the average less sensitive to transmission system transients [IWG96].

For multilevel LTC systems (see Fig. 2.22) emergency control actions at different levels should be coordinated. The main principles are: (i) keep voltage low at the distribution level, (ii) maintain normal (or even a little higher) voltages [CCM96] at the sub-transmission level, in order to keep reactive losses as low as possible, while getting the most out of the HV shunt compensation.

Another countermeasure, applicable when there is some reactive generation reserve, is the boosting of generator terminal voltages. By increasing generator voltage the transmission system voltages will also rise and the resulting reduction of reactive losses will contribute to stabilize the system. The boosting of generator voltages is particularly useful when it affects EHV transmission. This countermeasure can be applied through the secondary voltage control discussed in Section 3.2.3.

In the following subsections 8.6.2 and 8.6.3 we concentrate on two aspects of the long-term equilibrium restoration problem, while in subsection 8.6.4 we reintroduce the other two objectives into the picture.

8.6.2   Long-term equilibrium restoration: where to act

System picture in power space

The LT1 instability mechanism is best visualized in the power space introduced in Section 7.5.1 [VJM94, Ove94, VC95]. We consider a vector **p** of load demand

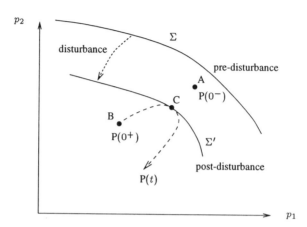

Figure 8.22   Critical point in the load space

parameters and we assume for simplicity that load restores to constant power, so that load demand is expressed in terms of power. A two-dimensional view of the power space is shown in Fig. 8.22. By Σ in this figure we denote the loadability or bifurcation surface defined in Section 7.5.1.

The system operates initially at the long-term equilibrium A, where the consumed power is $P(0^-)$, equal to the demand p. A disturbance occurring abruptly causes the loadability surface to shrink, so that point A is left outside the feasible operating region bounded by the post-disturbance loadability surface Σ'. This means that the load demand p cannot be met, and the long-term equilibrium is lost.

We can follow the response of the system in the power space by tracking the powers *consumed* by loads as a function of time, thus obtaining a trajectory $P(t)$. Immediately after the disturbance the power consumed by loads drops due to decreased voltages. Thus the initial post-disturbance operating point is B, where the power consumed is $P(0^+)$. Starting from point B the load restoration mechanism tries to bring power $P(t)$ back to point A, which is infeasible. At some point C the trajectory of consumed powers $P(t)$ will hit the loadability surface and will turn backwards, being unable to cross it. Since it lies on the bifurcation surface, point C is a point where all curves showing a dependent variable as a function of the parameters will fold. It is thus the *critical point* defined in Section 8.5.1, but seen this time in the power space.

Any point along the trajectory $P(t)$ can be thought of as a long-term equilibrium corresponding to a demand p equal to the present consumption $P(t)$. By assuming $p = P(t)$ we get a family of equilibrium points, which we will call *implicit equilibria*.

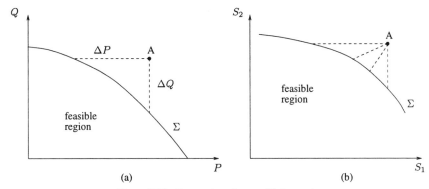

Figure 8.23 Restoration of an equilibrium point

For this family of equilibria the critical point C, lying on the bifurcation surface Σ, is a saddle-node bifurcation. One can thus compute along the system trajectory the unreduced long-term Jacobian $\mathbf{J}_\ell$ (or the reduced one $\mathbf{H}_z$) corresponding to each implicit equilibrium point. At the critical point C both the reduced and the unreduced Jacobians become singular, having a zero eigenvalue.

Assuming that no oscillatory instability is involved, the implicit equilibria lying on the trajectory $\mathbf{P}(t)$ before the critical point are *stable* having a real negative eigenvalue approaching zero, while those after the critical point are *unstable* with a real positive eigenvalue.

From the sensitivity formula (7.20) it is clear that as the critical point is crossed the sensitivities change sign through infinity following the change in sign of the eigenvalue passing through zero. The practical aspect of this property has been shown in Fig. 8.14: before the critical point C, the LTC restores both secondary voltage and power; after the critical point, the LTC reduces both primary and secondary voltages.

Equilibrium restoration

Thus the problem of restoring a long-term equilibrium is that of establishing through appropriate corrective actions an operating point inside the feasible region, bounded by the loadability surface Σ. Considering again load restoration to constant power, in the parameter space of load powers we can include active generation and reactive compensation as negative loads.

Figure 8.23 gives two examples of power spaces. Again, we show a two-dimensional view, of a possibly more complicated picture. In Fig 8.23a the parameters are one real and one reactive demand, whereas in Fig 8.23b the parameters are the apparent power demands at two different buses. In both figures point A corresponds to the load demand and Σ is the loadability, or bifurcation surface.

In the case of Fig 8.23a, a feasible operating point can be restored by introducing locally the active generation ΔP, which will effectively reduce the net active load, or by switching on reactive compensation that will effectively reduce the net reactive load by the amount ΔQ. In the case of Fig. 8.23b, the same goal can be achieved shedding load from bus 1 only, from bus 2 only, or from both buses according to a rule of shared participation. Thus, the objective of restoring an equilibrium can be achieved through a variety of corrective action combinations, which show up in parameter space as different directions.

Figure 8.23 clearly shows that along each direction there is a different *distance* to the loadability surface Σ. Defining an appropriate norm, we can look for a way of restoring equilibrium that minimizes the necessary amount of corrective action to be taken for equilibrium restoration.

This location aspect constitutes a fundamental difference of voltage with respect to frequency problems. In the latter case, underfrequency load shedding is independent of load location within the problem area (see also the discussion on voltage versus frequency stability in Chapter 1).

Corrective action evaluation based on the tangent hyperplane

An accurate determination of the needed corrective actions would require in principle to know the exact boundary surface Σ. Obviously this hypersurface is extremely complex to obtain in practice, due to its very nature as well as the high dimensionality of the power space to consider. It can only be envisaged to get some points of Σ, using techniques based on properties derived in Chapter 7 and further described in Chapter 9.

However, a linear approximation of Σ can be obtained in the form of the tangent hyperplane $\mathcal{H}$ introduced in Section 7.5.2 and reproduced in Fig. 8.24. Given point C of the hypersurface Σ, the tangent hyperplane is easily obtained by computing the left eigenvector $\mathbf{w}$ of the zero eigenvalue of $\mathbf{J}_\ell$ (or $\mathbf{H}_z$) and therefrom the normal vector $\mathbf{n}$, simply related to $\mathbf{w}$ through (7.64). From there the equilibrium restoration problem amounts to bringing the (pre-contingency operating) point A back on the feasible side of hyperplane $\mathcal{H}$, an elementary geometry problem. Obviously, $\mathcal{H}$ only provides a linear approximation, which is less accurate for large corrections (i.e. for less effective actions).

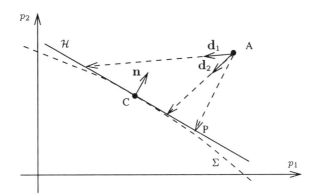

Figure 8.24   The tangent hyperplane for evaluating various corrective actions

Different criteria or norms have been proposed, yielding different optimum combinations of corrective actions.

Ref. [Ove95] considers the Euclidean distance between point A and the point P of $\mathcal{H}$ closest to A, as a measure of unsolvability. To compensate for the curvature of Σ, an iterative procedure is proposed in which point A is moved to point P, a new point C and tangent hyperplane are determined, and so on until the problem is again solvable. The best corrective actions are identified by computing the sensitivity of the above unsolvability measure to available controls, ending up in formulae somewhat similar to (7.72,7.73).

In [VJM94] a list of candidate corrective actions is considered, each of them corresponding to a known direction $\mathbf{d}$ of corrective action (see e.g. $\mathbf{d}_1$ and $\mathbf{d}_2$ in Fig. 8.24) and being characterized by the distance between point A and hyperplane $\mathcal{H}$. The various actions are then taken by increasing order of this distance, until point A ends up on the right side of the hyperplane.

8.6.3   Long-term equilibrium restoration : when to act

The time available for taking a corrective measure aimed at restoring a long-term equilibrium is limited by attraction considerations [XM94, Ove95]. If the control action is delayed too much, the system may exit the region of attraction of the post-control stable equilibrium, thereby leading to LT2 instability according to the terminology of Section 8.2.2. We illustrate this with the following example of capacitor switching.

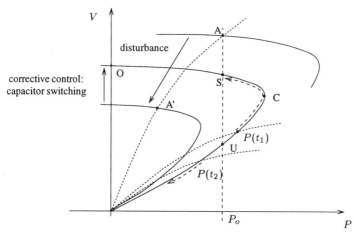

Figure 8.25 Capacitor switching

Consider the PV curves shown in Fig. 8.25. We assume a load with a constant admittance short-term characteristic restoring to constant power is the long term. The system is initially operating at the long-term equilibrium point A on the pre-disturbance network characteristic. A sudden disturbance results in a post-disturbance network PV curve that does not intersect the long-term load characteristic $P = P_o$. We have thus a case of LT1 instability. After the disturbance the operating point jumps to A' and subsequent load restoration takes the system along the post-disturbance PV curve. This is shown in Fig. 8.25 with a series of dotted short-term load characteristics. The switching of shunt compensation in the transmission system yields a new, post-control network PV characteristic, on which there is now a stable (S) and an unstable (U) long-term equilibrium point.

Let t_1 be the time instant, at which the capacitor switching takes place and let $P(t_1)$ be the consumed power right after this action (see Fig. 8.25). Since at this point the load power exceeds that of the long-term load characteristic P_o, the load restoration process will reduce the admittance and the system will be attracted to the stable equilibrium S. On the contrary, if the switching action it taken at time t_2, where the post control power is $P(t_2)$, the load admittance will go on increasing and an LT2 instability will result. In fact, as explained in Section 5.1.5, the region of attraction of the stable equilibrium point S is bounded by the unstable equilibrium U[4].

The following conclusion can be drawn from this example:

---

[4] Note that in multidimensional systems the boundary of the region of attraction is the stable manifold of the unstable equilibrium point

For each corrective action that is initially able to restore a long-term equilibrium, there is a time limit after which its implementation will no longer stabilize the system.

8.6.4 Corrective actions: the amount vs. time issue

The timing aspect is different for different control measures. When the corrective action to be taken refers to reactive compensation, or generator voltage boosting, there is no reason to delay this action, since there is no adverse effect on the consumers. By implementing such actions as soon as possible, it may be possible in some cases to restore a stable equilibrium point without any load reduction.

In contrast, when the corrective action to be taken is load shedding, or another form of load reduction, the control action should be delayed as much as possible, to make sure that it is absolutely necessary for the survival of the system. In the previous section we have seen that for a given amount of corrective action there is an upper time limit for its implementation. In a similar way we can determine (e.g. using numerical simulation) the minimum amount of corrective action ΔP_{min} necessary to meet an objective as a function of time [AAH97].

We will illustrate this idea of time dependent control requirement with a couple of examples. Our discussion is not restricted to long-term equilibrium restoration, but it incorporates as well the other two objectives listed in Section 8.6.1. Regarding the third objective, we will assume that system degradation is specified in terms of a transmission-side voltage, and thus the objective takes on the form of keeping this voltage above its lowest acceptable value.

Load reduction through LTC setpoint decrease

In this example we neglect LTC deadband effects and load self-restoration. We also assume that the LTC does not hit a limit. Thus, at long term equilibrium the load restores to constant power and the amount of this power depends upon the LTC voltage setpoint.

The pre- and post-contingency network PV curves are shown in Fig. 8.26 with solid lines. The pre-contingency operating point is A and the corresponding demand is P_A. The long-term load characteristic is shown with a dashed line and a few short-term ones are drawn with dotted lines. Once more, we have a case of LT1 instability, since the demand P_A is larger than the maximum power P_C corresponding to the critical point.

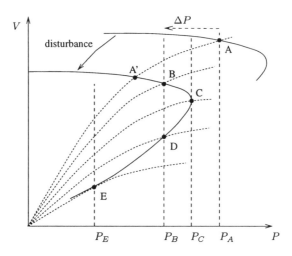

Figure 8.26   Load reduction through LTC setpoint

Consider now the time response of the system after the disturbance. At time $t = 0^+$ the operating point is A'. At time $t = t_C$ the critical point is crossed, and at time $t = t_E$ an S-LT1 instability is encountered and the system collapses.

Since the reduction of load is implemented by changing the LTC setpoint the short-term load characteristics are not affected by the control action, whereas the long-term characteristic moves to the left by an amount ΔP. We will now investigate how the minimum required amount of ΔP to restore a long-term equilibrium changes with time.

Before time t_C, at which the critical point C is crossed, this amount is constant and equal to that restoring a single (saddle-node) equilibrium, i.e. $\Delta P_{min} = P_A - P_C$. After this time, the minimum required amount of corrective action increases, because it must also guarantee attraction towards the stable post-control equilibrium. For instance, at time t_D, when the system has reached point D, the load reduction must be made at least equal to $P_A - P_B > P_A - P_C$, for the post control stable equilibrium to be attracting.

We plot the minimum required load reduction as a function of time in Fig. 8.27 with a dashed line. Note that there is an upper time limit for any control action: this is t_E, the time it takes to reach the S-LT1 instability. After this time objective no. 2 of Section 8.6.1 cannot be met and the system collapses. This limit is shown with a white dot in Fig. 8.27.

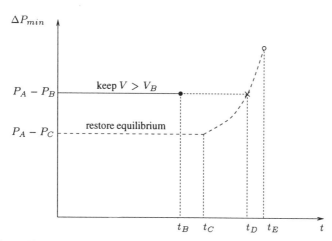

Figure 8.27   Minimum load reduction vs. time (specified voltage V_B above critical)

In the same figure we draw with a solid line the amount of load reduction necessary to achieve the objective of keeping the voltage above a specified value. In this figure the specified value V_B is *above* the critical one V_C. The minimum load reduction required to achieve this goal is constant and equal to $P_A - P_B$, larger than $P_A - P_C$, which is the minimum load reduction required to restore an equilibrium. If the control action is taken before t_B (the time needed to reach point B in Fig. 8.26), the voltage will continue to drop, but it will remain always above V_B. Thus t_B is the upper time limit for achieving the objective of not allowing voltage to decrease below V_B. This point is shown with a black dot in Fig. 8.27.

Note that if the control action is taken after t_B, but before t_D (the time required to reach the unstable point D in Fig. 8.26) the voltage will drop transiently below the specified value V_B, but it will eventually recover to V_B. The time t_D is in this case the upper time limit for obtaining a steady-state voltage higher than V_B. The time dependent requirement for this less stringent version of the objective ($V > V_B$ in steady state) is shown in Fig. 8.27 with a dotted line ending with an "x".

Let us examine now the consequence of choosing a specified voltage *lower* than the critical one, for instance V_D corresponding to point D in Fig. 8.26. In this case the objective of restoring an equilibrium is identical, up to time t_D, to that of keeping the voltage above V_D. After this time, the objective of keeping a voltage higher than V_D cannot be achieved. This case is shown graphically in Fig. 8.28. Clearly, by specifying a voltage lower than the critical one, the required amount of load power reduction remains the same, but the upper time limit for taking the corrective action becomes shorter.

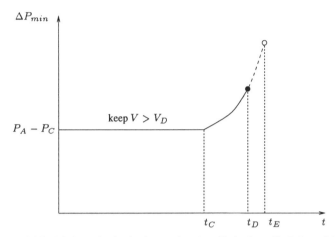

Figure 8.28 Minimum load reduction vs. time (specified voltage V_D below critical)

Load shedding

The case of load shedding is similar to the previous one, with one added complexity: when shedding load in order to move the long-term load characteristic to the left, the short-term load characteristic changes as well. This allows some more time for taking the control action, as we will see in the sequel. The analysis is restricted to the objective of restoring a stable equilibrium, but the conclusions drawn apply also to objective no. 2 of stopping system degradation. The effect of load shedding is illustrated in Fig. 8.29, where again we assume a load restoration to constant power.

As seen in this figure, when the amount of load shedding is the minimum one $P_A - P_C$, the time limit for restoring an equilibrium is equal to $t_{C'}$, the time required for the short-term characteristic to reach point C', such that the post-shedding short-term load characteristic passes through point C. After this time ($t_{C'}$), the amount of necessary load shedding increases. The same is true for the discrepancy between the pre-shedding and post-shedding short-term characteristics. Thus, when the load to be shed is $P_A - P_B$, as shown in Fig. 8.29, the corresponding time limit for restoring the stable equilibrium B is $t_{D'}$, where again the point D' corresponds to the pre-shedding short-term characteristic, such that the post-shedding one passes through point D.

The load shedding versus time diagrams are similar to those of Fig. 8.27 (when the specified voltage is above the critical) and Fig. 8.28 (when the specified voltage is below the critical one) with a slight translation to the right (more time available) to allow for the change in short-term characteristic discussed above. Note, however, that

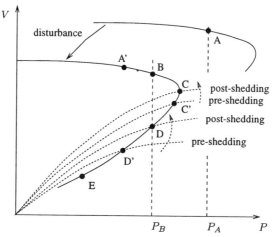

Figure 8.29   Effect of load shedding

t_E (the ultimate time limit for any control action to be taken) remains unchanged, since at point E the system collapses due to the S-LT1 instability.

8.7 PROBLEMS

8.1. Simulate the system of Fig. 8.1 with 3 circuits tripped and find the critical clearing time for a circuit to switch back in operation before the motor stalls. The data for the system are (per unit on motor rating):

$$R_s = 0 \quad X_s = 0.091 \quad X_r = 0.071 \quad X_m = 2.23 \quad R_r = 0.059 \quad H = 0.342$$

The mechanical torque is 0.9 pu, the transformer reactance 0.10 pu and each line has a reactance 0.4 pu, all in motor rating.

8.2. Obtain the limit cycle of Fig. 8.9 by imposing a transient torque pulse on a stable equilibrium point just before the Hopf bifurcation point A. By adjusting the duration of the pulse you can find a critical trajectory which is trapped in the stable manifold of the unstable limit cycle and converges to the limit cycle itself. Generator data:

$$X_d = 0.8958 \quad X_q = 0.8645 \quad X'_d = 0.1198 \quad T'_{do} = 6 \text{ s} \quad G = 30 \quad T = 0.5 \text{ s}$$

Motor data:

$$R_s = 0 \quad X_s = 0.12 \quad X_r = 0.051 \quad X_m = 3.904 \quad R_r = 0.0116 \quad H = 1.5$$

The line impedance is 0.1 pu and all per unit values are on a common base.

8.3. Using the data of the previous problem try to avoid the oscillatory behaviour by making the generator AVR faster.

8.4. Repeat the simulation runs of Cases 1, 2, and 3 of Section 8.5.

8.5. Reproduce the PV curves of Figs. 8.15, 8.16, 8.19. See the hint in Problem 4.3.

8.6. Draw the minimum load reduction vs. time diagram for Cases 1, 2, and 3 of Section 8.5 specifying an appropriate voltage criterion.

9

CRITERIA AND METHODS FOR VOLTAGE SECURITY ASSESSMENT

"To err is human, but to really foul things up requires a computer"
Arthur Bloch (Murphy's law, Vol. III)

Last but not least, this chapter is devoted to a description of computer methods that are used, or have been proposed, for voltage security analysis. For their presentation, we have chosen to start from voltage security criteria and group in the same section methods that basically address the same criterion. Three such categories are successively considered, namely contingency evaluation, loadability limit computation, and secure operation limit determination. We proceed with a presentation of eigenvector computation methods, for instability diagnosis purposes. Next, we illustrate some of the methods described in this chapter using examples from a real-life system. We conclude with some remarks dealing with real-time aspects.

Research efforts in voltage stability analysis have resulted in a huge amount of technical papers devoted to computer methods. Although we tried to provide a representative sample of them, our choice was inevitably subjective: we included methods with which we had some prior experience, as well as methods that naturally fit within the framework developed in previous chapters.

9.1 VOLTAGE SECURITY: DEFINITIONS AND CRITERIA

9.1.1 Operating states

Generally speaking, power system security can be defined as the absence of risk of system operation disruption. In practice, it is defined as the ability of the system to withstand without serious consequences any one of a list of "credible incidents".

A conceptual framework for security assessment and enhancement is provided by the notion of system operating states [DL74, FC78, DL78], which we briefly recall by way of introduction.

Power system operation is subjected to two sets of constraints: (i) the *load constraints* express that the load demand is met by the system; they take on the form of equality constraints which are basically the long-term equilibrium equations discussed in Chapter 6; (ii) the *operating constraints* impose minimum or maximum limits (or inequality constraints) on variables (e.g. maximum line currents, bus voltages within some range, etc.) associated with the system components.

If both the load and the operating constraints are satisfied, the system is said to be in a *normal (operating) state*. On the occurrence of a disturbance, the system may either settle down to a new normal state or it may enter: (i) an *emergency (operating) state*, if some operating constraints are not satisfied, or (ii) a *restorative state*, if the operating constraints are satisfied (for the operating part of the system), but not the load constraints (due to a blackout or a load shedding action).

It is of interest, especially within the context of voltage security, to consider the dynamic status of the system during an emergency, and distinguish between static and dynamic emergencies. A static emergency occurs when, following a disturbance, the system reaches a new steady state (or long-term equilibrium), but the latter violates some operating constraints, which cannot be tolerated for a long time. A dynamic emergency occurs when the disturbance causes the system to become unstable; this may result in a severe system disruption (e.g. loss of synchronism, motor stalling, etc.) in which neither load nor operating constraints are satisfied. Correspondingly, *static security assessment* deals with the ability of the system to survive disturbances without entering a static emergency, while *dynamic security assessment* deals with the system ability to reach a stable post-disturbance operating point. Dynamic security is thus a prerequisite.

Corrective control refers to an action taken after a disturbance to bring the system from emergency to normal state. Again, the type of emergency is of importance. Static emergency conditions may usually be tolerated for some time, allowing corrective actions to be taken by operators. This is especially true for low voltage conditions, which do not cause equipment stress. On the other hand, dynamic emergencies usually require fast, automatic corrective actions. The latter are taken by automatic system protection devices which rely on local or area-wide measurements.

Voltage stability is one aspect of dynamic security. It is sometimes referred to as a separate class of problems because, for a long time, dynamic security has been assimilated to transient (angle) stability only. Furthermore, the fact that static tools (i.e. tools based on algebraic equations) are often used to speed up computations should not lead to assimilating voltage security to static security, in a confusion of means and ends.

As already shown, short-term voltage instability develops over a few seconds while long-term voltage instability may take several minutes. Both are usually considered too fast to be corrected by system operators, and hence corrective actions are implemented through automatic controls, as discussed in Chapter 8. However, the time taken by the long-term instability to develop, while short for a human operator, would be ample for a modern computer executing efficient software to identify the problem, warn the operator and suggest or trigger corrective actions. Such protection, based upon on-line system analysis and adapting its decision to the disturbance of concern, deserves attention, although it is still beyond the state-of-the-art.

The overall objective of security control is to keep the system in the normal state. We may distinguish between three levels of assessment, that take place either in real-time or at the operational planning stage. These are security monitoring, security analysis and security margin determination. *Security monitoring* is the first level and consists in checking the above mentioned operating constraints. The other two levels are computationally more demanding; they are discussed in detail in the next two sections.

9.1.2   Security analysis

Security analysis consists in checking the system ability to undergo disturbances. A system is said to be *secure* if it can withstand each specified disturbance without getting into an emergency state. Otherwise it is *insecure* or in an *alert state*. *Preventive control* refers to actions taken in the normal state, to bring the system from an insecure to a secure state. Preventive controls usually require to operate the system at a higher cost. It is also possible that "normal" preventive controls are unable to bring the system to the secure state. In such extreme cases, load shedding may be envisaged as an alternative.

Obviously, if one were to consider all possible disturbances, it would be impossible to find a secure power system. In practice, system security is checked with reference to a set of credible disturbances, i.e. disturbances with a reasonable probability of occurrence, referred to as *(next-)contingencies*. *Contingency evaluation* thus consists in assessing the level of system security at a given operating point, by evaluating the impact of credible contingencies, taking into account post-contingency controls.

Contingencies. For long-term voltage stability analysis, the relevant contingencies are outages of transmission or generation facilities; in this time frame the sequence of events that lead to such outages does not really matter. On the contrary, for short-term voltage stability, the system response to short-circuits must be investigated, in addition to outages. A well-known criterion is the one of $N - 1$ security, according to which a system must be able to withstand any single transmission or generation outage without entering an emergency state. It may be additionally required that no generator operates under reactive power limit after a contingency. Multiple contingencies may also enter security criteria, but usually they have to relate to a common cause making their simultaneous occurrence a credible event. A typical example is a bus-bar fault, cleared by tripping all equipments connected to the bar of concern.

Post-contingency controls. System protection devices may contribute to stabilize the system in the post-contingency configuration and hence must be taken into account in contingency evaluation. It is however appropriate to distinguish between the various controls with respect to the impact on customers, in terms of interruption of service and power quality. Compensation switching, increase in generator voltage setpoints, and secondary voltage control typically have no impact on customers, as opposed to LTC blocking, LTC voltage reduction, and in the last resort, load shedding. A common practice is to assess the system ability to survive credible contingencies with the sole help of controls which have no impact on customers. Complementary to this, the adequacy of stronger controls is checked against more severe disturbances, not entering usual security criteria.

Techniques for contingency evaluation within the context of voltage stability will be detailed in Section 9.2.

9.1.3   Security margin determination

The main limitation of contingency evaluation lies in the fact that it focuses on a particular operating point. Even when the system state is voltage secure for a given operating condition, it is desirable to know how far the system can move away from its current operating point and still remain secure. This brings us to the third level of security assessment : the determination of *security margins*. Within this context, the objective of preventive control is not only to bring the system from an insecure to a secure state (i.e. from a negative to a zero margin) but also to *maintain adequate security margins*.

Security margins rely on the definition of a *system stress*, which corresponds to *large* deviations of parameters that system operators can either *observe* (e.g. the system load) or *control* (e.g. a generation rescheduling). This stress is also characterized by a "direction" in the parameter space, which takes on the form of bus participations to load increase and/or generation rescheduling. These margins are particularly needed in the transmission open access environment which is prevailing in an increasing number of countries. In this case, the operator wants to know how much load or transfer increase can be accepted without the system becoming insecure.

Note that security margins along a single direction can be generalized to security regions in the space of stress parameters. A graphical display of the security boundary is convenient for two parameters, with a third one used to produce a family of curves. The most interesting case is when the stress parameters interfere.

A security margin is a measure of proximity to: (i) either a *post-contingency loadability limit*, in which case we speak of loadability margin, or (ii) a *secure operation limit*, in which case we speak of secure operation margin. Generally speaking, in order to encounter a post-contingency loadability limit, we apply a given contingency and then stress the system along the specified direction, whereas to reach a secure operation limit we stress the system as far as it can withstand the given contingency.

The determination of both types of margins will be considered in detail in Sections 9.3 and 9.4.

9.1.4 Notation for static approaches

In this chapter, computational methods are classified according to the type of security assessment they can be used for. A more technical, commonly used distinction is between static and dynamic methods. Basically, dynamic approaches rely on the dynamic model (6.1 - 6.4) outlined in Chapter 6 while static approaches deal with the equilibrium equations of this model, also discussed in Chapter 6.

At many places in this chapter, the emphasis will be on long-term voltage stability. Reusing the notation of Section 7.1.2, the corresponding long-term equilibrium equations will be written for short as:

$$\varphi(\mathbf{u}, \mathbf{p}) = 0 \tag{9.1}$$

where $\mathbf{u}$ is a vector of variables and $\mathbf{p}$ a vector of parameters.

These equations have been derived in Section 6.5. Let us briefly recall that they take on the form of:

- either equations (6.39a,b,c) in which case **u** is made up of long-term variables **z** (e.g. LTC ratios), short-term variables **x̃** (relative to generators, motors, etc.) and network variables **y** (bus voltage magnitudes and phase angles);

- or a network-only model of the type (6.43) in which case **u** is made up of **y** and the system frequency ω_{sys}. In its simplest form, this model consist of load flow equations (see discussion in Section 6.5.4) involving **y** only.

9.2 CONTINGENCY EVALUATION

9.2.1 Post-contingency load flow

The simplest way of evaluating the impact of contingencies on long-term voltage stability is by computing the post-contingency long-term equilibrium.

When using the load flow equations as the long-term equilibrium model, the above equilibrium computation is simply referred to as a *post-contingency load flow*. The latter is routinely used in applications ranging from planning to real-time.

In Chapter 8 we have shown that an important mechanism of long-term instability following a contingency, is the absence of a long-term equilibrium in the post-contingency configuration (see LT1 mechanism in Section 8.5.1). Since (9.1) has no solution in such a case, any numerical algorithm trying to solve these equations will divergence. This provides a simple way to check whether an equilibrium exists. However, while very simple, this approach suffers from several drawbacks:

1. the divergence of the numerical method may result from purely numerical problems that do not relate to a physical voltage instability. This is particularly true in systems where many control adjustments are needed to get a feasible operating point;

2. in a truly unstable case, we are left without any information regarding the nature and location of the problem, as well as possible remedies;

3. as quoted at the end of Section 6.5.2, in some cases it may not be obvious to account for discrete-type devices whose final state depends on the system time evolution.

4. there are some situations where the existence of a post-contingency equilibrium does not guarantee a stable system behaviour: a typical example is the LT2

instability mechanism of Section 8.6.3, in which voltage stability results from a lack of attraction towards the existing equilibrium point.

Divergence problems become a real burden when contingency analysis is applied to systems that are vulnerable to voltage instability. The methods described in the next two sections may contribute to improving the aspects listed under items 1 and 2 above.

9.2.2 Load flow methods for unsolvable cases

Several modifications to the standard Newton-Raphson method have been proposed to deal with cases of difficult convergence and/or absence of solution. We briefly describe hereafter a method inspired of [IT81] and used in [RAU93, Ove94, Ove95]. Although developed for load flow calculations in rectangular coordinates, it could be applied to other models (9.1) of the network-only type.

The standard method to solve these equations is the Newton method, which consists of repeated iterations on the following linear system:

$$\varphi_{\mathbf{u}} \left[\mathbf{u}_{(k+1)} - \mathbf{u}_{(k)} \right] = -\varphi(\mathbf{u}_{(k)}, \mathbf{p}) \tag{9.2}$$

where subscripts in parentheses refer to iteration numbers ($k = 0, 1, 2, \ldots$). Consider the quadratic error function

$$E(\mathbf{u}_{(k)}) = \sum_{i=1}^{n} \varphi_i^2(\mathbf{u}_{(k)}, \mathbf{p}) \tag{9.3}$$

When convergence is normal, this function decreases towards zero over successive iterations, since each term (or at least a majority of terms) vanishes as $\mathbf{u}_{(k)}$ approaches a solution. This is no longer true in case of difficult convergence or divergence, and from some iteration one observes:

$$E(\mathbf{u}_{(k+1)}) > E(\mathbf{u}_{(k)}) \tag{9.4}$$

In the modified Newton method the value of E is checked over successive iterations and, when it is found to increase, a scaling factor $\alpha < 1$ is applied to the normal Newton correction:

$$\mathbf{u}_{(k+1)} = \mathbf{u}_{(k)} - \alpha \, \varphi_{\mathbf{u}}^{-1} \, \varphi(\mathbf{u}_{(k)}, \mathbf{p}) \tag{9.5}$$

The value of α is chosen so that $E[\mathbf{u}_{(k+1)}]$ is minimal. This can be done by testing successive values of α, computing $\mathbf{u}_{(k+1)}$ from (9.5) and E from (9.3), or more directly if rectangular coordinates are used. When the best value of α has been found, the corresponding $\mathbf{u}_{(k+1)}$ is adopted as the solution of the $(k+1)$-th step and the iterations proceed. The procedure stops at $\mathbf{u}^\star$ such that: (i) either all $|\varphi_i(\mathbf{u}^\star, \mathbf{p})|$'s are smaller than a tolerance, i.e. the equations (9.1) are solved, or (ii) the best found value of α is negligible, indicating that no further decrease in E can be obtained. In the latter case, $\mathbf{u}^\star$ does not satisfy (9.1), most likely because there is no solution.

If φ consists of power flow equations, $\varphi_i(\mathbf{u}^\star, \mathbf{p})$ is a real or reactive power mismatch of the form:

$$\varphi_i(\mathbf{u}^\star, \mathbf{p}) = S_i^{inj} - S_i(\mathbf{u}^\star, \mathbf{p}) \qquad i = 1, \ldots, n$$

where S_i^{inj} is a real or reactive bus power injection. In this case, if the method converges to a nonzero $E(\mathbf{u}^\star)$, an interesting by-product of the algorithm is an indication of how to restore solvability of the equations. Indeed, by considering that the injections are modified from S_i^{inj} to $S_i(\mathbf{u}^\star, \mathbf{p})$, one gets a modified load flow problem for which $\mathbf{u}^\star$ is a solution. Furthermore, it can be shown [Ove94, Ove95] that at $\mathbf{u}^\star$ the Jacobian $\varphi_\mathbf{u}$ is singular. In other words, in the parameter space of injections, the point corresponding to the modified injections lies on the bifurcation surface Σ (see Section 7.5.1). Although it has been observed that the largest components $|\varphi_i(\mathbf{u}^\star, \mathbf{p})|$ tend to concentrate in the voltage problem area [RAU93], the above mentioned change of injections does not necessarily correspond to the minimal load shedding. In [Ove94, Ove95], $\mathbf{u}^\star$ is used as the starting point for determining the minimum load shedding, using the procedure of equilibrium point restoration outlined in Section 8.6.2.

9.2.3 VQ curves

The VQ curves [CTF87, MJP88, MAR94] introduced in Section 2.7 can be used for three purposes: (i) obtain a solution of (9.1) in cases of difficult convergence (caused for instance by proximity to instability); (ii) restore solvability in unstable cases for which (9.1) has no solution, and (iii) obtain a reactive power margin to stability or instability.

Let us recall the simple technique for producing VQ curves: a fictitious generator with zero active power is added to the system and its voltage V is varied in order to record its reactive power production Q_c as a function of voltage V, yielding the curves shown in Fig. 2.18. Note that the VQ curve technique is a special case of the continuation method to be described in Section 9.3.2.

To illustrate point (i) above, assume that for the power transfer corresponding to curve 2 in Fig. 2.18 we have convergence problems when solving (9.1). We first add the fictitious generator, taking for instance $V/E = 1$, which allows the equations to be solved normally. Then we decrease V progressively, up to meeting point O' where

$Q_c = 0$ (and $dQ_c/dV > 0$). At this point, the generator does not produce anything and hence the current system state **u** is a solution of the original equations (9.1).

Continuing the curve up to its minimum (the left branch need not be determined far below this minimum) yields the reactive power margin Q_2 with respect to the loss of an operating point. Curve 3 in the same figure corresponds to an unstable situation, and the same technique yields the Q_3 margin to system operability.

In fact, the above Q margin is the distance to a particular loadability limit, obtained by varying the reactive power injection at a single bus. It must be emphasized, however, that by using such a peculiar loading pattern, the resulting mode of instability (as characterized by the eigenvectors of the zero eigenvalue at the loadability limit) is very different from those corresponding to realistic directions of stress (including active power as well).

Beside security margins, VQ curves also provide information on the amount of shunt compensation needed to either restore an operating point or reach a specified voltage, as shown in Figs. 2.20 and 2.21. In this application as well as in margin calculation, the location of the fictitious generator requires some judgment and some knowledge of the system. Expectedly, different locations yield different curves and different margins. A technique to obtain the best location for additional reactive power support is by driving the system to the limit along a realistic direction of stress and identifying the largest reactive power components of the eigenvector corresponding to the zero eigenvalue [MAR94].

Application to weak area identification

An interesting problem, with some practical applications, is that of identifying the areas within a power system that are most prone to voltage instability. In systems with a special structure, such as radial systems, the voltage stability limited areas can be easily identified using engineering judgment and operator experience. In meshed power systems, however, the problem of locating weak areas is not trivial.

Reference [Sch98] proposes an approach to identify areas prone to voltage instability, where coherent bus groups are defined imposing a reactive power stress on each bus (similar to obtaining VQ curves) and grouping together the buses which depend upon the same reactive reserves. One advantage of this method is that it does not depend upon linearization.

9.2.4   Multi-time-scale simulation methods

Advantages of time-domain methods

Within the context of voltage stability analysis, time-domain methods in general offer several advantages, for instance:

- higher modelling accuracy: time simulation is often used as the benchmark, e.g. to validate simplified, faster methods

- possibility to study other instability mechanisms than the loss of equilibrium captured by static methods (refer to the classification in Chapter 8);

- higher interpretability of results (in terms of sequence of events leading to instability or collapse), possibility to obtain information on remedial actions, higher educational value, etc.

On the other hand, time simulation is generally much more demanding in terms of computing times and data gathering, unless specific methods are devised. Nevertheless, the more careful modelling forced by time simulation is often a good opportunity to update or correct models used in routine computations.

Multi-time-scale modelling

Multi-time-scale simulation of voltage phenomena requires the numerical integration of the large set of differential-algebraic, continuous-discrete time equations (6.1 - 6.4), which we recall here for convenience:

$$\dot{\mathbf{x}} = \mathbf{f}(\mathbf{x}, \mathbf{y}, \mathbf{z}_c, \mathbf{z}_d) \tag{9.6a}$$

$$0 = \mathbf{g}(\mathbf{x}, \mathbf{y}, \mathbf{z}_c, \mathbf{z}_d) \tag{9.6b}$$

$$\dot{\mathbf{z}}_c = \mathbf{h}_c(\mathbf{x}, \mathbf{y}, \mathbf{z}_c, \mathbf{z}_d) \tag{9.6c}$$

$$\mathbf{z}_d(k+1) = \mathbf{h}_d(\mathbf{x}, \mathbf{y}, \mathbf{z}_c, \mathbf{z}_d(k)) \tag{9.6d}$$

This model has been thoroughly discussed in Chapter 6. Short-term voltage stability simulation relies on (9.6a,b). As discussed in Chapter 6, this is basically the model used in transient (angle) stability studies, with proper account for load behaviour. Thus, in principle, the numerical integration methods used for transient stability [Sto79] apply equally well to short-term voltage stability studies. Long-term voltage stability simulation, on the other hand, raises problems that require specific solutions.

In the sequel we first review some fundamentals of numerical integration before dealing with the specific aspects of multi-time-scale simulation.

A quick overview of numerical integration methods

We consider provisionally the simpler problem of integrating a scalar Ordinary Differential Equation (ODE) of the type:

$$\dot{x} = f(x) \tag{9.7}$$

Numerical integration consists in discretizing the ODE to end up with a recursive, algebraic formula that can be processed by a computer. Numerous methods have been proposed to this purpose [KMN88, Gea71]. For many of them the algebraized relationship takes on the form:

$$x^{j+1} = \sum_{i=1}^{k} \alpha_i \, x^{j+1-i} + h \sum_{i=0}^{\ell} \beta_i \, \dot{x}^{j+1-i} \tag{9.8}$$

where j denotes the time step and h is the integration step size[1]. Integration formulae with $k = \ell = 1$ are called *single-step* since, at each point in time, they use results from the previous point only. Table 9.1 gives three well-known examples of single-step methods. Methods such that $k > 1$ or $\ell > 1$ are called *multi-step*.

An important distinction is between implicit and explicit methods. An integration method such that $\beta_0 \neq 0$ is called *implicit*: the value of x at a given time step depends on the derivative of x at the same step. For such a method it is convenient to rewrite (9.8) as:

$$x^{j+1} = h \, \beta_0 \, f(x^{j+1}) + c^{j+1} \tag{9.9}$$

where c^{j+1} is a known constant at the $(j+1)$-th step. An *explicit* method, on the other hand, is such that $\beta_0 = 0$. Among the methods of Table 9.1, the Euler method is explicit while the Backward Euler and the Trapezoidal Rule are implicit.

Consider the solution of (9.7) over the time interval $[t^j \quad t^{j+1}] = [t^j \quad t^j + h]$. Using a Taylor series expansion (including the remainder), the *exact* values of x at both ends of the interval are related through:

$$x(t^{j+1}) = x(t^j) + h\dot{x}(t^j) + \frac{h^2}{2}\ddot{x}(\xi) = x(t^j) + hf(x(t^j)) + \frac{h^2}{2}\ddot{x}(\xi) \tag{9.10}$$

---

[1] Runge-Kutta methods do not follow the above formulation. They are not considered here, because their poor numerical stability [KMN88, Gea71] does not make them good candidates for multi-time-scale simulation

Table 9.1 Examples of single-step integration methods

method	formula (9.8)	LTE	amplification factor
Euler	$x^{j+1} = x^j + h\,\dot{x}^j$ $(\alpha_1 = 1, \beta_0 = 0, \beta_1 = 1)$	$\dfrac{h^2}{2}\ddot{x}(\xi)$	$1 + h\,J(\zeta)$
Backward Euler	$x^{j+1} = x^j + h\,\dot{x}^{j+1}$ $(\alpha_1 = 1, \beta_0 = 1, \beta_1 = 0)$	$\dfrac{h^2}{2}\ddot{x}(\xi)$	$\dfrac{1}{1 - h\,J(\zeta)}$
Trapezoidal Rule	$x^{j+1} = x^j + \dfrac{h}{2}\left(\dot{x}^j + \dot{x}^{j+1}\right)$ $(\alpha_1 = 1, \beta_0 = 0.5, \beta_1 = 0.5)$	$\dfrac{h^3}{12}\dddot{x}(\xi)$	$\dfrac{1 + 0.5\,h\,J(\zeta)}{1 - 0.5\,h\,J(\zeta)}$

where $t^j < \xi < t^{j+1}$. Assume for simplicity that we use the Euler method (see Table 9.1) to solve (9.7). For the same time steps we have:

$$x_{em}(t^{j+1}) = x_{em}(t^j) + h\,f(x_{em}(t^j)) \tag{9.11}$$

where the lowerscript em denotes the numerical approximation corresponding to Euler method. Subtracting (9.11) from (9.10), we obtain:

$$x(t^{j+1}) - x_{em}(t^{j+1}) = x(t^j) - x_{em}(t^j) + h\left[f(x(t^j)) - f(x_{em}(t^j))\right] + \frac{h^2}{2}\ddot{x}(\xi) \tag{9.12}$$

Using the Mean Value Theorem, the term between brackets becomes:

$$f(x(t^j)) - f(x_{em}(t^j)) = J(\zeta)(x(t^j) - x_{em}(t^j))$$

where $J(\zeta)$ denotes the derivative of f with respect to x evaluated at a specific (but unknown) point ζ. Introducing this result in (9.12) and rearranging terms we obtain:

$$x(t^{j+1}) - x_{em}(t^{j+1}) = [1 + h\,J(\zeta)]\left[x(t^j) - x_{em}(t^j)\right] + \frac{h^2}{2}\ddot{x}(\xi) \tag{9.13}$$

The left-hand side of this equation represents the *global error* at time t^{j+1}, i.e. the difference at the $(j + 1)$-the step between the exact and the computed solutions. Consider now the right-hand side. The last term is the *Local Truncation Error* (LTE). This error is introduced at each time step by the truncation of the Taylor series expansion in the numerical integration formula. Equivalently, this would be the integration error at time t^{j+1} if the solution at time t^j was perfectly accurate. The first term stems from the errors accumulated at previous time steps. The expression $1 + h\,J(\zeta)$ is the *amplification factor* by which $x(t^j) - x_{em}(t^j)$, the global error at the previous step, is multiplied.

A fundamental requirement for any method is that the amplification factor remains always less than 1, otherwise the global error will grow with time and the integration will be *numerically unstable*. To preserve accuracy, the amplification factor should be made as small as possible. The practical consequences of this are best seen on the simple test equation:

$$\dot{x} = \lambda\ x \qquad (9.14)$$

for which the amplification factor in Euler method is $1 + h\,\lambda$. For numerical stability it is required to have $|1 + h\,\lambda| < 1$, or equivalently $-2 < h\lambda < 0$. We conclude that, for a stable system ($\lambda < 0$) the numerical response response by Euler method is stable if h is smaller than $|2/\lambda|$. Setting in this case $\lambda = -1/T$, instability occurs if the integration step size is larger than twice the time constant T. In a simplistic integration scheme like the Euler method, there is thus a stringent requirement to use a small integration step size h in order to keep both the LTE and the amplification factor small.

The LTE and amplification factor are indicated for all three methods in Table 9.1. As can be seen, the Trapezoidal Rule is the most accurate in terms of LTE [2]. However, the most salient result is that, for $\lambda < 0$, the amplification factor of the two implicit methods remains smaller than 1 (or even decreases to zero for Backward Euler) when h tends to infinity. This illustrates the general property that implicit methods have much better numerical stability.

What has been said for the simple system (9.14) with a single, real λ can be generalized to the linear system $\dot{x} = \mathbf{A}\mathbf{x}$. In this case, the numerical stability condition keeps on the same form with λ being the largest eigenvalue (in magnitude) of $\mathbf{A}$. Consequently, a region of numerical stability in the complex $h\lambda$ plane (generalizing the stability interval on the real $h\lambda$ axis) can be determined. An integration method is called *A-stable* when this region is the half left complex plane: in this case, if the system is stable, the numerical response will be stable for any value of h. The Backward Euler method and the Trapezoidal Rule are both A-stable. The A-stability, the simplicity and the lower LTE [3] of the Trapezoidal Rule have contributed to the popularity of this method [BCS77].

---

[2] at least for small h, in which case h^3 is much smaller than h^2

[3] Note however that the amplification factor of the Trapezoidal Rule tends to -1 as h increases, which corresponds to a risk of numerical oscillations

Let us concentrate now on the way to compute x^{j+1}. If the method is explicit ($\beta_0 = 0$), this amounts to introducing the past values of x and $\dot{x}$ in the right-hand side of (9.8). If the method is implicit, the nonlinear equation (9.9) has to be solved with respect to x^{j+1}. This can be done using *functional iterations* or *Newton method*.

In functional iterations, the value of x^{j+1} is first predicted, using an explicit formula. This predicted value and the known past values of x and $\dot{x}$ are introduced in the right-hand side of (9.9), which provides a new value for x^{j+1}, and the procedure is repeated until convergence. For this reason, the functional iteration scheme belongs to the family of *predictor-corrector* methods[4]. Although simple, functional iterations have a major limitation: in order for the x^j sequence to converge, it is required that [Gea71]:

$$|h\beta_0 \, J(x^{j+1})| < 1$$

or equivalently

$$h < \frac{1}{|\beta_0 \, J(x^{j+1})|}$$

which puts a limit on the maximal allowed step size. This limit may be constraining even for a numerically stable method. This difficulty is circumvented by using Newton method to solve (9.9), which allows to use large time step sizes. The potential risk of divergence of Newton iterations can be avoided by using a predictor that provides a good initial guess.

Let us now consider differential-algebraic systems of the type (9.6a,b). They can be solved using either a *partitioned* or a *simultaneous* scheme. In a partitioned scheme, the ODEs are solved for **x**, and the algebraic constraints are solved for **y**, alternatively to avoid interface errors, and in a way that avoids fully iterating on each subproblem. In simultaneous schemes, both the algebraized ODEs and the algebraic constraints are solved at the same time. A partitioned scheme is better suited to explicit methods or implicit methods using functional iterations, while a simultaneous scheme is more attractive when the implicit integration formula is handled by Newton method, in which case this method is used to solve the two sets of equations:

$$\mathbf{f}(\mathbf{x}^{j+1}, \mathbf{y}^{j+1}) - \frac{1}{h\beta_0}\mathbf{x}^{j+1} + \mathbf{d}^{j+1} \;=\; 0 \qquad (9.15a)$$

$$\mathbf{g}(\mathbf{x}^{j+1}, \mathbf{y}^{j+1}) \;=\; 0 \qquad (9.15b)$$

where (9.15a) comes from (9.9) and $\mathbf{d}^{j+1}$ is a known constant. The Newton method involves iterations on the following linear system (which in the case of power systems

---

[4] It has been shown that the corrector must be repeated until convergence (within some tolerance) in order to preserve the numerical stability of implicit methods [Gea71, KMN88]

is sparse):

$$
\begin{bmatrix} \mathbf{f_x} - \dfrac{1}{h\beta_0}\mathbf{I} & \mathbf{f_y} \\[2mm] \mathbf{g_x} & \mathbf{g_y} \end{bmatrix} \begin{bmatrix} \mathbf{x}_{(k+1)}^{j+1} \\[2mm] \mathbf{y}_{(k+1)}^{j+1} \end{bmatrix} = \begin{bmatrix} -\mathbf{f}(\mathbf{x}_{(k)}^{j+1},\mathbf{y}_{(k)}^{j+1}) + \dfrac{1}{h\beta_0}\mathbf{x}_{(k)}^{j+1} - \mathbf{d}^{j+1} \\[2mm] -\mathbf{g}(\mathbf{x}_{(k)}^{j+1},\mathbf{y}_{(k)}^{j+1}) \end{bmatrix}
$$
(9.16)

where $\mathbf{I}$ is a unity diagonal matrix and parenthesized subscripts refer to iteration numbers ($k = 0, 1, 2, \ldots$). In the so-called *dishonest Newton method*, the Jacobian matrix of the above equation is kept constant over several time steps, as long as convergence remains satisfactory and no discontinuity occurs.

Multi-time-scale simulation aspects

The handling of the long-term equations (9.6c,d) makes the model to be integrated *stiff*. A stiff problem is one in which the underlying physical process contains components operating on widely separated time scales, or equivalently the dynamics of some part of the process (in our case the short-term dynamics) are very fast compared to the interval over which the whole process is studied (in our case, the long-term time frame). In linear systems, stiffness is measured by the ratio of the largest to the smallest eigenvalue.

A number of approaches are in use for long-term simulation [KOO93, SDP93, DS93, MGK93, Kun94, KF93]:

■ computation of the whole system response with the small time step size needed for accurate simulation of the short-term dynamics. This is a simple but brute force approach, resulting in large computing times and huge amounts of output points to process afterwards. To fix ideas, using a 0.01 s step size, it takes a matter of 30,000 time steps to simulate 5 minutes of the system response, which is common in voltage stability studies;

■ increase in time step size after the short-term transients have died out. The idea is to subsequently filter out the fast transients, if any, that are not significant in the system response. Provision must be made to switch back to a smaller step size, upon detection of an S-LT instability. A criterion is to observe the rate of change of the fast variables;

■ automatic adjustment of step size to the system behaviour, i.e. shorter when the fast dynamics is excited, and longer when only slow transients are present.

In the last two cases, the increase in step size requires to use a numerically stable integration method, even if fast transients have completely vanished in the system response. A robust solution scheme is also needed. For previously explained reasons, implicit simultaneous integration using the Newton method appears as the most appropriate to this purpose.

Step size control is already an old idea and the interested reader may refer for instance to [Gea71] for more details. It may rely on heuristic criteria (such as the convergence behaviour of the Newton iterations) [Sto79] but a well-known strategy is by control of the LTE. In practice, at a given time step of the numerical integration, the global error (i.e. the total error with respect to the exact solution) cannot be controlled. The only term that can be acted on is the LTE (i.e. the error introduced at each new step). The latter can be estimated and compared to some threshold ϵ. If the error is found to decrease below (resp. increase above) ϵ, the time step size h is increased (resp. decreased), based on the known relationship between LTE and h.

We illustrate this strategy using the Trapezoidal Rule as an example. Assume that at the j-th step the value of x^j and $\dot{x}^j$ are known exactly. Substituting a Taylor series expansion for the first-order derivative $\dot{x}^{j+1}$ in the Trapezoidal Rule formula of Table 9.1 yields:

$$x^{j+1} = x^j + h\,\dot{x}^j + \frac{h^2}{2}\ddot{x}^j + \frac{h^3}{4}\,\dddot{x}^j + O(h^4) \qquad (9.17)$$

Comparing with the Taylor series expansion of x^{j+1} up to the third order we get:

$$\text{LTE} = \frac{h^3}{4}\dddot{x}^j - \frac{h^3}{6}\dddot{x}^j = \frac{h^3}{12}\dddot{x}^j$$

as indicated in Table 9.1. To estimate the LTE, the third derivative can be evaluated by finite differences. An alternative consists in using the difference between predicted and corrected values of x. Using the quadratic prediction:

$$x_{(0)}^{j+1} = x^j + h\dot{x}^j + \frac{h^2}{2}\ddot{x}^j \qquad (9.18)$$

and subtracting (9.18) from (9.17) we get:

$$x^{j+1} - x_{(0)}^{j+1} = \frac{h^3}{4}\dddot{x}^j = 3 \cdot \text{LTE}$$

Therefrom we deduce that in order to keep the magnitude of LTE equal to ϵ, the step size must be changed from the old value h_{old} to to the new value h_{new} according to:

$$h_{new} = h_{old}\left[\frac{3\,\epsilon}{|x^{j+1} - x_{(0)}^{j+1}|}\right]^{1/3}$$

In practice, this formula can be used to check that the last step size was acceptable and provide the next step length. Also, several equal length steps are made before a change is allowed.

Some final remarks regarding the discrete-type equations (9.6d) are in order. When changes in z_d occur at a given time instant t, the continuous variables $\mathbf{x}$ and $\mathbf{z}_c$ do not change while both the algebraic variables $\mathbf{y}$ and the derivatives $\dot{\mathbf{x}}$ and $\dot{\mathbf{z}}_c$ undergo a discontinuity. In principle, the numerical integration has to be reinitialized by computing $\mathbf{y}(t^+)$ and therefrom $\dot{\mathbf{x}}(t^+)$ and $\dot{\mathbf{z}}_c(t^+)$. A single-step method is needed to restart the integration.

In long-term voltage stability simulation of large systems, the numerous discrete-type devices such as LTCs give rise to frequent discrete transitions. The latter are a natural obstacle to large increases in step size, because the transition times have to be identified with some reasonable accuracy and too large time steps would eventually require some time consuming interpolation. In large systems the time step size can hardly be increased beyond several tenths of a second (which, however, yields a substantial gain). The above discontinuities can be avoided by using the continuous-time approximate LTC model (4.35) and/or aggregate load recovery models of the type (4.51a,b).

When the time step size is kept below one second, it is possible to use a (simpler) partitioned explicit method to integrate the long-term ODEs (9.6c) with time constants in the order of one minute, keeping a simultaneous implicit scheme for integration of $\mathbf{x}$ and $\mathbf{y}$ and possibly some faster $\mathbf{z}_c$ variables.

Illustrative example

In the following example, we show simulation results obtained with the EUROSTAG software [DS93], one of the first production-grade program to implement the variable step size concept for a unified treatment of short and long-term dynamics. Figures 9.1 to 9.4 show typical outputs for a 33-bus test system used by CIGRE task forces on long-term and voltage stability [CTF94a, CTF95].

At time $t = 50$ s machine M2 is tripped. After some delay, the LTCs start trying to restore their secondary voltages (the positions of two of them are shown in Fig. 9.3) but without success. As a result, transmission voltages sag as shown in Fig. 9.1. This is aggravated by the switching of machines M1 and M3 under constant field voltage by their rotor protection, at $t = 85$ and 220 s respectively, as shown in Fig. 9.2. The limitation of M3 causes the transmission voltage to drop fast and machine M1, whose field voltage was reduced to an unduly low value, loses synchronism. This is observed in the final voltage oscillations. This instability falls in the S-LT1 category described

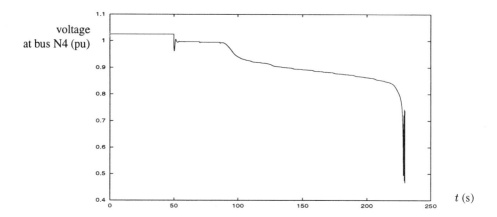

Figure 9.1   CIGRE system : EUROSTAG output of voltage at a transmission bus

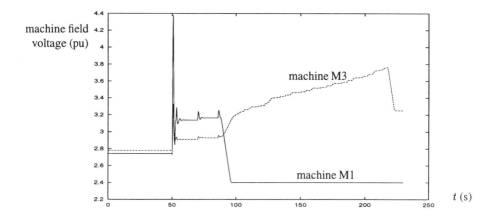

Figure 9.2   CIGRE system : EUROSTAG output of two machine field voltages

in Chapter 8. The tripping of the unstable machine by its out-of-step relay results in a blackout. A similar scenario was studied in [VCV96].

The variable step size h used by EUROSTAG is shown in the semi-logarithmic plot of Fig. 9.4. Its range of possible variation was set to $[10^{-4} \quad 10^1]$ seconds. The initiating disturbance has been applied at $t = 50$ s in order to demonstrate the capability to

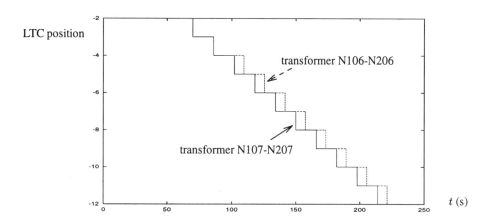

Figure 9.3   CIGRE system : EUROSTAG output of two LTC positions

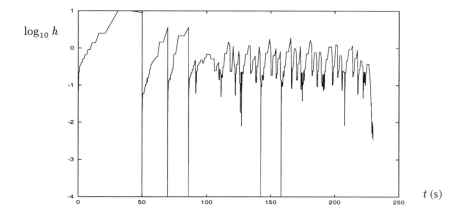

Figure 9.4   CIGRE system : variable time step size used by EUROSTAG (semi-log scale)

perform large, stable step sizes when no dynamics are excited (the step sizes should be compared to the subtransient time constants of the 4-winding Park model used for machines). Apart from the period following the disturbance as well as the final short-term instability, the step size is reduced after the LTC and OXL switchings, because the latter excite the short-term dynamics to an extent that is considered significant by the LTE control logic.

Another example of multi-time scale simulation is given in Section 9.6.3.

9.2.5   QSS long-term simulation

In spite of the increase in computer power and the development of efficient variable step size algorithms, multi-time-scale simulation remains a heavy approach in terms of computing times, data handling and output processing. These constraints are even more stringent within the context of real-time voltage security assessment.

When dealing with long-term instability mechanisms, it is desirable to speed up calculations by neglecting the short-term dynamics, at least up to some point. In the previous section, it was suggested to use a larger time step size to filter out the fast dynamics. Another approach consists in merely replacing the short-term dynamics with their equilibrium equations and concentrating on long-term phenomena. This is nothing but the Quasi Steady-State (QSS) approximation of long-term dynamics introduced in Fig. 6.9 and used in our analyses of Chapters 6 and 7. This idea is rather old [Lac79] but in the recent years it has been developed either as one mode of operation of a dynamic simulation package [KOO93] or as a separate time-domain simulation program [STI82, VC93, VJM95, VJM94]. It also generalizes the empirical "step-by-step load flow" technique described in [IWG93].

Modelling

Formally, the equations of the QSS approximation are (9.6a - 9.6d) with the short-term dynamics (9.6a) replaced by the equilibrium equations:

$$0 = \mathbf{f}(\mathbf{x}, \mathbf{y}, \mathbf{z}_c, \mathbf{z}_d) \qquad (9.19)$$

In practice, a reduced set of the above equations, involving a reduced state vector $\mathbf{x}$, is preferred. The corresponding *reduced equilibrium model* is easier to handle, although equivalent to the original set of equations. Its derivation has been discussed in detail in Section 6.5.1.

Computational aspects

The QSS calculation procedure is outlined in Fig. 9.5. The time step size h is in the order of 1 to 10 seconds in practice. Each dot represents a short-term equilibrium point, i.e. a solution of (9.6b,9.19) with $\mathbf{z}_c$ and $\mathbf{z}_d$ fixed at their current values.

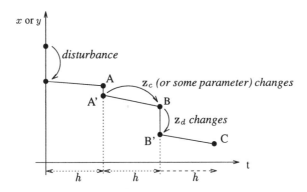

x or y

disturbance

A

z_c *(or some parameter) changes*

A'

B

z_d *changes*

B'

C

t

h h h

Figure 9.5 principle of QSS simulation

The transitions from A to A', B to B', etc. come from the discrete dynamics (9.6d) of OXLs, LTCs, switched shunt elements, secondary controls, etc. (see description in Section 6.1). These devices undergo a transition once a condition has been fulfilled for some time. This delay may be constant, obey an inverse-time characteristic or even be zero (e.g. for SVC susceptance limitation, considered infinitely fast in the QSS approximation). In QSS simulation there is no point in identifying very accurately the time of each transition, considering that the short-term dynamics have been neglected anyway. Rather, the various discrete devices are checked at multiples of the time step h and switched as soon as their internal delays are overstepped. The discrete devices are thus more or less "synchronized" depending on the value of h.

Points A', B', etc. are obtained by solving (9.6b,9.19) with respect to x and y using the Newton method. This requires to iterate on the following linear, sparse system:

$$\begin{bmatrix} \mathbf{f_x} & \mathbf{f_y} \\ \mathbf{g_x} & \mathbf{g_y} \end{bmatrix} \begin{bmatrix} \mathbf{x}^{j+1}_{(k+1)} \\ \mathbf{y}^{j+1}_{(k+1)} \end{bmatrix} = \begin{bmatrix} -\mathbf{f}(\mathbf{x}^{j+1}_{(k)}, \mathbf{y}^{j+1}_{(k)}) \\ -\mathbf{g}(\mathbf{x}^{j+1}_{(k)}, \mathbf{y}^{j+1}_{(k)}) \end{bmatrix} \tag{9.20}$$

where parenthesized subscripts refer to iteration numbers ($k = 0, 1, 2, \ldots$). Prediction from previous values can hardly be envisaged due to the discontinuous nature of the time evolution (see Fig. 9.5); thus the iterations start from the x and y values of the last time step.

The transitions from A' to B, B' to C, etc. correspond to the differential equations (9.6c) and/or smooth variations of some parameters with time (e.g. during load increase). If there is no differential equation (9.6c), the system evolution is a mere succession of short-term equilibrium points in which B is obtained from A', C from B', etc. by iterating on (9.20) with the parameters updated. To deal with differential equations, an explicit integration scheme is sufficient since the time step size h is small compared to the time constants of (9.6c). Moreover, it is common to have discrete transitions at almost all time steps, and hence (i) a single-step integration method is needed; (ii) a partitioned solution scheme allows the linear system (9.20) to be used for both integrating and solving the discontinuities.

The sparse Jacobian in (9.20) is updated and factorized following events like equipment trippings or OXL activation. On the contrary, the numerous changes in transformer ratios or shunt susceptances as well as the smooth change in system parameters do not require a Jacobian update, unless some components of the right-hand-side of (9.20) become large or a slow convergence rate is experienced.

Limitations of the method and illustrative examples

The QSS approximation is based on the assumption that the short-term dynamics are stable. Therefore, with reference to the instability mechanisms discussed in Chapter 8:

- it cannot deal with the ST1, ST2 and ST3 instabilities of short-term dynamics (which is not the purpose of the method);

- it cannot reproduce S-LT1, S-LT2 and S-LT3 instabilities where the long-term unstable evolution triggers a short-term instability. Note however that S-LT1 instability is detected in the form of a loss of the short-term equilibrium. This is revealed in practice by diverging Newton iterations (9.20).

For illustrative purposes, Figs. 9.6 and 9.7 show how the QSS approximation performs in two scenarios of the case study of Section 8.5. The solid line curves (reproduced from Figs. 8.10 and 8.17) have been obtained by variable step size integration using the Trapezoidal Rule, while the dotted lines correspond to QSS simulation.

Figure 9.6 corresponds to Case 1 of LT1 instability. The two curves can hardly be distinguished, showing that the approximation is excellent. Figure 9.7 corresponds to Case 2 of S-LT1 instability. The loss of a short-term equilibrium emphasized by the PV curves of Fig. 8.16 is detected by the QSS simulation, where the time evolution cannot be determined beyond the black dot of Fig. 9.7. Up to this point the approximation is quite good. The accuracy somewhat deteriorates close to the saddle-node bifurcation of short-term dynamics, as discussed in Section 7.3.6. Nevertheless, the QSS simulation shows clearly an unacceptable long-term evolution, which is the *cause* of system trouble and it does predict the final collapse (albeit with a time error).

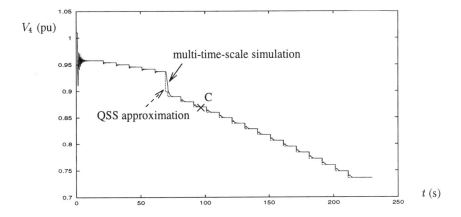

Figure 9.6   Case 1 of Section 8.5.1 : detailed vs. QSS simulation

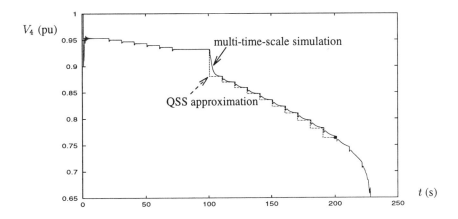

Figure 9.7   Case 2 of Section 8.5.2 : detailed vs. QSS simulation

Coming back to Fig. 9.6, the point labeled C is the critical point shown in Fig. 8.15. Note that the QSS simulation proceeds through this point without encountering any numerical difficulty.

QSS approximation offers an interesting compromise between the computational efficiency of static methods and the advantages of time-domain methods. Multi-time-scale simulation should be used to deal with short-term instability as well as to simulate the final collapse in S-LT scenarios (e.g. for protection design). For long-term voltage security assessment, QSS simulation is quite accurate in most cases, as will be evidenced in Section 9.6.3.

9.3   LOADABILITY LIMIT COMPUTATION

9.3.1   Definition and problem statement

Given a direction of system stress, the loadability limit indicates how much the system must be stressed in order to reach instability.

Loadability limits can be computed on the current configuration, in order to assess the system ability to face a forecasted load increase. However, *post-contingency loadability limits* are more useful in practice. They aim at characterizing the security margin which would remain after a contingency. They correspond to an experiment where the post-contingency situation is stressed to evaluate its robustness.

The concept is illustrated in Fig. 9.8 using PV curves, where load is assumed to restore to constant power. Note that the post-contingency operating point A may incorporate the effect of post-contingency controls, as discussed in Section 9.1.2. Similarly, controls may be effective while stressing the system up to the loadability limit.

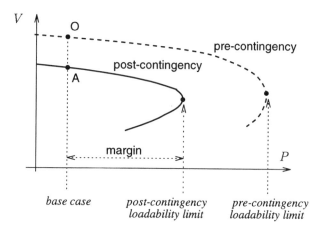

Figure 9.8   PV-curve sketch of a post-contingency loadability limit

Loadability limits have been defined and their properties studied in Chapter 7. We briefly recall hereafter some fundamentals needed for the description of computational methods.

Loadability limits correspond to the loss of equilibrium of the system subject to smooth changes in parameters $\mathbf{p}$. The relevant system model is thus made up of n equilibrium equations (9.1) on n variables $\mathbf{u}$. The system is considered to be at equilibrium with respect to both the short and long-term dynamics. Hence, (9.1) corresponds to the long-term equilibrium equations discussed in Section 6.5.2 (see Section 9.2.1 for a short description). We focus on changes in parameters along a direction $\mathbf{d}$ associated with the specified system stress (see Fig. 7.18):

$$\mathbf{p} = \mathbf{p}^o + \mu\,\mathbf{d} \tag{9.21}$$

so that (9.1) can be rewritten for short as:

$$\varphi(\mathbf{u}, \mu) = 0 \tag{9.22}$$

The loadability limit corresponds to the maximum value μ^* such that (9.22) has a solution.

In Section 3.6 we have emphasized the impact of generator reactive power limits on the maximum deliverable power. To handle these limits, an inequality-constraint formulation has been introduced in Section 7.6. For an n_g-generator system, it consists of replacing the n equations (9.22) by:

$$\tilde{\varphi}(\mathbf{u}, \mu) \;=\; 0 \tag{9.23a}$$
$$\text{subject to} \quad \psi_i^{avr}(\mathbf{u}, \mu) \;\geq\; 0 \qquad i = 1, \ldots, n_g \tag{9.23b}$$
$$\psi_i^{lim}(\mathbf{u}, \mu) \;\geq\; 0 \qquad i = 1, \ldots, n_g \tag{9.23c}$$
$$\psi_i^{avr}(\mathbf{u}, \mu).\psi_i^{lim}(\mathbf{u}, \mu) \;=\; 0 \qquad i = 1, \ldots, n_g \tag{9.23d}$$

where $\psi_i^{avr} = 0$ characterizes the i-th generator under AVR control (see Tables 6.4 and 6.7) and $\psi_i^{oxl} = 0$ characterizes the i-th generator under (field or armature current) limit (see same tables). Note that (9.23d) is aimed at forcing (at least) one of the inequality constraints (9.23b, 9.23c) to be active, expressing that the generator is either under AVR control or limited. Finally (9.23a) consists of the other $n - n_g$ equilibrium equations, not relative to generator excitation control.

For a given value of μ, knowing which generators are under AVR control and which ones are under limit, we may rewrite (9.23a-d) as the following set of n equations:

$$\tilde{\varphi}(\mathbf{u}, \mu) \;=\; 0 \tag{9.24a}$$
$$\psi_i^{avr}(\mathbf{u}, \mu) \;=\; 0 \quad \text{for each generator under AVR control} \tag{9.24b}$$
$$\psi_i^{lim}(\mathbf{u}, \mu) \;=\; 0 \quad \text{for each generator under limit} \tag{9.24c}$$

Note that (9.24a,b,c) make up a set of n equations on the $n + 1$ unknowns $\mathbf{u}$ and μ. It thus defines a family of solutions with a single degree of freedom. If, as μ varies, a generator gets limited, there is a point where both (9.24b) and (9.24c) hold for the generator of concern. This will be called a *breaking-point* in the sequel. After this point, (9.24b) is replaced by (9.24c) for the limited generator and the total number of equations is again n.

Let us recall finally that loadability limits are encountered at either Saddle-Node Bifurcation (SNB) points or breaking-points.

9.3.2 Continuation methods

The purpose of continuation methods is to compute a solution path of equations of the type (9.22), corresponding to a range of values of the parameter μ [Sey88].

Early applications of continuation methods to power systems were reported in [FG83, Pri84]. In voltage stability analysis, they are used more specifically to compute the solution path from a base case to a loadability limit, taking into account that the path folds at the limit point (see e.g. Fig. 7.5b or 7.19) [ISE91, CMT91, AC92, CA93, IWG93]. The method is also called *continuation power flow* with reference to the use of power flow equations for the model (9.22).

For simplicity we will refer to (9.22) when describing the method but in practice the set of equations to be considered is (9.24a-c), with the above mentioned switching from (9.24b) to (9.24c). In this respect, continuation methods can provide loadability limits of both the SNB and the breaking-point type.

Numerically, a sampling of points on the solution path is computed. We first consider passing from point $(\mathbf{u}^j, \mu^j)$ to point $(\mathbf{u}^{j+1}, \mu^{j+1})$ $(j = 0, 1, 2, \ldots)$, where upperscripts refer to point numbers.

A simple scheme consists in choosing a new value μ^{j+1} for the parameter μ and solving (9.22) for $\mathbf{u}$. In this scheme, μ is the *continuation parameter*. Using the Newton method to solve (9.22), we fix μ at the value μ^{j+1} and we iterate on the linear system:

$$\varphi_{\mathbf{u}} \left[\mathbf{u}_{(k+1)}^{j+1} - \mathbf{u}_{(k)}^{j+1} \right] = -\varphi(\mathbf{u}_{(k)}, \mu^{j+1}) \tag{9.25}$$

where parenthesized subscripts refer to iteration numbers ($k = 0, 1, 2, \ldots$). Note that this equation can be rewritten as:

$$\begin{bmatrix} \varphi_u & \varphi_\mu \\ 0^T & 1 \end{bmatrix} \begin{bmatrix} \mathbf{u}_{(k+1)}^{j+1} - \mathbf{u}_{(k)}^{j+1} \\ \mu_{(k+1)}^{j+1} - \mu_{(k)}^{j+1} \end{bmatrix} = \begin{bmatrix} -\varphi(\mathbf{u}_{(k)}^{j+1}, \mu^{j+1}) \\ 0 \end{bmatrix} \qquad (9.26)$$

where φ_μ is used, for the sake of uniformity, to denote the gradient $\nabla_\mu \varphi$ of φ with respect to μ.

While simple, the above scheme faces two problems: (i) if $\mu^{j+1} > \mu^\star$, there is no solution of (9.22) and the algorithm will diverge; (ii) at the SNB point ($\mu = \mu^\star$), we know from Chapter 7 that φ_u is singular; hence, for μ^{j+1} smaller but close to $\mu^\star$, (9.25) suffers from ill-conditioning and the iterations may diverge [AHI78]. In practice, for some systems it is possible to approach the SNB point very closely, while for others it is difficult.

In order to obtain the solution path around the SNB point, another variable than μ must be used as the continuation parameter. In the sequel, we restrict ourselves to the *local parameterization* technique, which consists in choosing one component of $\mathbf{u}$ as the new continuation parameter [Sey88, AC92]. Assume that we choose the i-th component. Thus we compute the point of the solution path corresponding to u_i taking on the value u_i^{j+1}. This can be written in matrix form as:

$$\begin{bmatrix} \mathbf{e}_i^T & 0 \end{bmatrix} \begin{bmatrix} \mathbf{u} \\ \mu \end{bmatrix} - u_i^{j+1} = 0 \qquad (9.27)$$

where $\mathbf{e}_i$ is an n-dimensional vector with all elements equal to 0 except for the i-th element being equal to 1. Using the Newton method to solve (9.22) and (9.27) with respect to $\mathbf{u}$ and μ, we fix u_i at the value u_i^{j+1} and we iterate on the linear system:

$$\begin{bmatrix} \varphi_u & \varphi_\mu \\ \mathbf{e}_i^T & 0 \end{bmatrix} \begin{bmatrix} \mathbf{u}_{(k+1)}^{j+1} - \mathbf{u}_{(k)}^{j+1} \\ \mu_{(k+1)}^{j+1} - \mu_{(k)}^{j+1} \end{bmatrix} = \begin{bmatrix} -\varphi(\mathbf{u}_{(k)}^{j+1}, \mu_{(k)}^{j+1}) \\ 0 \end{bmatrix} \qquad (9.28)$$

As can be seen, the Jacobian matrix in (9.28) is obtained by merely adding one row and one column to the usual Jacobian matrix φ_u. Note also its similarity with the Jacobian of (9.26): the only difference lies in the nonzero component of the last row. Because of its additional row and column, the Jacobian in (9.28) can be made nonsingular at the SNB point and the solution path around this point can be obtained without numerical difficulty.

In fact we have already met an example of local parameterization in the VQ curve technique of Section 9.2.3. In this case $\mu = Q_c$ and the fictitious generator voltage V was the continuation parameter.

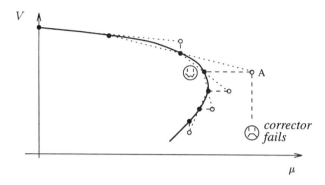

Figure 9.9   principle of continuation methods with local parameterization

Continuation methods usually employ a predictor-corrector scheme and involve a logic to change the continuation parameter as appropriate. This is sketched in Fig. 9.9. At each step, the white dots represent the predicted values and the black dots the corresponding corrected values. The solution path is first obtained using μ as the continuation parameter and iterating on (9.25) (or equivalently (9.26)). At point A, however, there is no correction corresponding to the prediction and the Newton iterations will diverge. Upon detection of convergence problems, the continuation parameter is changed to a component of $\mathbf{u}$. A reasonable choice is the lowest voltage magnitude or the voltage magnitude with the largest rate of decrease. In Fig. 9.9 we assume that the shown voltage is the new continuation parameter; hence, the predicted and corrected values are at the same ordinate. The former parameter μ is now a variable and is computed together with the other unknowns. The reaching of the loadability limit is detected by the fact that μ starts decreasing, indicating that the lower solution branch starts being computed. In practice it is of little interest to determine a large portion of this branch. It is enough to ascertain (e.g. visually) that the loadability limit has been reached.

Several predictors are possible. Let us quote: the tangent to the solution path, the secant, or simply no predictor. Using the tangent vector predictor, the predicted values $\mathbf{u}_{(0)}^{j+1}$ and $\mu_{(0)}^{j+1}$ are obtained from the current corrected values $\mathbf{u}^j$ and μ^j by solving:

$$\begin{bmatrix} \varphi_{\mathbf{u}} & \varphi_\mu \\ \mathbf{0}^T & 1 \end{bmatrix} \begin{bmatrix} \mathbf{u}_{(0)}^{j+1} - \mathbf{u}^j \\ \mu_{(0)}^{j+1} - \mu^j \end{bmatrix} = \begin{bmatrix} \mathbf{0} \\ \mu^{j+1} - \mu^j \end{bmatrix} \qquad (9.29)$$

when the continuation parameter is μ, and

$$\begin{bmatrix} \varphi_{\mathbf{u}} & \varphi_\mu \\ \mathbf{e}_i^T & 0 \end{bmatrix} \begin{bmatrix} \mathbf{u}_{(0)}^{j+1} - \mathbf{u}^j \\ \mu_{(0)}^{j+1} - \mu^j \end{bmatrix} = \begin{bmatrix} \mathbf{0} \\ u_i^{j+1} - u_i^j \end{bmatrix} \qquad (9.30)$$

when the continuation parameter is u_i. The Jacobians in (9.29) and (9.30) must be updated at each step, which is computationally demanding.

A simple predictor is the secant, or first-order polynomial extrapolation [Sey88, CFS95], corresponding to:

$$\mathbf{u}_{(0)}^{j+1} - \mathbf{u}^j = \mathbf{u}^j - \mathbf{u}^{j-1}$$
$$\mu_{(0)}^{j+1} - \mu^j = \mu^j - \mu^{j-1}$$

and shown with dotted lines in Fig. 9.9. It does not require a Jacobian update for prediction purposes (only when generators get limited along the solution path and when the continuation parameter is changed, as in any continuation method).

As regards correctors, other techniques than the above described local parameterization have been proposed. Let us mention for instance the intersection between the solution path and the perpendicular plane to the tangent vector [ISE91, CA93]. The local parameterization, however, yields a simple algorithm and is sufficient for most voltage stability computations.

Step-length control has been also proposed to speed-up calculations, especially on the low-curvature parts of the solution path (e.g. [CFS95]).

Variants

The method used in [FJC85, LPT90, FOC90, FFC92], approximates the solution path by a collection of linear segments corresponding to the tangent vectors given by (9.29,9.30). Each point corresponds to one additional generator reaching its reactive limit, i.e. switching from (9.24b) to (9.24c). The step size is thus variable. The method can be seen as a repeated tangent predictor without corrector. Because of this, there is some discrepancy between the computed and the exact loadability limit.

In [HC96] the authors compute the successive breaking-points of the solution path. A tangent predictor is used as in [FJC85] to identify the next generator to reach its limit. At the corresponding breaking-point, both (9.24b) and (9.24c) hold for the switching generator. This brings the additional equation needed to solve the equations with respect to $\mathbf{u}$ and μ in the correction step. The method thus extends the one described in the previous paragraph, by adding a corrector. The loadability limit is taken as corresponding to the breaking-point with the largest value of μ. The method cannot determine whether an SNB exists around this point, which leaves some uncertainty on the limit.

9.3.3 Optimization methods

Earlier in this section we have recalled that the loadability limit corresponds to the maximum value of μ such that (9.22) has a solution. This suggests that the loadability limit can be determined as the solution of the following optimization problem:

$$\max \quad \mu \tag{9.31a}$$
$$\text{subject to} \quad \varphi(\mathbf{u}, \mu) = 0 \tag{9.31b}$$

Methods based on the above formulation, or a close one, have been proposed in [VC88b, OB88, VC91a, PMS96, IWT97]. The optimization method may be called *direct* in the sense that the loadability limit is computed directly, without determining the solution path between the base case and limit points. Compared to continuation methods, the optimization method is thus expected to be computationally more efficient. On the other hand, continuation methods are more attractive when it is of interest to obtain the solution path explicitly, or when the effect of controls acting along the solution path must be incorporated.

The derivation of a loadability limit as the solution of an optimization problem has been introduced in Chapter 7 in order to establish properties of SNB points and loadability limits. Note that in this derivation the objective function $\zeta(\mathbf{p})$ was general, while in this chapter we focus on determining loadability limits in a given direction $\mathbf{d}$ of parameter change.

We first concentrate on solving the equality-constrained problem (9.31a,b), letting aside the inequalities corresponding to generator reactive limits. We will come back to this important point later on. We thus define the Lagrangian

$$\mathcal{L} = \mu + \mathbf{w}^T \varphi(\mathbf{u}, \mu) = \mu + \sum_{i=1}^{n} w_i \, \varphi_i(\mathbf{u}, \mu)$$

where $\mathbf{w}$ is a vector of Lagrange multipliers. The first-order (necessary) optimality conditions are obtained by setting to zero the derivatives of $\mathcal{L}$ with respect to $\mathbf{u}$, $\mathbf{w}$ and μ:

$$\nabla_{\mathbf{u}}\mathcal{L} = 0 \quad \Leftrightarrow \quad \varphi_{\mathbf{u}}^T \mathbf{w} = 0 \tag{9.32a}$$
$$\nabla_{\mathbf{w}}\mathcal{L} = 0 \quad \Leftrightarrow \quad \varphi(\mathbf{u}, \mu) = 0 \tag{9.32b}$$
$$\frac{\partial \mathcal{L}}{\partial \mu} = 0 \quad \Leftrightarrow \quad 1 + \mathbf{w}^T \varphi_{\mu} = 0 \tag{9.32c}$$

where (9.32c) has been obtained using the chain rule and taking (9.21) into account. Equation (9.32a), identical to (7.7c), confirms a result obtained in Chapter 7: $\varphi_{\mathbf{u}}$ is singular at the solution and the vector $\mathbf{w}$ of Lagrange multipliers is the left eigenvector of the corresponding zero eigenvalue. Thus, a by-product of the optimization is the left eigenvector $\mathbf{w}$, from which useful information on the instability mode can be obtained, as explained in Sections 7.5.2 and 7.5.3. Being Lagrange multipliers, the components of $\mathbf{w}$ can be also interpreted as the sensitivities of the objective function, i.e. the margin μ^*, to the various constraints (9.22) [GMW81].

The set of nonlinear equations (9.32a,b,c) can be solved using the Newton method [SAB84]. This involves iterations on the linear system:

$$\underbrace{\begin{bmatrix} \sum_{i=1}^{n} w_i \dfrac{\partial^2 \varphi_i}{\partial \mathbf{u}^2} & \varphi_{\mathbf{u}}^T & 0 \\ \varphi_{\mathbf{u}} & 0 & \varphi_{\mu}^T \\ 0^T & \varphi_{\mu} & 0 \end{bmatrix}}_{\mathbf{H}} \begin{bmatrix} \Delta \mathbf{u} \\ \Delta \mathbf{w} \\ \Delta \mu \end{bmatrix} = - \begin{bmatrix} \varphi_{\mathbf{u}}^T \mathbf{w} \\ \varphi(\mathbf{u}, \mu) \\ 1 + \mathbf{w}^T \varphi_{\mu} \end{bmatrix} \qquad (9.33)$$

In this equation, $\mathbf{H}$ is the matrix of first derivatives, or Jacobian, of (9.32a,b,c) with respect to $(\mathbf{u}, \mathbf{w}, \mu)$. It is thus the matrix of second derivatives, or Hessian, of $\mathcal{L}$ with respect to the same variables. As can be seen the Hessian $\mathbf{H}$ involves $\varphi_{\mathbf{u}}$ as a sub-matrix. While the latter is singular at the solution, the former is not. $\mathbf{H}$ is about twice as large as $\varphi_{\mathbf{u}}$ but it is symmetric, which speeds up its LU decomposition. Also, $\mathbf{H}$ is very sparse and is advantageously stored and manipulated in blocked form [SAB84]. Finally, it can be kept constant from one iteration to the next as the solution is approached within some tolerance.

With proper initialization of the $\mathbf{w}$ variables, the Newton method converges quickly and reliably to the loadability limit: typically 2 to 4 iterations (9.33) are sufficient to reach the accuracy required for practical applications [VC91a]. The key-point, however, is the handling of the inequality constraints (9.23b,c,d). As in any constrained optimization, the problem is to identify which constraints are active at the solution: in our case, which generators obey (9.24b) and which ones obey (9.24c) at the loadability limit. In continuation methods, generators are switched from (9.24b) to (9.24c) at breaking-points of the solution path. In the optimization approach, this sequence of generator switchings is not known a priori. Moreover, in order to preserve the computational efficiency which makes the direct method attractive, it is essential to identify the set of limited generators with reasonable computational effort.

In the procedure described in [VC91a], the optimization problem (9.31a,b) is first solved with generators controlling voltages as in the base case. This yields a margin $\tilde{\mu}$, which obviously overestimates the real one ($\tilde{\mu} > \mu^*$). Based on the reactive power generations computed at respectively $\mu = 0$ (base case) and $\mu = \tilde{\mu}$ (unconstrained limit), a linear interpolation is used to estimate the fraction r_i ($0 < r_i < 1$) of the margin $\tilde{\mu}$ at which the i-th generator meets its limit. All generators such that $r_i < s.\max_i r_i$, where $s < 1$ is a security coefficient, are switched from (9.24b) to (9.24c). If all r_i's are larger than $s.\max_i r_i$, the generator with the smallest r_i is switched alone. The optimization problem with the new set of constraints (9.24a,b,c) is then solved, starting from the solution of the previous optimization. This yields a new margin $\tilde{\mu}$ and hence new r_i's. The procedure is repeated until all constraints are satisfied, thereby providing a sequence of decreasing $\tilde{\mu}$'s which converges to $\mu^\star$. In [VC91a] a total number of at most 10 Newton iterations (9.33) was reported from tests performed on a real-life system. This makes the optimization method much faster than continuation methods.

One must keep in mind the possibility for a loadability limit to correspond to a breaking-point. Referring to Figs. 7.22.a and 7.22.b, when the generator inequality (9.23c) is enforced, the optimization method converges to the SNB point B. In the case of Fig. 7.22.b, however, at point B the relaxed constraint (9.23b) is violated, as indicated by B being outside the corresponding feasible region. As suggested by the figure, the generator voltage will rise instead of falling after the limit is enforced, which does not correspond to anything physical. This limit violation can be used in practice as an indication that the loadability limit is at the breaking-point. In this case, both equalities are enforced for the generator of concern (thus leaving no degree of freedom) and the method converges to point A. Some logic is needed to deal with the case where the limit violation follows the switching of several generators.

As an alternative to the above inequality handling procedure, it has been proposed to use the Interior Point method of optimization [PMS96, IWT97], successfully applied to other power system optimization problems. This approach consists in adding to the objective "barrier functions" penalizing operating points that approach inequality constraint limits. A multiplier controls the barrier "height". For a given value of the multiplier, the Newton method is used to solve the optimization problem with barrier functions, and a step-length correction is applied to preserve feasibility at each iteration of the method. The multiplier is progressively reduced as convergence is achieved. In our case, this causes the generators which would normally be limited at the solution, to move closer to their limits. The reader is referred to the quoted references for more details. It must be noted that the method requires to solve a linear system of much larger size than (9.33) (due to the introduction of slack variables in addition to $\mathbf{u}$, $\mathbf{w}$ and μ). The use of very efficient sparsity techniques is essential, in order to preserve the computational advantages of optimization over the simpler continuation methods.

Other applications of optimization methods

Incorporation of controls. In the formulation (9.31a,b), our optimization problem has a single degree of freedom. Optimization methods offer the possibility to take into account adjustable controls and compute the (normally larger) loadability limit which can be achieved thanks to the additional degrees of freedom brought by these controls [PMS96, IWT97]. For instance, generator voltages may be let free to vary within specified ranges: provided that some generators remain under voltage control at the loadability limit, the latter will be somewhat larger. Other controls usually considered in optimal power flow may be used as well.

Other objective functions. In some applications it is possible to obtain a loadability limit as the solution of an optimal power flow with a more "traditional" objective than (9.31a). For instance the maximal power transfer from one area to another can be obtained as the solution of an economic dispatch problem with fictitious cheap (resp. expensive) production costs assigned to generators in the sending (resp. receiving) area. As another example, reactive power margins to stability can be computed, in unstable cases, by minimizing the total shunt reactive power additions [ARU91].

Solvability restoration using optimization. When computing a loadability limit, it is assumed that, starting from a feasible base case, the system is stressed up to losing equilibrium, i.e. up to being unsolvable. The dual problem is to determine, when starting from an infeasible point, the minimum control actions needed to restore solvability. This problem has been described in Section 8.6.2, and a solution using a linear approximation of the bifurcation surface Σ has been outlined. A direct approach [GMM96] consists in solving a minimization problem whose objective is the total amount of control actions, subject to the system equations (9.1) as well as inequalities on control and state variables. The objective considered in [GMM96] is minimum load shedding and the Interior Point method is used to solve the optimization problem.

Point of Collapse method

The method named Point of Collapse by its authors has been initially proposed in [AJ88] and subsequently adapted in [CAD92, CA93], where tests on AC/DC systems are reported. In this method the loadability limit is defined a priori as a solution of (9.22) at which Jacobian $\varphi_{\mathbf{u}}$ is singular. In [AJ88], the singularity condition is expressed using the right eigenvector $\mathbf{v}$, which results in:

$$\varphi(\mathbf{u}, \mu) = 0 \tag{9.34a}$$
$$\varphi_{\mathbf{u}} \mathbf{v} = 0 \tag{9.34b}$$

$$\sum_i v_i^2 \;=\; 1 \qquad\qquad\qquad (9.34c)$$

where (9.34c) ensures a nonzero vector $\mathbf{v}$. Alternatively, one could impose a particular component v_k to be equal to one, provided that this component of the right eigenvector is nonzero at the loadability limit. From the analytical viewpoint, the set of equations (9.34a,b,c) is strictly equivalent to (9.32a,b,c): the latter involves the left eigenvector $\mathbf{w}$ of the zero eigenvalue and imposes this vector to be nonzero through (9.32c), while the former uses the right eigenvector and condition (9.34c). From a practical viewpoint, however, the formulation (9.32a,b,c) is more advantageous because it yields a symmetric matrix $\mathbf{H}$ in the Newton method used to solve the equations. This merely comes from the fact that $\mathbf{H}$ is the Hessian of $\mathcal{L}$. The left eigenvector formulation was adopted in [CAD92, CA93].

Note finally that (9.34a,b,c) provide only SNB points and not loadability limits that occur at breaking-points.

9.3.4   Time simulation coupled with sensitivity analysis

Since loadability limits correspond to smooth changes of parameters driving the system to instability, one way of obtaining these limits is by computing the time response of the system to a progressive increase in system stress μ. The practical advantage of this approach is that the same simulation tool can be used for both evaluating contingencies and computing loadability limits, although in the latter case simulation must be complemented by a stability diagnosis, as explained hereafter [VJM95, VJM94].

The best way to depict the situation is again in the load power space, assuming for simplicity that loads recover to constant power. Consider Fig. 9.10, where $\mathbf{p}$ moves along direction $\mathbf{d}$, starting from $\mathbf{p}_0$. We plot with dotted lines the vector $\mathbf{P}(t)$ of load powers at successive times. While $\mathbf{p}$ is moving smoothly between $\mathbf{p}_0$ and $\mathbf{p}^\star$, both $\mathbf{p}$ and $\mathbf{P}(t)$ coincide. On the other hand, as $\mathbf{p}$ continues beyond $\mathbf{p}^\star$, thus leaving the feasible region bounded by the bifurcation surface Σ, $\mathbf{P}(t)$ separates from $\mathbf{p}$ and after touching Σ it goes backward in the feasible region. This part of the system trajectory is similar to the one shown in Fig. 8.22.

Note that in a continuation method, after reaching $\mathbf{p}^\star$, $\mathbf{p}$ is brought back to $\mathbf{p}_0$ by the method itself, moving in the opposite direction to $\mathbf{d}$, and producing another solution path. The corresponding decrease in μ indicates that the loadability limit has been reached.

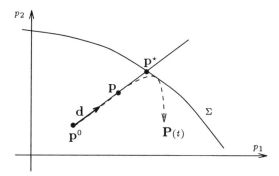

Figure 9.10   System behaviour in the load space during a stress increase

In time simulation we need a criterion to identify the critical point, where $\mathbf{P}(t)$ hits Σ. This criterion cannot be provided by the maximum of load power $\sum_i P_i(t)$, especially in large systems. Indeed, if the mode of instability is local, it is possible that after touching Σ, the total load power goes on increasing because the decrease in the unstable area is masked by a continuing increase in the remaining of the system.

Referring to Section 8.6.2, a sound criterion for identifying the critical point is the dominant eigenvalue of the long-term dynamics, which becomes zero at any point on Σ. A simpler criterion involves the sensitivities (7.17), which change sign through infinity at the same points.

Since we deal with long-term instability, the Jacobian $\varphi_{\mathbf{u}}$ to be used in (7.17) is either $\mathbf{J}_\ell$ or $\mathbf{G_y}$, as discussed in Section 7.3.4. Although this is not the Jacobian used in (multi-time-scale or QSS) time simulation, the computational effort of the sensitivity calculation is quite moderate. Indeed $\varphi_{\mathbf{u}}$ is sparse, as mentioned in Section 7.4, and the sensitivities can be computed at a sample of points along the simulated trajectory.

In practice, $\varphi_{\mathbf{u}}$ is first computed and factorized. Then, the sparse linear system

$$\varphi_{\mathbf{u}}^T \, \mathbf{a} = \nabla_{\mathbf{u}} \eta$$

is solved with respect to $\mathbf{a}$ and the sensitivities (7.17) are obtained as:

$$S_{\eta \, \mathbf{P}} = \nabla_{\mathbf{P}} \eta - \varphi_{\mathbf{P}}^T \, \mathbf{a}$$

A numerical example is given in Section 9.6.6.

9.3.5   Methods based on multiple load flow solutions

Based on the original research reported in [TMI83] a school of voltage stability analysis focuses not directly on the loadability limit, but rather on the distance separating a stable equilibrium form the unstable one closest to it [TMI88, DO90, ISE90, YHS94, OK96]. Referring for instance to Fig. 7.12, close to the loadability limit C the stable equilibrium S and the unstable one U will coalesce and disappear. Thus the distance between S and U provides a measure of the proximity to instability. Note that this measure can be interpreted in terms of system response, since the unstable equilibrium determines the region of attraction of the stable one.

One problem in applying this method is how to obtain the closest unstable equilibrium. Note that a large number of candidate solutions exist, especially in large systems. One has also to consider the possibility that the closest unstable equilibrium at a given operating point may be different from that coalescing with the stable one at the SNB.

This was graphically illustrated in Fig. 7.19 for a system consisting of only two load buses. At point O the closest unstable equilibrium is the one coalescing with it at the SNB point A. For a demand lower than Σ'', however, this unstable equilibrium does not even exist. This gives an idea of the complexity of the problem. In [OD91] it is suggested that the search for unstable solutions should be restricted only to type-1 unstable equilibria, i.e. equilibria with only one positive eigenvalue. Using again Fig. 7.19 as an example, the unstable equilibria lying on the curve between points A and A' are type-1, whereas the unstable equilibria below point A' are type-2 with two positive eigenvalues.

9.4   SECURE OPERATION LIMIT DETERMINATION

9.4.1   Definition and problem statement

Given a direction of system stress, a secure operation limit corresponds to the most stressed among a set of operating points, such that the system can withstand a specified contingency [GMK96, VCM98]. Unlike a post-contingency loadability limit, a secure operation limit refers to the present, i.e. pre-contingency, system configuration and provides a security margins that is easier to interpret in system operation.

A secure operation limit encompasses three types of information : direction of system stress, operator/controller actions while the system is stressed, and post-contingency controls. We comment hereafter on the last two aspects.

Operator/controller actions. Prior to any contingency, operators or controllers react to the stress imposed on the system. Most often their rôle is to keep the voltage profile within limits and to maximize the reactive reserves readily available to face incidents. The following are typical examples of such actions are:

- shunt capacitor/reactor switching

- operator adjustment of generator terminal voltage in order to keep a constant voltage at the network terminal (high voltage side of the step-up transformer)

- secondary voltage control, aimed at regulating pilot-node voltages and maximizing reactive reserves

- operator adjustment of LTCs on transformers connecting two transmission levels (see Fig. 2.22).

Post-contingency controls. As mentioned earlier in this chapter, post-contingency controls are typically automatic and faster than human operators, who play a rôle in the pre-contingency situation only. In the definition of secure operation limits, there is thus a clear decomposition between the pre-contingency situation, where slow controls/actions have time to act, and the post-contingency situation, where only fast controls can act.

The above notions are illustrated hereafter on a simple example. Consider the system of Fig. 9.11.a. In the long-term, the load is restored to constant power by the transformer LTC, which may cause voltage instability. Shunt compensation is available at bus A. It is used in two ways: (i) in normal operating conditions, an operator or controller adjusts the compensation at bus A to keep its voltage within some narrow limits; (ii) following an incident, upon detection of a low voltage condition, pre-determined amounts of compensation are switched on, as a corrective control. The system stress is an increase in demand at the load bus while the contingency of concern is some line tripping causing the reactance X to increase suddenly.

Figure 9.11.b describes the system behaviour in terms of PV curves, relating the voltage V_A to the load power P. Consider first the operation at point B. Right after the contingency, the PV characteristic is the curve shown with dotted line, while after the corrective compensation switching it becomes the solid line curve. The resulting operating point is B'. The system being able to withstand the contingency[5], we move towards the security limit by increasing the pre-contingency load up to the level

---

[5]Note that the system may fail to reach B' by lack of attraction, e.g. because the post-contingency controls are triggered too late (see Section 8.6.3). This case is not considered in Fig. 9.11.b for simplicity.

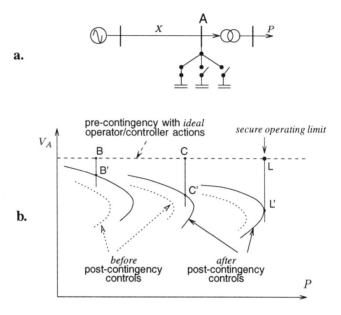

Figure 9.11   PV-curve sketch of a secure operation limit

corresponding to point C. This system stress is accompanied by an operator/controller action, which manages in this example to keep the voltage at bus A perfectly flat (see dashed line in Fig. 9.11.b). Note that the post-contingency curves change accordingly. Similarly, the post-contingency controls may be different. In other words, there is not a single post-contingency PV characteristic but a family of them. Eventually the sought limit corresponds to point L, the ultimate pre-contingency situation such that there is a resulting operating point L'.

The notion of secure operation limit is general and applies to other aspects of power system security: for instance, transient (angle) stability aspects are discussed in [MMG93, MSS96] and thermal limits are straightforwardly incorporated as well.

9.4.2   Binary search

Handling of a single contingency

A simple and robust method to determine a secure operation limit is the *binary search* (also referred to as the *dichotomic* search or *bisection method*). Simply stated, it

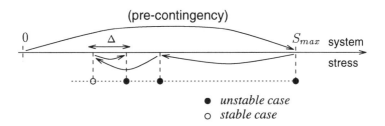

Figure 9.12 Binary search for secure operation limit determination

consists in building progressively an interval of stress $[S_\ell \ S_u]$ such that S_ℓ corresponds to a stable scenario, S_u to an unstable one and $(S_u - S_\ell)$ is smaller than a specified tolerance Δ. Starting with $S_\ell = 0$ (base case) and $S_u = S_{max}$, the interval is divided in two equal parts at each step; if the mid-point is found stable (resp. unstable) it is taken as the new lower (resp. upper) bound. S_{max} is the largest stress of interest.

Clearly, if the system responds in a stable way at S_{max}, the search stops. In case it is found unstable in the base case, the search is carried out in the reverse direction.

The procedure is illustrated in Fig. 9.12, where the black (resp. white) dots indicate unstable (resp. stable) scenarios and the arrows show the sequence of calculations. It is similar to the one used for critical clearing time determination in transient angle stability.

There are basically two computational tasks involved in a binary search:

■ contingency evaluation: the various methods described in Section 9.2 may be used. Please refer to this section for a discussion of their relative merits;

■ generation of pre-contingency operating points corresponding to various levels of stress: this is usually performed by an "external" load flow or optimal power flow program. An alternative implementation will be described in Section 9.6.5.

Handling of several contingencies

When a secure operation limit has to be computed with respect to several contingencies, it would be a waste of time to compute the individual limit for each contingency to finally keep the smallest value as the global limit. It is much more efficient to process the various contingencies together [VCM98].

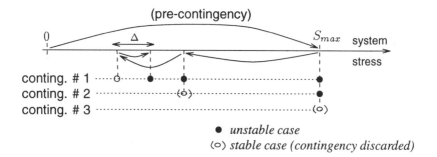

Figure 9.13   Handling of multiple contingencies in binary search

The procedure is outlined in Fig. 9.13. At a given step of the binary search, the various contingencies stemming from the previous step are simulated. If there is at least one unstable contingency, the stable ones (shown as parenthesized white dots in Fig. 9.13) can be discarded since the corresponding limits are higher than the current stress level. The unstable ones make up the reduced list of contingencies to be processed at the next step. The saving in computing time depends obviously on how the individual limits are distributed in the $[0 \ S_{max}]$ interval. As a by-product, the stress level at which a contingency has been discarded provides a lower (pessimistic) bound on the corresponding individual security limit.

Acceleration of the limit search

When contingency evaluation takes much time, it may be desirable to speed up the limit search process by decreasing the number of tested stress levels. In the binary search, the information extracted from each time simulation is merely an acceptable/unacceptable verdict. References [MMG93, MSS96, VCM98] report on methods that extract more information from the simulated time responses in order to extrapolate the limit.

9.5   EIGENANALYSIS FOR INSTABILITY DIAGNOSIS

9.5.1   Eigenvector analysis

When a system is brought to instability due to smooth parameter increase (in loadability limit computation) or large-disturbance (in contingency evaluation and secure operation limit determination), it is of interest to identify which control actions are most adequate to prevent or correct the instability. Such a diagnosis can be obtained from eigenvector analysis. Before dealing with calculation aspects, we will briefly recall the significance of eigenvectors, concentrating once more on long-term instability.

Consider first the case of a smooth parameter increase bringing the system to a saddle-node bifurcation. As shown in Section 7.3.4, at this point, the state matrix $\mathbf{H_z}$ of the long-term dynamics has a zero eigenvalue, together with the unreduced Jacobian $\mathbf{J}_\ell$. In Section 7.4.1 we have recalled how (left and right) eigenvectors relative to the unstable eigenvalue of a state matrix like $\mathbf{H_z}$ provide information on the *instability mode*, its cause and its remedies. We have also shown that for a zero eigenvalue, the left eigenvector of $\mathbf{H_z}$ is included in the corresponding left eigenvector $\mathbf{w}_c$ of $\mathbf{J}_\ell$, and similarly for the right eigenvector. The $\mathbf{w}_c$ vector, which relates to both state and algebraic variables, plays an important rôle in saddle-node bifurcation analysis, as explained in Sections 7.5.2 and 7.5.3. Let us quote in particular the possibility to compute sensitivities of the power margin to parameters, using (7.68, 7.72 or 7.73).

In case of a large disturbance leading to loss of equilibrium, both $\mathbf{H_z}$ and $\mathbf{J}_\ell$ have a zero eigenvalue at the critical point (see Section 8.6.2). In the power space, corrective actions to restore an equilibrium point may rely on the tangent hyperplane approximation of the bifurcation surface Σ. Here too, the $\mathbf{w}_c$ eigenvector is required in order to obtain the normal vector $\mathbf{n}$ of Σ, using (7.64).

Note that similar results are obtained using the network-only formulation of [GMK92, IWG93], $\mathbf{J}_\ell$ being replaced by $\mathbf{G_y}$, as explained in Section 7.3.4.

Thus, for instability diagnosis purposes, $\mathbf{w}_c$ must be computed at either the bifurcation or the critical point provided by one of the methods previously described in this chapter[6]. In continuation methods, the bifurcation point is obtained when μ reaches a maximum. Optimization methods, on the other hand, directly yield $\mathbf{w}_c$ as the vector of Lagrange multipliers. When using time simulation (see Section 9.3.4), sensitivity analysis performed at a sample of points along the system trajectory provides either the bifurcation point or the critical point. The modified load flow method of Section 9.2.2 ends up with a point belonging to Σ.

Up to now we have focused on the zero eigenvalue. It has been also proposed to compute at normal operating points the eigenvectors relative to a subset of dominant, real eigenvalues of either the state matrix $\mathbf{H_z}$ or the $\mathbf{J}_{QV}$ matrix (see Section 7.3.4). The objective is to identify possible instability modes of the system. It must be kept in mind, however, that *the dominant mode at a normal point may be quite different from the one encountered at the loadability limit obtained by stressing the system*. In

---

[6]beside diagnosis aspects, the tracking of the largest real eigenvalue has been also proposed as a voltage stability index: please refer to the general discussion of Section 9.7.2

applying small-disturbance methods one should keep in mind the very nonlinear nature of the original problem.

9.5.2 Eigenvector computation methods

Inverse Iteration method

Consider first the problem of computing some eigenvalue(s) of $\mathbf{H_z}$. Inverse Iteration is a well-known method to compute an eigenvalue λ and the corresponding eigenvector, starting from an estimate $\hat{\lambda}$ close to λ [Jen77]. The algorithm, written hereafter for the left eigenvector, is very simple:

1. Factorize $\mathbf{H_z} - \hat{\lambda}\mathbf{I}$;
    initialize $\mathbf{r}^{(0)}$; $k := 1$
2. Solve:
$$(\mathbf{H_z} - \hat{\lambda}\mathbf{I})^T \mathbf{w}^{(k)} = \mathbf{r}^{(k-1)} \tag{9.35}$$
3. Normalize:
$$\mathbf{r}^{(k)} = \frac{\mathbf{w}^{(k)}}{w_{max}^{(k)}} \tag{9.36}$$
4. If $\|\mathbf{r}^{(k)} - \mathbf{r}^{(k-1)}\| < \epsilon$
    then $\lambda = \hat{\lambda} + 1/w_{max}^{(k)}$; stop
    else $k := k+1$; go to 2

Note that $\mathbf{w}$ is the eigenvector of $\mathbf{H_z}$ (we drop subscript $_z$ to avoid heavy notation) and $w_{max}^{(k)}$ is the component of $\mathbf{w}^{(k)}$ of largest magnitude.

As already mentioned in Section 7.4.1, for computational efficiency, the dense linear system (9.35) is replaced by the equivalent but sparse system :

$$\left[\begin{array}{c|cc} \mathbf{h_z} - \hat{\lambda}\mathbf{I} & \mathbf{h_x} & \mathbf{h_y} \\ \hline \mathbf{f_z} & & \\ \mathbf{g_z} & & \mathbf{J}_s \end{array} \right]^T \left[\begin{array}{c} \mathbf{w}^{(k)} \\ \mathbf{w}_x^{(k)} \\ \mathbf{w}_y^{(k)} \end{array} \right] = \left[\begin{array}{c} \mathbf{r}^{(k-1)} \\ 0 \\ 0 \end{array} \right] \tag{9.37}$$

Once an (almost) zero eigenvalue of $\mathbf{H_z}$ has been located, the next problem is to compute the left eigenvector $\mathbf{w}_c$ of $\mathbf{J}_\ell$. To this purpose we could apply (9.37) to $\mathbf{J}_\ell$, subtracting $\hat{\lambda}\mathbf{I}$ from its whole diagonal and factorizing the so obtained matrix. This step, however, can be avoided, as shown hereafter. Introducing (9.36) into (9.37) yields:

$$
\left[\begin{array}{c|cc} \mathbf{h_z} - \hat{\lambda}\mathbf{I} & \mathbf{h_x} & \mathbf{h_y} \\ \hline \mathbf{f_z} & & \\ \mathbf{g_z} & & \mathbf{J}_s \end{array} \right]^T \left[\begin{array}{c} \mathbf{w}^{(k)} \\ \mathbf{w}_x^{(k)} \\ \mathbf{w}_y^{(k)} \end{array} \right] = \frac{1}{w_{max}^{(k-1)}} \left[\begin{array}{c} \mathbf{w}^{(k-1)} \\ 0 \\ 0 \end{array} \right] \tag{9.38}
$$

As convergence is achieved:

$$
\frac{1}{w_{max}^{(k-1)}} \rightarrow \lambda - \hat{\lambda}
$$

Now, if $\mathbf{H_z}$ has a zero eigenvalue and a zero estimate $\hat{\lambda}$ is used, (9.38) becomes:

$$
\left[\begin{array}{c|cc} \mathbf{h_z} & \mathbf{h_x} & \mathbf{h_y} \\ \hline \mathbf{f_z} & & \\ \mathbf{g_z} & & \mathbf{J}_s \end{array} \right]^T \left[\begin{array}{c} \mathbf{w}^{(k)} \\ \mathbf{w}_x^{(k)} \\ \mathbf{w}_y^{(k)} \end{array} \right] = 0
$$

which means that, as convergence is achieved, $\mathbf{w}_c$ is nothing but the vector which would be used for an additional unreduced Inverse Iteration. This vector is exactly equal to $\mathbf{w}_c$ only if $\lambda = \hat{\lambda} = 0$. In practice λ is never exactly zero and a small nonzero $\hat{\lambda}$ may be used. However, experience has shown that the obtained vector is accurate for λ and $\hat{\lambda}$ reasonably close to zero [RVC96].

Simultaneous Iteration method

While simple, the Inverse Iteration method is prone to convergence difficulties if $\hat{\lambda}$ is not close enough the sought eigenvalue λ [Jen77]. In fact, if λ and λ' are the two eigenvalues closest to $\hat{\lambda}$, good convergence towards λ requires that the ratio $|(\lambda - \hat{\lambda})/(\lambda' - \hat{\lambda})|$ is as small as possible, and in any case smaller than unity.

Figure 9.14 illustrates two common situations where such difficulties are met. $\hat{\lambda}$ has been taken equal to zero, at least in the absence of a better estimate. In Fig. 9.14.a, the operating point is stable but there is a cluster of eigenvalues close to the dominant λ, a very common situation in systems with many LTCs having the same characteristics. In Fig. 9.14.b the operating point is unstable but the eigenvalue λ' closest to $\hat{\lambda} = 0$ is a stable one. Obviously, the information obtained from the left eigenvector of λ' is irrelevant.

These drawbacks are circumvented by the Simultaneous Iterations method [Jen77, GMK92, RVC96]. Simply stated, the latter consists of Inverse Iterations performed in parallel on m different eigenvectors, the latter being re-oriented to prevent convergence towards repeated eigenvalues:

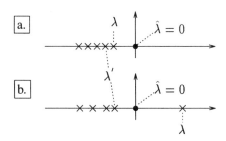

Figure 9.14   Two common eigenvalue patterns

1.   Factorize $\mathbf{H_Z} - \hat{\lambda}\mathbf{I}$;
      initialize $\mathbf{r}_i^{(0)}$ $(i = 1,\ldots,m)$; $\quad k := 1$
2.   For $i = 1,\ldots,m$, solve:

$$(\mathbf{H_Z} - \hat{\lambda}\mathbf{I})^T \mathbf{w}_i^{(k)} = \mathbf{r}_i^{(k-1)}$$

      form $\mathbf{W}^{(k)}$ = $\left[\mathbf{w}_1^{(k)}, \ldots, \mathbf{w}_m^{(k)} \right]$

             $\mathbf{R}^{(k)}$ = $\left[\mathbf{r}_1^{(k)}, \ldots, \mathbf{r}_m^{(k)} \right]$

3.   Solve for $\mathbf{B}^{(k)}$:

$$\left(\mathbf{R}^{T(k)} \mathbf{R}^{(k)} \right) \mathbf{B}^{(k)} = \mathbf{R}^{T(k)} \mathbf{W}^{(k)}$$

4.   Perform QR decomposition:

$$\mathbf{B}^{(k)} \mathbf{Q}^{(k)} = \mathbf{Q}^{(k)} \, diag(\lambda_1^{(k)}, \ldots, \lambda_m^{(k)})$$

5.   Re-orientate the $\mathbf{r}_i$ vectors:

$$\left[\tilde{\mathbf{r}}_1^{(k)}, \ldots, \tilde{\mathbf{r}}_m^{(k)} \right] = \mathbf{W}^{(k)} \mathbf{Q}^{(k)}$$

6.   Normalize the $\tilde{\mathbf{r}}_i$ vectors:

$$\mathbf{r}_i^{(k)} = \frac{\tilde{\mathbf{r}}_i^{(k)}}{r_{i\,max}^{(k)}}$$

7.   If $\|\lambda_i^{(k)} - \lambda_i^{(k-1)}\| < \epsilon'$, $i = 1,\ldots,d$
      then $\lambda_i = \hat{\lambda} + 1/\lambda_i^{(k)}$; stop
      else $k := k+1$; go to 2

The larger the ratio $|\lambda_i / \lambda_{m+1}|$, the better the convergence rate for λ_i. Hence, if d eigenvalues are desired, the number k of iterations can be decreased by applying the above algorithm with m larger than d (the convergence test of step 7 being restricted to the d eigenvalues of interest). Of course, the decrease in k must be balanced against the computational cost of handling the $m - d$ additional "guard" vectors $\mathbf{w}_i$. When the dominant eigenvalue is sought, $d = 1$ and a typical choice is $m = 4$ [RVC96].

9.6 EXAMPLES FROM A REAL-LIFE SYSTEM

In this section we illustrate some of the methods previously described in this chapter, using examples from a real-life system. Most of them have been obtained with the ASTRE[7] software developed at the University of Liège. The heart of this program is a fast QSS simulation of long-term voltage phenomena. It has been tested on various systems and, at the time of writing this book, it is used by Electricité De France and Hydro-Québec. The results shown hereafter deal with the Hydro-Québec system [VCM97, VCM98]. Additional results obtained with the French system may be found in [VJM95, VJM94].

9.6.1 Voltage stability of the Hydro-Québec system

The Hydro-Québec system (see Fig. 9.15) is characterized by long distances (up to 1000 km) between the northern main generation centers and the southern main load area. The results given hereafter refer to a 1996 situation.

The long EHV transmission lines have high series reactances and shunt susceptances. Therefore, power transfer fluctuations result in large variations of the reactive losses. At low power transfers, the reactive power generation of EHV lines is compensated by connecting 330-Mvar shunt reactors at the 735 kV substations. At peak load (which is around 35,000 MW), most of the shunt reactors are disconnected while sub-transmission voltage control requires connection of shunt capacitors. Both effects contribute to a very capacitive characteristic of the system. Due to the remote location of power plants, voltage control along the transmission system is mainly insured by 20 SVCs and Synchronous Condensers (SCs). These compensators offer a total reactive reserve of about 5000 Mvar in production and 3000 Mvar in absorption. An operating range of 2000 Mvar is used for voltage regulation purposes.

In the northern regions, power transfer limits are mainly dictated by transient (angle) stability. The installed series compensation has significantly improved the system robustness over the recent years, allowing for instance stable operation after a 3-phase fault normally cleared. The stability is maintained with SVCs contribution but without automatic devices such as generation shedding whose intervention is limited to extreme contingencies.

In the southern region, voltage stability is the main concern. At peak load, the total reactive reserve of compensators is not sufficient to insure voltage stability following the most severe contingencies. For this reason, automatic Shunt Reactor Tripping

---

[7]French acronym for "Analyse de la Sécurité de Tension des Réseaux Electriques"

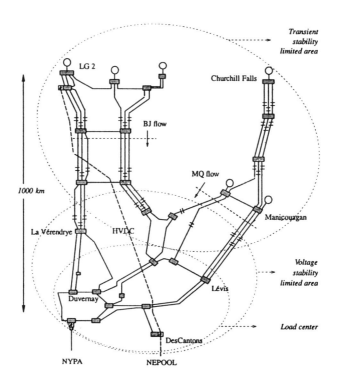

Figure 9.15   One-line diagram of the 735-kV Hydro-Québec system

(SRT) devices were introduced in 1990, providing an additional 2300 Mvar support near the load centers. This amount has almost tripled since 1996. These devices are implemented in different substations to insure redundancy and reliability. They are activated by either low bus voltage or high compensator reactive output [BTS96]. Additional countermeasures to instability involve the compensation of SVC and SC step-up transformers, as well as an SC voltage boosting device (described in the sequel).

9.6.2   System modelling

Following the 1988 overvoltage instability incident [Mai94], a multi-time-scale version of the Hydro-Québec ST600 stability program was developed. Numerical integration relies on the Trapezoidal Rule with fixed time step size. Since 1989, it has been extensively used to determine the adequate settings of SRT devices and revise the

Table 9.2   Salient data of Hydro-Québec system model

	multi-time-scale simulation (ST600)	*QSS simulation (ASTRE)*
long-term dynamics	$\simeq 250$ load tap changers under- and overexcitation limiters shunt reactor tripping devices synchronous condenser voltage boosters	
short-term dynamics	88 generators + 9 synchronous condensers	
	3-winding Park model each with saturation effects + AVR + PSS + governor + turbine	3 $\tilde{f}$ equations each (see Table 6.4) + system frequency ω_{sys}
	6 SVCs	
	each modeled in detail	no $\tilde{f}$ equation (see Table 6.6)
	1650 differential eqs.	*292 algebraic eqs.*
instantaneous	voltage magnitudes and phase angles at 514 buses	

operating strategies, in particular power transfer limits. This is the benchmark needed for detailed studies but it is computationally very demanding.

ASTRE is used for faster QSS simulation. It has been provided with an interface picking up the relevant information from the load flow and ST600 data files, so that the user can easily switch from one method to the other.

An overall description of the model is given in Table 9.2, which follows the time decomposition adopted in this book.

The network model consists of 514 buses (at various levels from 120 to 735 kV) and 681 branches. For QSS simulation, the short-term dynamics are replaced by their equilibrium equations, reduced as explained in Section 6.5.1. Loads are represented by the exponential model (4.2a,b) with various values of α and β. The long-term dynamics consist of the discrete-type devices listed in Table 9.2, all acting after internal time delays (except SVC susceptance limitation, which takes place without delay). Load power restoration is due to several levels of LTCs in cascade; no self-restoration has been considered.

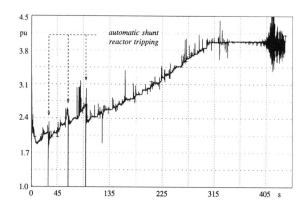

Figure 9.16   ST600 output: field current of an SC, in an unstable scenario

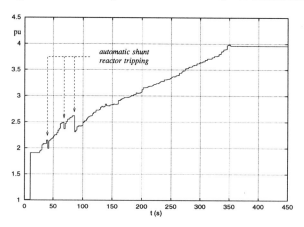

Figure 9.17   Field current of Fig. 9.16 obtained with ASTRE

9.6.3   Validation of QSS simulation

Illustrative examples of multi-time-scale and QSS simulation are given in Figs. 9.16 and 9.17 respectively, which show the field current of an SC in an unstable scenario. It is clearly seen how QSS simulation filters out the short-term dynamics, while reproducing the long-term evolution, in particular the field current limitation. The final oscillations in Fig. 9.16 come from a loss of synchronism when the system eventually collapses (S-LT1 instability). A loss of short-equilibrium is correspondingly detected in ASTRE at t=456 s. At that time, many transmission voltages are below 0.8 pu, and long-term instability is evident.

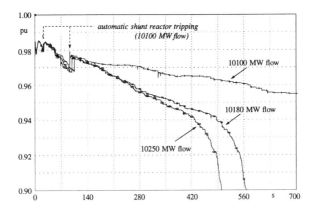

Figure 9.18   ST600 output: voltage at Duvernay bus following a line tripping

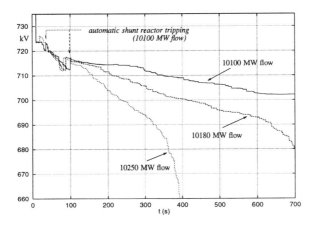

Figure 9.19   Voltages of Fig. 9.18 obtained with ASTRE

The QSS and multi-time-scale simulations have been compared by considering the system response to contingencies, for a few pre-contingency operating points close to the corresponding stability limit.

Simulation outputs provided by ST600 are shown in Fig. 9.18. Each curve corresponds to a different pre-contingency power flow in the Manicouagan - Québec corridor (see "MQ flow" in Fig. 9.15). The contingency of concern is the loss of a series-compensated line between Manicouagan and Lévis, with 330-Mvar shunt reactor tripping. To be more realistic, a 6-cycle 1-phase fault at Lévis as been simulated, cleared by line opening. The 10100 and 10180 MW cases are marginally stable and unstable, respectively. Incidentally, the figure illustrates a well-known property of saddle-node bifurcation: the closer to the stability limit, the longer it takes to either collapse or settle down at a long-term equilibrium.

The corresponding curves obtained with ASTRE are shown in Fig. 9.19. As there is no long-term differential equation (see Table 9.2), each curve is a succession of transient equilibrium points. The short-circuit is obviously ignored while an (arbitrary) 10 s delay before line tripping accounts for the non-simulated short-term transients. The time step size used is 2 s. Although time step sizes of up to 5 s were found equally good (and obviously more efficient) [VCM97], a value of 1 s is now adopted to deal with the recently increased number of SRT devices[8].

As can be seen, from the security limit viewpoint, both methods are in very good agreement. In fact, the discrepancies between the ASTRE and ST600 unstable curves are attributable to:
– the marginal nature of the test cases considered: close to the stability limit, small initial changes have large final effects;
– the neglected short-term dynamics in ASTRE. This causes the discrete LTC and SRT transitions to be somewhat shifted in time;
– the "synchronization" of discrete events at multiples of the time step size.

Similar tests have been performed for the Baie James power flow, denoted "BJ flow" in Fig. 9.15. The contingency of concern is the loss of the LaVérendrye - Duvernay line along with 330 Mvar shunt reactor. The following results have been found:
– highest stable flow: ST600: 12450 MW   ASTRE: 12500 MW
– lowest unstable flow: ST600: 12475 MW   ASTRE: 12525 MW.
Taking the highest stable flow as the security limit, the discrepancy is 50 MW, i.e. 0.4 % of the limit itself. It is very small taking into account that the stability limit is not very sensitive to this particular pre-contingency stress. In comparison, the system is operated at typically 500 MW below the (pre-computed) limit.

Another example of validation, dealing with SRT is provided in the next section.

QSS simulations as shown above typically take between 1 and 6 s CPU on a 300-MHz SUN UltraSparc-2x workstation and about 2.5 times more on a 166-MHz Pentium PC. Longer computing times correspond to cases closer to the stability limit. In very unstable scenarios, simulation is stopped early while in stable ones, there is almost no computational effort once long-term equilibrium has been reached. A comparison

---

[8]For a system driven by LTCs and secondary voltage control, a time step size of 10 s was used in [VJM95, VJM94]

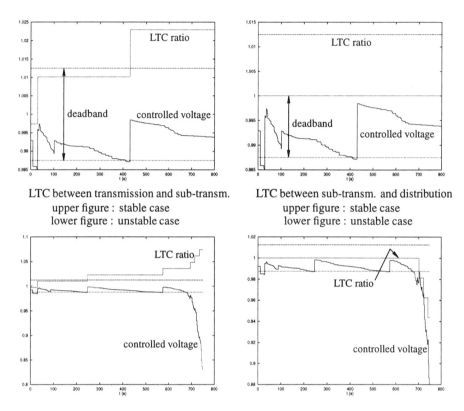

Figure 9.20 Responses of two LTC in stable and unstable cases

between QSS and multi-time-scale simulation has shown that the former is *more than 1000 times faster* than the latter [VCM97].

9.6.4 Examples of instability mechanisms and countermeasures

This section provides QSS simulation results illustrating some long-term instability mechanisms and countermeasures discussed in Chapter 8.

LTC instability

Figure 9.20 shows the response of two LTCs operating in cascade. In all plots, the horizontal lines represent the deadband of the corresponding LTC. The LTC ratio and the controlled (secondary) voltage are also shown, in per unit. The upper two plots

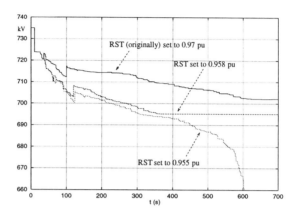

Figure 9.21   Influence of SRT settings on stability

show the stable response to a contingency occurring at time $t = 10$ s. LTC operation is normal with the upper level acting first, so that the lower level does not need to move (see Section 4.4.4). The lower two plots correspond to the same contingency but a slightly higher initial system stress, for which the system response is unstable. The upper-level LTC moves repetitively to bring the controlled voltage back within the deadband. Although the first attempts are successful, the secondary voltage eventually leaves the deadband under the combined effects of all LTCs. This causes the lower-level to start acting, once its internal delay has elapsed, but with no more success. The simulation stops due to loss of short-term equilibrium. This figure is typical of the LT1 or LT2 instability mechanisms in which LTCs are unable to restore load voltages.

SRT device settings

The action of SRT devices is clearly seen in Figs. 9.16 to 9.19. The tuning of these controls must take into account several contingencies as well as equipment and operating constraints. One aspect of the problem is to choose adequate settings so that reactors are tripped neither unduly nor too late. Indeed we have shown in Section 8.6.3 (see Fig. 8.25) that the time available to take an action is limited by attraction considerations (LT2 instability).

As an example, consider the marginally stable 10100 MW scenario of Fig. 9.19. Each SRT device is set to trigger a reactor once the monitored EHV voltage falls below 0.97 pu for some time. By decreasing this threshold, the tripping is delayed. Figure 9.21 shows the system response for the original settings as well as for the lowest stable and the highest unstable settings.

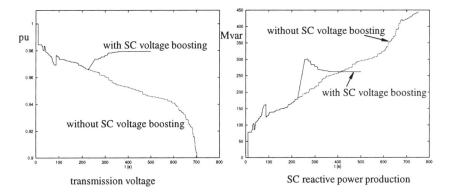

Figure 9.22 Examples of SC voltage boosting

LT2 instability is a typical situation requiring time simulation for proper analysis. Further comparison between multi-time-scale and QSS simulation gave the following results:
– lowest stable setting : ST600 : 0.963 pu   ASTRE : 0.958 pu
– highest unstable setting : ST600 : 0.960 pu   ASTRE : 0.955 pu
The discrepancy amounts to 0.005 pu, which is quite satisfactory compared to the 0.003 pu accuracy expected for voltage measurements.

SC voltage boosting

As already mentioned, another automatic corrective action used by Hydro-Québec consists in acting on SC voltage setpoints. SCs are normally operated at low reactive power output (i.e. with high reactive reserve) to face incidents. Hence, a large enough reactive production may be the indicator of a stressed situation. Based on this idea, the SC reactive power output is monitored and if it remains for some time above a fraction f_1 of the SC capability, the AVR voltage setpoint is increased (at a rate of typically 1 % per 10 s). This voltage boosting stops when either the reactive output reaches a fraction f_2 ($> f_1$) of the SC capability, or the setpoint has increased by more than a specified amount (typically a few %).

Figure 9.22 shows a case where this action saves the system from instability. Without SC voltage boosting, the system is unstable due to LT1 or LT2 instability. The SC produces more and more reactive power and its field current is limited at time $t =$ 680 s. With voltage boosting, the system is attracted towards an acceptable long-term equilibrium. During voltage boosting, the SC reactive power increases faster than in the unstable case, but at the end it stops well below the field current limit. The final decrease is due to LTCs moving in the opposite direction under the effect of the increased transmission grid voltages.

9.6.5   Secure operation limit determination

As mentioned in Section 9.4.2, the determination of a secure operation limit requires to generate pre-contingency operating points corresponding to various levels of system stress and to determine the subsequent system response to a contingency. In ASTRE, both tasks are performed using QSS simulation. The advantages are a higher computational efficiency and a perfect coherency in system modelling. In practice, the parameters which characterize the system stress (active power setpoint of turbines, active and reactive demand of voltage dependent loads) are increased linearly with time up to the level corresponding to S_{max}. The rate of change is taken low enough to consider that the system evolves through a succession of long-term equilibria, and high enough to save computing time. The system time response to this parameter change, together with the pre-contingency controller / operator actions, is determined by QSS simulation and stored in memory. During the binary search, for a specified level of system stress, the relevant time interval of the pre-contingency simulation is simply retrieved from memory and the contingency is applied at the end of this time interval.

We provide hereafter a detailed example. Starting from a base case, the system is stressed by rescheduling generation between the Baie James and Montréal areas, under constant load level, so that power flow denoted "BJ flow" in Fig. 9.15 is increased. The maximum stress of interest S_{max} is set to 3300 MW (measured at the sending end) above the base case. Without control action, this additional power transfer would severely depress grid voltages and exhaust system reactive reserves. However, the operators intervene to avoid this by switching off shunt reactors at the 735-kV level and adjusting SVCs and SCs. This pre-contingency action is modelled by connecting to the buses of concern fictitious SCs which maintain 1 pu voltages while the system is stressed and are replaced by shunt admittances for contingency simulation. This is an acceptable approximation of the real, discrete-type actions of operators.

The contingency studied is the loss of the 735-kV line between La Vérendrye and Duvernay. Referring to the discussion of Section 9.1.2, the secure operation limits are evaluated taking into account the benefit of normal post-contingency controls such as shunt reactor tripping and SC voltage boosting. Stronger controls such as load (or even generation) shedding are envisaged for more severe disturbances.

Consider the curves in Figs. 9.23 and 9.24 relative to 100 % stress. The time interval $[0\ 150\ s]$ corresponds to the pre-contingency loading. Figure 9.23 shows the active

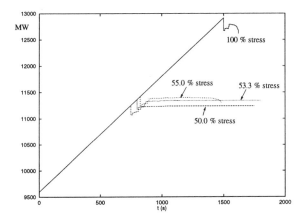

Figure 9.23 Active power generation of La Grande area (4 simulations of the binary search are shown)

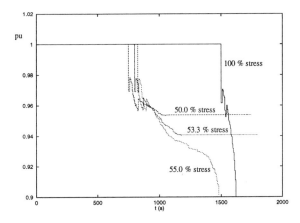

Figure 9.24 Voltage at bus Duvernay (4 simulations of the binary search are shown)

power production in the sending area while the flat voltage profile of Fig. 9.24 corresponds to the above mentioned operator actions. The contingency is simulated at time $t = 1500$ s and clearly leads to instability.

According to the binary search strategy, the 50 % stress level is considered next. To this purpose the system is "brought back" at time $t = 1500/2 = 750$ s and the same contingency is applied, leading now to a stable behaviour (see Fig. 9.24). The procedure is repeated until the stress interval bounded by the highest stable and the

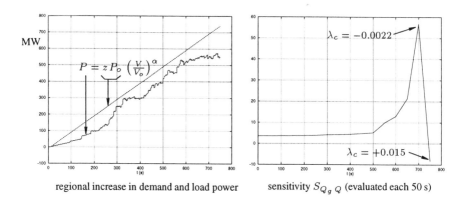

regional increase in demand and load power   sensitivity $S_{Q_g\,Q}$ (evaluated each 50 s)

Figure 9.25   Example of loadability limit determination

lowest unstable cases falls below the tolerance Δ, set in this example to 60 MW. This has led to testing eight levels of stress. The marginally stable and unstable cases, corresponding to 53.3 and 55.0 % stress, respectively, are shown in Figs. 9.23 and 9.24.

In Figure 9.23, the active power variations following the contingency (e.g. after $t = 1500$ s for 100 % stress) correspond to governor control. In the stable cases, the generation recovers to almost its pre-disturbance level, except for a small difference that comes from active power losses in the post-contingency network as well as sensitivity of loads to voltage within the LTC deadbands.

Note finally that a large step size of 5 s is used to simulate the smooth pre-contingency variations while it is reduced to 2 s for contingency simulation.

9.6.6   Loadability limit determination

Although secure operation limits are the Hydro-Québec choice for security assessment, we briefly illustrate on this system the determination of a loadability limit using the technique of Section 9.3.4.

The case shown in Fig. 9.25 corresponds to a load increase in the Montréal area. To this purpose, the scaling factor z of the exponential load model (4.2a,b) has been increased linearly with time at the participating buses. The left plot in Fig. 9.25 shows the total increase in demand and power, respectively. The difference between the two curves comes from LTC deadband effects.

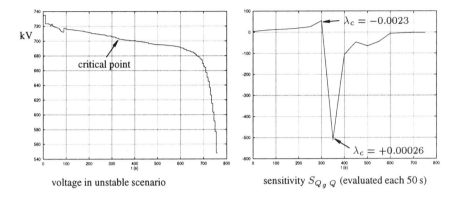

Figure 9.26 (left) voltage in unstable scenario

(right) sensitivity $S_{Q_g Q}$ (evaluated each 50 s)

Figure 9.26   Example of critical point identification

To detect the crossing of the bifurcation point, we use the sensitivities of the total reactive power production[9] to reactive power loads, as defined in the simple example of Section 7.2.2. They are computed as explained in Section 9.3.4, each 50 s during the load increase, and the simulation is stopped as soon as a change in sign "through infinity" is detected. The sensitivity relative to a major 735-kV bus is shown in the right plot of Fig. 9.25. The loadability limit is reached sometime between $t = 700$ and 750 s; referring to the left plot, the loadability limit is about 550 MW.

The dominant eigenvalue of $\mathbf{H}_z$ has been computed using Simultaneous Iterations. The values before and after bifurcation crossing are given in Fig. 9.25 and confirm the result given by sensitivities. The eigenvector $\mathbf{w}_c$ of $\mathbf{J}_\ell$, computed at $t = 750$ s, can be used for diagnosis purposes.

9.6.7   Critical point identification

The technique of the previous section can be used to locate the critical point along the trajectory of a system made unstable by a contingency (see Section 8.6.2). An example is given in Fig. 9.26. Sensitivities computed again at 50 s intervals indicate the crossing of the critical point in between times $t = 300$ and 350 s. The corresponding eigenvalues, as computed by the Simultaneous Iteration method are shown in the right plot and the critical point is shown in the left one. The eigenvector $\mathbf{w}_c$ of $\mathbf{J}_\ell$, computed at $t = 350$ s, can be used for corrective action determination.

---

[9] this includes all components that control voltages: generators, SVCs and SCs

9.7   REAL-TIME ISSUES

In stability limited systems, the traditional approach to security assessment has been
to establish off-line security limits to be monitored on-line by operators. The determination of such limits requires extensive simulation and engineering judgment based on
a good knowledge of the system. The information from the numerous off-line studies
has to be gathered in a set of rather simple operating rules, which typically take on the
form of bounds set on key parameters, or simple diagrams showing regions of secure
operation. The variables used for this purpose should be quantities that operators
can either monitor or control. The main difficulties are the large number of network
operating conditions, as well as the influence of various "secondary" parameters on
the security problem of concern. This may lead to deriving security limits for a limited
number of plausible system configurations and/or adopting conservative limits to cover
uncertainties in secondary parameters.

Some utilities now envisage to perform voltage security analysis in the real-time
environment of their control centers. Among the expected advantages, one should
count the ability to face operating conditions that could not be forecasted or analyzed
off-line. Moreover, in a real system situation, the uncertainty on secondary parameters
becomes less important.

9.7.1   Contingency selection

Security margin calculations become heavy when dealing with a large number of contingencies, especially within the context of real-time applications. Thus *Contingency
selection* (or filtering) becomes essential.

This important topic has been much less investigated within the context of voltage
security margins than for static security analysis [RAU93, EIM95, CWF97, VCM98,
Sch98]. It must be emphasized that the objectives are different. In static security
analysis, contingency filtering aims at quickly identifying the contingencies that will
lead to some operating constraint violations. As regards voltage security margins, the
objective is to quickly identify the contingency which yields the lowest margin. A less
ambitious objective would be to identify a group of severe contingencies containing the
most dangerous one. The simultaneous binary search of Fig. 9.13 could, for instance,
be applied to identify the contingency giving the lowest margin.

Note that a reliable contingency selection should take into account the impact of
disturbances *as well as the direction of system stress*. Indeed, the most constraining
contingency may change when the direction of stress changes.

9.7.2 Voltage stability indices

Many voltage stability indices (or indicators) have been proposed in the literature, whose aim is to evaluate how far the system is from instability. Most of them are intended to be used in a real-time environment. Following [CTF94a], they may be roughly classified into:

- indices based on the current system state only, which we call for short *state-based indices*. They range from simple voltage drops to sensitivities, eigenvalues and singular values. Most of them are deduced from properties derived in Chapters 2 and 7;

- *large-deviation* indices, involving changes in parameters associated with a system stress. These are typically the security margins previously considered in Sections 9.3 and 9.4.

State-based indices are usually simpler to compute or measure than large-deviation indices. However, the latter provide a more explicit measure of the system security margin.

Note that state-based indices may be embedded in the determination of large-deviation indices: an example is the coupling of sensitivity information with time simulation, as described in Section 9.3.4.

Indices based on system linearization around a given operating point generally suffer from poor prediction capabilities because the system behaviour between this point and a bifurcation is nonlinear, in particular due to generator limitation effects.

Practical indices should be expressed in terms of parameters that operators can observe or control. Even more important, they should have the capability to anticipate the effects of contingencies.

9.7.3 Automatic Learning methods

In parallel with the development of real-time voltage security assessment, research efforts have been devoted in the recent years towards applying Automatic Learning methods to obtain elaborate stability indicators or rules for secure system operation. The overall approach can be summarized as follows:

1. Off-line: a large data base of pre-analyzed scenarios is built. To this purpose, system configurations and operating points are generated randomly. Secondary parameters known with uncertainty are also varied randomly. It is essential that human experts are involved in this generation step.

Each operating point of this data base is analyzed with respect to contingencies. As a result, it is labeled secure or insecure or, even better, a security / insecurity margin is computed. Each point is also characterized by "attributes", i.e. quantities relevant to the security problem of concern (e.g. load levels, power flows in corridor, voltages, power imports/exports, etc.)

This step is computationally very demanding. Contingency evaluation is however parallel by nature and can be distributed on several computers.

2. Off-line: Automatic Learning methods analyze the so obtained data base to extract the complex relationships between the attributes and the classification or margin. This information can be expressed in the form of decision trees, artificial neural networks, etc. Although it may vary from one method to the other, this step is usually much less computationally demanding than the previous one.

It is essential that human experts interpret and validate the outputs of these methods, and iterations between steps 1 and 2 are usually needed.

3. On-line: the so obtained synthetic information is very simple and fast to use: for instance, data relative to the current operating conditions are passed on to the decision tree or neural network to obtain the security classification or margin.

The success of these approaches strongly depends upon the representativity of the off-line data base, whose elaboration requires special care.

Applications to voltage stability problems are reported in [VWP93, WCP94], where the above approaches are used to obtain both preventive and emergency voltage indicators. A detailed presentation of the methods and their application to power systems is given in [Weh97].

REFERENCES

[AAH97]   S. Arnborg, G. Andersson, D. J. Hill, and I. A. Hiskens. On undervoltage load shedding in power systems. *International Journal of Electrical Power and Energy Systems*, 19:141–149, 1997.

[AAH98]   S. Arnborg, G. Andersson, D. J. Hill, and I. A. Hiskens. The influence of load modelling for undervoltage load shedding studies. *IEEE Transactions on Power Systems*, 1998. To appear.

[AC92]   V. Ajjarapu and C. Christy. The continuation power flow: a tool for steady state voltage stability analysis. *IEEE Transactions on Power Systems*, 7:416–423, 1992.

[ACN90]   V. Archidiacono, S. Corsi, A. Natale, C. Raffaelli, and V. Menditto. New developments in the application of enel transmission system voltage and reactive power automatic control. In *CIGRE Proceedings*, 1990. Paper 38/39-06.

[ADH94]   F. L. Alvarado, I. Dobson, and Y. Hu. Computation of closest bifurcations in power systems. *IEEE Transactions on Power Systems*, 9:918–928, 1994.

[AFI82]   S. Abe, Y. Fukunaga, A. Isono, and B. Kondo. Power systems voltage stability. *IEEE Transactions on Power Apparatus and Systems*, 101:3830–3840, 1982.

[AHI78]   S. Abe, N. Hamada, A. Isono, and K. Okuda. Load flow convergence in the vicinity of a voltage stability limit. *IEEE Transactions on Power Apparatus and Systems*, 97:1983–1993, 1978.

[AJ88]   F. L. Alvarado and T. H. Jung. Direct detection of voltage collapse conditions. In Fink [Fin88], pages 5.23–5.38.

[AL92]   V. Ajjarapu and B. Lee. Bifurcation theory and its application to nonlinear dynamic phenomena in an electric power system. *IEEE Transactions on Power Systems*, 7:424–431, 1992.

[AM94]   M. M. Adibi and D. P. Milanicz. Reactive capability limitation of synchronous machines. *IEEE Transactions on Power Systems*, 9:29–40, 1994.

[Arn86]   Vladimir I. Arnol'd. *Catastrophe Theory.* Springer Verlag, 1986.

[ARU91]   R. R. Austria, N. D. Reppen, J. A. Uhrin, M. C. Patel, and A. Galatic. Applications of the optimal power flow to analysis of voltage collapse limited power transfer. In Fink [Fin91], pages 311–319.

[AZT91]   S. Ahmed-Zaid and M. Taleb. Structural modeling of small and large induction machines using integral manifolds. *IEEE Transactions on Energy Conversion*, 6:529–535, 1991.

[BB80]   C. Barbier and J.-P. Barret. An analysis of phenomena of voltage collapse on a transmission system. *Revue Générale de l'Electricité*, pages 3–21, 1980. Special issue.

[BBM96]   A. Berizzi, P. Bresesti, P. Marannino, G.P. Granelli, and M. Montagna. System-area operating margin assessment and security enhancement against voltage collapse. *IEEE Transactions on Power Systems*, 11:1451–1462, 1996.

[BCR84]   P. Borremans, A. J. Calvaer, J.-P. De Reuck, J. Goossens, E. Van Geert, J. Van Hecke, and A. Van Ranst. Voltage stability - fundamental aspects and comparison of practical criteria. In *CIGRE Proceedings*, 1984. Paper 38-11.

[BCS77]   Boeing Computer Services. Power system dynamic analysis - phase i. *EPRI report EL-484*, 1977.

[Ber86]   A. R. Bergen. *Power systems analysis.* Electrical and Computer Engineering Series. Prentice Hall, 1986.

[BP92]   M. Begovic and A. Phadke. Control of voltage stability using sensitivity analysis. *IEEE Transactions on Power Systems*, 7:114–123, 1992.

[BTS96]   S. Bernard, G. Trudel, and G. Scott. A 735-kv shunt reactors automatic switching system for Hydro-Quebec network. *IEEE Transactions on Power Systems*, 11:2024–2030, 1996.

[CA93]   C. A. Cañizares and F. L. Alvarado. Point of collapse and continuation methods for large ac/dc systems. *IEEE Transactions on Power Systems*, 8:1–8, 1993.

[CAD92]   C. A. Cañizares, F. L. Alvarado, C. L. DeMarco, I. Dobson, and W. F. Long. Point of collapse methods applied to ac/dc power systems. *IEEE Transactions on Power Systems*, 7:673–683, 1992.

[Cal83]   A. J. Calvaer. On the maximum loading of active linear electric multiports. *Proceedings of the IEEE*, 71:282–283, 1983.

[Cal84]   M. S. Calović. Modeling and analysis of under-load tap-changing transformer control systems. *IEEE Transactions on Power Apparatus and Systems*, 103:1909–1915, 1984.

[Cal86]   A. J. Calvaer. Voltage stability and collapses : a simple theory based on real and reactive currents. *Revue Générale de l'Electricité*, 8:1–17, 1986.

[CCM96]   F. Carbone, G. Castellano, and G. Moreschini. Coordination and control of tap changers under load at different voltage level transformers. In *Proceedings Melecon*, pages 1–3, Bari (Italy), 1996.

[CDK87]   L. O. Chua, C. A. Desoer, and E. S. Kuh. *Linear and nonlinear circuits*. McGraw Hill, 1987.

[CFS95]   H.D. Chiang, A.J. Flueck, K.S. Shah, and N. Balu. Cpflow: a practical tool for tracing power system steady-state stationary behavior due to load and generation variations. *IEEE Transactions on Power Systems*, 10:623–630, 1995.

[CG84]   A. J. Calvaer and E. Van Geert. Quasi steady-state synchronous machine linearization around an operating point and applications. *IEEE Transactions on Power Apparatus and Systems*, 103:1466–1472, 1984.

[CGS84]   J. Carpentier, R. Girard, and E. Scano. Voltage collapse proximity indicators computed from an optimal power flow. In *8th PSCC Proceedings*, pages 671–678, Helsinki, 1984.

[CIE95]   CIGRE & IEEE FACTS Working Groups. Facts overview. *Special Publication 95 TP 108*, 1995.

[Cla87]   H.K. Clark. Voltage control and reactive supply problems. In G.B. Sheblé, editor, *IEEE Tutorial on Reactive Power: Basics, Problems and Solutions*, pages 17–27, Publication 87EH0262-6-PWR, 1987.

[CMT91]   H. D. Chiang, W. Ma, R. J. Thomas, and J. S. Thorp. A tool for analyzing voltage collapse in electric power systems. *Proc. 10th Power System Computation Conference, Graz (Austria)*, pages 1210–1217, 1991.

[Con91]   C. Concordia. Voltage instability. *International Journal of Electrical Power and Energy Systems*, 13:14–20, 1991.

[Cot74]   R. W. Cottle. Manifestations of the Schur complement. *Linear Algebra and its Applications*, 8:189–211, 1974.

[CTF87]   CIGRE Task Force 38-01-03. Planning against voltage collapse. *Electra*, 111:55–75, 1987.

[CTF93]   CIGRE Task Force 38-02-10. Modelling of voltage collapse including dynamic phenomena. *CIGRE Publication*, 1993.

[CTF94a]   CIGRE Task Force 38-02-11. Indices predicting voltage collapse including dynamic phenomena. *CIGRE Publication*, 1994.

[CTF94b]   CIGRE Task Force 38-02-12. Criteria and countermeasures for voltage collapse. *CIGRE Publication*, 1994.

[CTF95]   CIGRE Task Force 38-02-08. Long term dynamics - phase ii. final report. *CIGRE Publication*, 1995.

[CTF96]   CIGRE Task Force 38-01-06. Load flow control in high voltage power systems using facts controllers. *CIGRE Publication*, 1996.

[CWF97]   H.D. Chiang, C.S. Wang, and A.J. Flueck. Look-ahead voltage and load margin contingency selection functions for large-scale power systems. *IEEE Transactions on Power Systems*, 12:173–180, 1997.

[CWG98]   CIGRE Working Group 34-08. Protection against voltage collapse. *CIGRE Publication*, 1998. To appear.

[DB84]   C. L. DeMarco and A. R. Bergen. Application of singular perturbation techniques to power system transient stability analysis. In *ISCAS Proceedings*, pages 597–601, Montreal, 1984.

[DC89]   I. Dobson and H. D. Chiang. Towards a theory of voltage collapse in electric power systems. *Systems and Control Letters*, 13:253–262, 1989.

[DDF97]   J. Deuse, J. Dubois, R. Fanna, and I. Hanza. Undervoltage load shedding scheme. *IEEE Transactions on Power Systems*, 12:1446–1454, 1997.

[DL74]   T. E. Dy Liacco. Real-time computer control of power systems. *Proceedings of the IEEE*, 62:884–891, 1974.

[DL78]   T. E. Dy Liacco. System security : the computer's role. *IEEE Spectrum*, pages 43–50, June 1978.

[DL91]   I. Dobson and L. Lu. Immediate change in stability and voltage collapse when generator reactive power limits are encountered. In Fink [Fin91], pages 65–73.

[DL93]   I. Dobson and L. Lu. New methods for computing a closest saddle-node bifurcation and worst case load power margin for voltage collapse. *IEEE Transactions on Power Systems*, 8:905–913, August 1993.

[dMF96]   F. P. de Mello and J. W. Feltes. Voltage oscillatory instability caused by induction motor loads. *IEEE Transactions on Power Systems*, 11:1279–1285, 1996.

[DO90]   C. L. DeMarco and T. J. Overbye. An energy based security measure for assessing vulnerability to voltage collapse. *IEEE Transactions on Power Systems*, 5:419–427, 1990.

[Dob92]   I. Dobson. Observations on the geometry of saddle node bifurcation and voltage collapse in electric power systems. *IEEE Transactions on Circuits and Systems – I*, 39(3):240–243, 1992.

[Dob94]   I. Dobson. The irrelevance of load dynamics for the loading margin to voltage collapse and its sensitivities. In Fink [Fin94], pages 509–518.

[DS93]   J. Deuse and M. Stubbe. Dynamic simulation of voltage collapses. *IEEE Transactions on Power Systems*, 8:894–904, 1993.

[EIM95]   G.C. Ejebe, G.D. Irisarri, S. Mokhtari, O. Obadina, P. Ristanovic, and J. Tong. Methods for contingency screening and ranking for voltage stability analysis of power systems. In *Proceedings PICA conference*, pages 249–255, 1995.

[Elg71]   O. I. Elgerd. *Electric Energy Systems Theory*. McGraw Hill, 1971.

[FC78]   L. H. Fink and K. Carlsen. Operating under stress and strain. *IEEE Spectrum*, pages 48–53, 1978.

[FFC92]   O. B. Fosso, N. Flatabø, T. Carlsen, O. Gjerde, and M. Jostad. Margins to voltage instability calculated for normal and outage conditions. In *CIGRE Proceedings*, Paris, 1992. Paper 38-209.

[FG83]   A. Fahmideh-Vojdani and F. D. Galiana. The continuation method and its application in system planning and operations. *Proc. of the CIGRE-IFAC Symposium on Control applications for power system security, Florence (Italy)*, pages 318–326, 1983.

[Fin88]   L. H. Fink, editor. *Bulk Power System Voltage Phenomena I – Voltage Stability and Security*, Potosi, Missouri, 1988.

[Fin91]   L. H. Fink, editor. *Bulk Power System Voltage Phenomena II – Voltage Stability and Security*, Deep Creek Lake, Maryland, 1991.

[Fin94]   L. H. Fink, editor. *Bulk Power System Voltage Phenomena III – Voltage Stability and Security*, Davos, Switzerland, 1994.

[FJC85]   N. Flatabø, N. Johannesen, T. Carlsen, and L. Holten. Evaluation of reactive power reserves in transmission systems. In *IFAC Conference Proceedings*, Rio de Janeiro, 1985.

[FKU83]   A. E. Fitzgerald, C. Kingsley, Jr., and S. D. Umans. *Electric Machinery*. McGraw-Hill, 4th edition, 1983.

[FM94]   D. C. Franklin and A. Morelato. Improving dynamic aggregation of induction motor models. *IEEE Transactions on Power Systems*, 9:1934–1941, 1994.

[FOC90]   N. Flatabø, R. Ognedal, and T. Carlsen. Voltage stability condition in a power system calculated by sensitivity methods. *IEEE Transactions on Power Systems*, 5:1286–1293, 1990.

[GDA97]   S. Greene, I. Dobson, and F. L. Alvarado. Sensitivity of the loading margin to voltage collapse with respect to arbitrary parameters. *IEEE Transactions on Power Systems*, 12:262–272, 1997.

[Gea71]   C. W. Gear. *Numerical initial value problems in ordinary differential equations*. Prentice Hall, 1971.

[GH83]   John Guckenheimer and Philip Holmes. *Nonlinear Oscillations, Dynamical Systems and Bifurcations of Vector Fields*. Springer Verlag, 1983.

[GMK92]   B. Gao, G. K. Morison, and P. Kundur. Voltage stability evaluation using modal analysis. *IEEE Transactions on Power Systems*, 7:1529–1542, 1992.

[GMK96]   B. Gao, G. K. Morison, and P. Kundur. Towards the development of a systematic approach for voltage stability assessment of large-scale power systems. *IEEE Transactions on Power Systems*, 11:1314–1324, 1996.

[GMM96]   S. Granville, J.C.O. Mello, and A.C.G. Melo. Application of interior point methods to power flow unsolvability. *IEEE Transactions on Power Systems*, 11:1096–1104, 1996.

[GMW81]   P. E. Gill, W. Murray, and M. H. Wright. *Practical Optimization*. Academic Press, London, 1981.

[Gra88]   K.-M. Graf. Dynamic simulation of voltage collapse processes in EHV systems. In Fink [Fin88], pages 6.45–6.54.

[Gyu94]   L. Gyugi. Dynamic compensation of ac transmission lines by solid-state synchronous voltage sources. *IEEE Transactions on Power Delivery*, 9:904–911, 1994.

[HC96]   I. A. Hiskens and B. B. Chakrabarti. Direct calculation of reactive power limits. *International Journal of Electrical Power and Energy Systems*, 18:121–129, 1996.

[HD97]   L. M. Hajagos and B. Danai. Laboratory measurements and models of modern loads and their effect on voltage stability studies. *IEEE Transactions on Power Systems*, 1997.

[HEZ89]   A. E. Hammad and M. Z. El-Zadek. Prevention of transient voltage instabilities due to induction motor loads by static var compensators. *IEEE Transactions on Power Systems*, 4:1182–1190, 1989.

[HH89]   I. A. Hiskens and D. J. Hill. Energy functions transient stability, and voltage behaviour in power systems with nonlinear loads. *IEEE Transactions on Power Systems*, 4:1525–1533, 1989.

[HH91]   I. A. Hiskens and D. J. Hill. Failure modes in a collapsing power system. In Fink [Fin91], pages 53–63.

[HH93]   I. A. Hiskens and D. J. Hill. Dynamic interaction between tapping transformers. In *11th PSCC Proceedings*, pages 1027–1034, Avignon, 1993.

[Hil93]   D. J. Hill. Nonlinear dynamic load models with recovery for voltage stability studies. *IEEE Transactions on Power Systems*, 8:166–176, 1993.

[Hil95]   D. J. Hill (editor). Special issue on nonlinear phenomena in power systems. *IEEE Proceedings*, 83, 1995.

[HK92]   J. Hale and H. Koçak. *Dynamics and Bifurcations*. Springer Verlag, 1992.

[HTL90]   Y. Harmand, M. Trotignon, J.F. Lesigne, J.-M. Tesseron, C. Lemaître, and F. Bourgin. Analysis of a voltage collapse incident and proposal for a time-based hierarchical containment scheme. In *CIGRE Proceedings*, 1990. Paper 38/39-02.

[ISE90]   K. Iba, H. Suzuki, M. Egawa, and T. Watanabe. A method for finding a pair of multiple load flow solutions in bulk power systems. *IEEE Transactions on Power Systems*, 5:582–591, 1990.

[ISE91]   K. Iba, H. Suzuki, M. Egawa, and T. Watanabe. Calculation of critical loading condition with nose curve using homotopy method. *IEEE Transactions on Power Systems*, 6:584–593, 1991.

[IT81]   S. Iwamoto and Y. Tamura. A load flow calculation method for ill-conditioned power systems. *IEEE Transactions on Power Apparatus and Systems*, 100:1736–1743, 1981.

[ITF93]   IEEE Task Force. Load representation for dynamic performance analysis. *IEEE Transactions on Power Systems*, 8:472–482, 1993.

[ITF95a]   IEEE Task Force. Bibliography on load models for power flow and dynamic performance simulation. *IEEE Transactions on Power Systems*, 10:523–538, 1995.

[ITF95b]   IEEE Task Force. Standard load models for power flow and dynamic performance simulation. *IEEE Transactions on Power Systems*, 10:1302–1313, 1995.

[ITF96a]   IEEE Digital Excitation Task Force. Computer models for representation of digital-based excitation systems. *IEEE paper 96 WM 031-5 EC*, 1996.

[ITF96b]   IEEE Task Force on Excitation Limiters. Recommended models for overexcitation limiting devices. *IEEE Transactions on Energy Conversion*, 1996.

[ITF96c]   IEEE Task Force on Excitation Limiters. Underexcitation limiter models for power system stability studies. *IEEE Transactions on Energy Conversion*, 1996.

[IWG81]   IEEE Working Group on Computer Modelling of Excitation Systems. Excitation system models for power system stability studies. *IEEE Transactions on Power Apparatus and Systems*, 100:494–509, 1981.

[IWG90]   IEEE Working Group on Voltage Stability. Voltage stability of power systems: concepts, analytical tools, and industry experience. *IEEE Special Publication 90TH0358-2-PWR*, 1990.

[IWG93]   IEEE Working Group on Voltage Stability. Suggested techniques for voltage stability analysis. *IEEE Publication 93TH0620-5PWR*, 1993.

[IWG96]   IEEE Working Group K12. Voltage collapse mitigation. *IEEE Power System relaying Committee*, 1996.

[IWT97]   G. D. Irisarri, X. Wang, J. Tong, and S. Mokhtari. Maximum loadability of power systems using interior point non-linear optimisation method. *IEEE Transactions on Power Systems*, 12:162–172, 1997.

[Jen77]   A. Jennings. *Matrix Computation for Engineers and Scientists*. John Wiley & Sons, 1977.

[JG81]   J. Jarjis and F. D. Galiana. Quantitative analysis of steady-state stability in power systems. *IEEE Transactions on Power Apparatus and Systems*, 100:318–326, January 1981.

[JSK94]   S. G. Johansson, F. G. Sjögren, D. Karlsson, and J. E. Daalder. Voltage stability studies with PSS/E. In Fink [Fin94], pages 651–661.

[JVS96]   X. Jiang, V. Venkatasubramanian, H. Schättler, and J. Zaborszky. On the dynamics of the large power system with hard limits. In *12th PSCC Proceedings*, pages 440–449, Dresden, 1996.

[Kar94]   D. Karlsson. Discussion for session 6. In Fink [Fin94], page 519.

[KF93]   R. J. Koessler and J.W. Feltes. Time-domain simulation investigates voltage collapse. *Computer applications in Power*, 5:18–22, 1993.

[KG86]   P. Kessel and H. Glavitsch. Estimating the voltage stability of a power system. *IEEE Transactions on Power Delivery*, 1:–, 1986.

[KH94]   D. Karlsson and D. J. Hill. Modelling and identification of nonlinear dynamic loads in power systems. *IEEE Transactions on Power Systems*, 9:157–166, 1994.

[KKO86]   P. Kokotović, H. K. Khalil, and J. O'Reilly. *Singular Perturbation Methods in Control: Analysis and Design*. Academic Press, 1986.

[KMN88]   D. Kahaner, C. Moler, and S. Nash. *Numerical methods and software*. Prentice Hall, 1988.

[KOO93]   A. Kurita, H. Okubo, K. Oki, S. Agematsu, D.B. Klapper, N.W. Miller, W.W. Price, J.J. Sanchez-Gasca, K. A. Wirgau, and T.D. Younkins. Multiple time-scale power system dynamic simulation. *IEEE Transactions on Power Systems*, 8:216–223, 1993.

[KPB86]   H. G. Kwatny, A. K. Pasrija, and L. H. Bahar. Static bifurcations in electric power networks: Loss of steady-state stability and voltage collapse. *IEEE Transactions on Circuits and Systems – I*, 33:981–991, 1986.

[KT51]   H. W. Kuhn and A. Tucker. Nonlinear programming. In *Proceedings of the 2nd Berkeley Symposium on Mathematical Statistics and Probability*, University of California, Berkeley, 1951.

[Kun94]   Prahba Kundur. *Power System Stability and Control*. EPRI Power system Engineering Series. McGraw Hill, 1994.

[LA93]   B. Lee and V. Ajjarapu. Period doubling route to chaos in an electric power system. *IEE Proceedings*, 140:490–496, 1993.

[Lac78]   W. R. Lachs. Voltage collapse in EHV power systems. In *IEEE PES Winter Meeting*, 1978. Paper A 78 057-22.

[Lac79]   W. R. Lachs. System reactive power limitations. In *IEEE PES Winter Meeting*, 1979. Paper A 79 015-9.

[LAH93]   P.-A. Löf, G. Andersson, and D. J. Hill. Voltage stability indices for stressed power systems. *IEEE Transactions on Power Systems*, 8:326–335, 1993.

[LAH95]   P.-A. Löf, G. Andersson, and D. J. Hill. Voltage dependent reactive power limits for voltage stability studies. *IEEE Transactions on Power Systems*, 10:220–228, 1995.

[Lia66]   A. M. Liapunov. *Stability of Motion*. Academic Press, 1966.

[LPT90]   C. Lemaitre, J.-P. Paul, J.-M. Tesseron, Y. Harmand, and Y. Zhao. An indicator of the risk of voltage profile instability for real-time control applications. *IEEE Transactions on Power Systems*, 5:154–161, 1990.

[LSA92]   P.-A. Löf, T. Smed, G. Andersson, and D. J. Hill. Fast calculation of a voltage stability index. *IEEE Transactions on Power Systems*, 7:54–64, 1992.

[LSI85]   M. Lotfalian, R. Schlueter, D. Idizior, P. Rusche, S. Tedeschi, L. Shu, and A. Yazdankhah. Inertial, governor, and agc/economic dispatch load flow simulations of loss of generation contingencies. *IEEE Transactions on Power Apparatus and Systems*, 104:3020–3028, 1985.

[LSP92]   B. C. Lesieutre, P. W. Sauer, and M. A. Pai. Sufficient conditions on static load models for network solvability. *Proceedings of the North American Power Symposium*, pages 262–271, 1992.

[LSP98]   B. C. Lesieutre, P. W. Sauer, and M. A. Pai. Existence of solutions for the network/load equations in power systems. *IEEE Transactions on Circuits and Systems – I*, 1998. To appear.

[Mai94]   R. Mailhot. Voltage stability: impacts on hydro-québec operations and simulation tools development. In Fink [Fin94], pages 207–213.

[Mar86]   N. Martins. Efficient eigenvalue and frequency response methods applied to power system small-signal stability studies. *IEEE Transactions on Power Systems*, 1:217–226, 1986.

[MAR94]   Y. Mansour, F. Alvarado, C. Rinzin, and W. Xu. SVC placement using critical modes of voltage instability. *IEEE Transactions on Power Systems*, 9:757–763, 1994.

[MGK93]   G.K. Morison, B. Gao, and P. Kundur. Voltage stability analysis using static and dynamic approaches. *IEEE Transactions on Power Systems*, 8:1159–1171, 1993.

[MIC87]   J. Medanić, M. Ilić, and J. Christensen. Discrete models of slow voltage dynamics for under load tap-changing transformer coordination. *IEEE Transactions on Power Systems*, 2:873–882, 1987.

[Mil82]   T. J. E. Miller, editor. *Reactive Power Control in Electric Power Systems*. John Wiley & Sons, 1982.

[MJP88]   Y. Mansour, C. D. James, and D. N. Pettet. Voltage stability and security - B.C. Hydro's operating practice. In Fink [Fin88], pages 2.9–2.25.

[MMG93]   R. J. Marceau, R. Mailhot, and F. D. Galiana. A generalized shell for dynamic security analysis in operations planning. *IEEE Transactions on Power Systems*, 8:1098–1106, 1993.

[MSS96]   R. J. Marceau, M. Sirandi, S. Soumare, X.D. Do, F. Galiana, and R. Mailhot. A review of signal energy analysis for the rapid determination of dynamic security limits. *Canadian Journal of Electrical & Computer Engineering*, 21:125–132, 1996.

[MXA94]   Y. Mansour, W. Xu, F. Alvarado, and C. Rinzin. Svc placement and transmission line reinforcement using critical modes. *IEEE Transactions on Power Systems*, 9:757–763, 1994.

[Nag75]   T. Nagao. Voltage collapse at load ends of power systems. *Electrical Engineering in Japan*, 95:62–70, 1975.

[NKP87]   F. Nozari, M. D. Kankam, and W. W. Price. Aggregation of induction motors for transient stability load modelling. *IEEE Transactions on Power Systems*, 2:1096–1102, 1987.

[NM92]   S. A. Nirenberg and D. A. McInnis. Fast acting load shedding. *IEEE Transactions on Power Systems*, 7:873–877, 1992.

[OB88]   O. O. Obadina and G. J. Berg. Determination of voltage stability limit in multimachine power systems. *IEEE Transactions on Power Systems*, 3:1545–1554, 1988.

[OD91]   T. J. Overbye and C. L. DeMarco. Improved techniques for power system voltage stability assessment using energy methods. *IEEE Transactions on Power Systems*, 6:1446–1452, 1991.

[OK96]   T. J. Overbye and R. P. Klump. Effective calculations of power system low-voltage solutions. *IEEE Transactions on Power Systems*, 11:75–82, 1996.

[O'M74]   R. E. O'Malley. *Introduction to Singular Perturbation*. Academic Press, 1974.

[Ost93]   T. Ostrup. Reactive power production capability chart of a generator. In *IEEE/NTUA Athens Power Tech Proceedings*, pages 537–541, 1993.

[Ove94]   T. J. Overbye. A power flow solvability measure for unsolvable cases. *IEEE Transactions on Power Systems*, 9:1359–1365, 1994.

[Ove95]   T. J. Overbye. Computation of a practical method to restore power flow solvability. *IEEE Transactions on Power Systems*, 10:280–287, 1995.

[OYS91]   H. Ohtsuki, A. Yokoyama, and Y. Sekine. Reverse action of on-load tap changer in association with voltage collapse. *IEEE Transactions on Power Systems*, 6:300–306, 1991.

[Pal92]   M. K. Pal. Voltage stability conditions considering load characteristics. *IEEE Transactions on Power Systems*, 7:243–249, 1992.

[Pal93]   M. K. Pal. Voltage stability: Analysis needs, modelling requirement and modelling adequacy. *IEE Proceedings - C*, 140:279–286, 1993.

[Par29]   R. H. Park. Two-reaction theory of synchronous machines - generalized method of analysis - part i. *AIEE Transactions*, 48:716–727, 1929.

[Par33]   R. H. Park. Two-reaction theory of synchronous machines - generalized method of analysis - part ii. *AIEE Transactions*, 52:352–355, 1933.

[PCJ90]   J.-P. Paul, C. Corroyer, P. Jeannel, J.-M. Tesseron, F. Maury, and A. Torra. Improvements in the organization of secondary voltage control in france. In *CIGRE Proceedings*, 1990. Paper 38/39-03.

[PLT87]   J.-P. Paul, J.-Y. Leost, and J.-M. Tesseron. Survey of the secondary voltage control in france: present realization and investigations. *IEEE Transactions on Power Systems*, 2:505–511, 1987.

[PM94]   M. Pavella and P.G. Murthy. *Transient stability of powers systems: theory and practice*. John Wiley & Sons, 1994.

[PMS96]   C. J. Parker, I. F. Morrison, and D. Sutanto. Application of an optimisation method for determining the reactive margin from voltage collapse in reactive power planning. *IEEE Transactions on Power Systems*, 11:1473–1481, 1996.

[Poi81]   H. Poincaré. Mémoire sur les courbes définies par une équation diffé-
          rentielle. *J. Math. Pures Appl.*, 7(3):375–422, 1881.

[Pri84]   G. B. Price. A generalized circle diagram approach for global analysis of
          transmission system performance. *IEEE Transactions on Power Apparatus
          and Systems*, 103:2881–2890, 1984.

[PSH92]   L. A. S. Pilotto, M. Szechtman, and A. E. Hammad. Transient ac voltage
          related phenomena for hvdc schemes connected to weak ac systems. *IEEE
          Transactions on Power Delivery*, 7:1396–1404, 1992.

[PSL95a] M. A. Pai, P. W. Sauer, and B. C. Lesieutre. Static and dynamic nonlin-
          ear loads and structural stability in power systems. *IEEE Proceedings*,
          83:1562–1572, 1995. Special Issue on Nonlinear Phenomena in Power
          Systems.

[PSL95b] M. A. Pai, P. W. Sauer, B. C. Lesieutre, and R. Adapa. Structural stability
          in power systems - effect of load models. *IEEE Transactions on Power
          Systems*, 10:609–615, 1995.

[PWM88]   W. W. Price, K. A. Wirgau, A. Murdoch, J. V. Mitsche, E. Vaahedi, and
          M. A. El-Kady. Load modeling for power flow and transient stability
          computer studies. *IEEE Transactions on Power Systems*, 3:180–187, 1988.

[Rao78]   S. S. Rao. *Optimization: Theory and Practice*. Wiley Eastern Ltd, New
          Delhi, 1978.

[RAU93]   N. D. Reppen, R. R. Austria, J. A. Uhrin, M. C. Patel, and A. Galatic.
          Performance of methods for ranking and evaluation of voltage collapse
          contingencies applied to a large-scale network. In *IEEE/NTUA Athens
          Power Tech Proceedings*, pages 337–343, 1993.

[RDM84]   G. Rogers, J. Di Mano, and R. T. H. Alden. An aggregate induction motor
          model for industrial plants. *IEEE Transactions on Power Apparatus and
          Systems*, 103:683–690, 1984.

[RLS92]   C. Rajagopalan, B. Lesieutre, P. W. Sauer, and M. A. Pai. Dynamic aspects
          of voltage/power characteristics. *IEEE Transactions on Power Systems*,
          7:990–1000, 1992.

[RVC96]   P. Rousseaux and T. Van Cutsem. Fast small disturbance analysis of
          long-term voltage stability. In *12th PSCC Proceedings*, pages 295–302,
          Dresden, 1996.

[SAB84] D. I. Sun, N. Ashley, B. Brewer, A. Hughes, and W. F. Tinney. Optimal power flow by newton approach. *IEEE Transactions on Power Apparatus and Systems*, 103:2864–2880, 1984.

[SAZ84] P. W. Sauer, S. Ahmed-Zaid, and M. A. Pai. Systematic inclusion of stator transients in reduced order synchronous machine models. *IEEE Transactions on Power Apparatus and Systems*, 103:1348–1354, 1984.

[Sch17] I. Schur. Potenzreihen im innern des einheitskreises. *J. Reine Angew. Math.*, 147:205–232, 1917.

[Sch98] R. A. Schlueter. A voltage stability security assessment method. *IEEE Transactions on Power Systems*, 1998. To appear.

[SCS88] R. A. Schlueter, A. G. Costi, J. E. Sekerke, and H. L. Forgey. Voltage stability and security assessment. Technical Report EL-5967, EPRI, 1988.

[SDP93] J.J. Sanchez-Gasca, R. D'Aquila, J.J. Paserba, W.W. Price, D.B. Klapper, and I. Hu. Extended-term dynamic simulation using variable time step integration. *Computer applications in Power*, 5:23–28, 1993.

[Sey88] R. Seydel. *From Equilibrium to Chaos: Practical Bifurcation and Stability Analysis*. Elsevier, 1988.

[SH79] G. Shackshaft and P. B. Henser. Model of generator saturation for use in power system studies. *Proceedings IEE*, 126:759–763, 1979.

[SL97] A. M. Stanković and B. C. Lesieutre. Parametric variations in dynamic models of induction machine clusters. *IEEE Transactions on Power Systems*, 1997.

[SO90] Y. Sekine and H. Ohtsuki. Cascaded voltage collapse. *IEEE Transactions on Power Systems*, 5:250–256, 1990.

[SP90] P. W. Sauer and M. A. Pai. Power system steady-state stability and the load flow Jacobian. *IEEE Transactions on Power Systems*, 5:1374–1383, 1990.

[SP94] P. W. Sauer and M. A. Pai. A comparison of discrete vs. continuous dynamic models of tap-changing-under-load transformers. In Fink [Fin94], pages 643–650.

[SP97] P. W. Sauer and M. A. Pai. *Power Systems Dynamics and Stability*. Prentice-Hall, 1997.

[SSH77] G. Shackshaft, O. C. Symons, and J. G. Hadwick. General-purpose model of power-system loads. *IEE Proceedings*, 124:715–723, August 1977.

[STI82]   Y. Sekine, K. Takahashi, Y. Ichida, Y. Ohura, and N. Tsuchimori. Method of analysis and assessment of power system voltage phenomena, and improvements including control strategies for greater voltage stability margins. In *CIGRE Proceedings*, 1982. Paper 38-206.

[Sto79]   B.J. Stott. Power system dynamic response calculation. *Proceedings of the IEEE*, 67:219–241, 1979.

[Tay92]   C. W. Taylor. Concepts of undervoltage load shedding for voltage stability. *IEEE Transactions on Power Delivery*, 7:480–488, 1992.

[Tay94]   C. W. Taylor. *Power System Voltage Stability*. EPRI Power System Engineering Series. McGraw Hill, 1994.

[TCN94]   C. W. Taylor, G. L. Comegys, F. R. Nassief, D. M. Elwood, and P. Kundur. Simulation and implementation of undervoltage load shedding for pacific northwest voltage stability. In *CIGRE '94*, Paris, 1994. Paper 38-103.

[TMI83]   Y. Tamura, H. Mory, and S. Iwamoto. Relationship between voltage instability and multiple load flow solutions. *IEEE Transactions on Power Apparatus and Systems*, 102:1115–1125, 1983.

[TMI88]   Y. Tamura, H. Mory, and S. Iwamoto. Voltage instability proximity index (vipi) based on multiple load flow solutions in ill-conditioned power systems. In *27th CDC Proceedings*, Austin, TX, 1988.

[TT87]   R. J. Thomas and A. Tiranuchit. Dynamic voltage instability. In *26th CDC Proceedings*, pages 53–58, Los Angeles, 1987.

[TT88]   A. Tiranuchit and R. J. Thomas. A posturing strategy against voltage instabilities in electric power systems. *IEEE Transactions on Power Systems*, 3:87–93, 1988.

[VC88a]   T. Van Cutsem. Dynamic and static aspects of voltage collapse. In Fink [Fin88], pages 6.55–6.79.

[VC88b]   T. Van Cutsem. Network-optimization based reactive power margin calculation. In *Proceedings IFAC Symposium on Power Systems modelling and control applications*, Brussels, 1988. Paper No 7.1.1.

[VC91a]   T. Van Cutsem. A method to compute reactive power margins with respect to voltage collapse. *IEEE Transactions on Power Systems*, 6:145–156, February 1991.

[VC91b]   T. Van Cutsem. Voltage collapse mechanisms : a case study. In Fink [Fin91], pages 85–101.

[VC93]   T. Van Cutsem. Analysis of emergency voltage situations. In *11th PSCC Proceedings*, pages 323–330, Avignon, 1993.

[VC95]   T. Van Cutsem. An approach to corrective control of voltage instability using simulation and sensitivity. *IEEE Transactions on Power Systems*, 10:616–622, 1995.

[VCM97]   T. Van Cutsem and R. Mailhot. Validation of a fast voltage stability analysis method on the hydro-québec system. *IEEE Transactions on Power Systems*, 12:282–292, 1997.

[VCM98]   T. Van Cutsem, C. Moisse, and R. Mailhot. Determination of secure operating limits with respect to voltage collapse. *IEEE Transactions on Power Systems*, 1998. Paper SM 97-942 PWRS, to appear.

[VCV96]   T. Van Cutsem and C. D. Vournas. Voltage stability analysis in transient and mid-term time scales. *IEEE Transactions on Power Systems*, 11:146–154, 1996.

[VJM94]   T. Van Cutsem, Y. Jacquemart, J.-N. Marquet, and P. Pruvot. Extensions and applications of a mid-term voltage stability analysis method. In Fink [Fin94], pages 251–270.

[VJM95]   T. Van Cutsem, Y. Jacquemart, J.-N. Marquet, and P. Pruvot. A comprehensive analysis of mid-term voltage stability. *IEEE Transactions on Power Systems*, 10:1173–1182, 1995.

[VL88]   T.K. Vu and C.C. Liu. Analysis of tap-changer dynamics and construction of voltage stability regions. In *Proceedings IEEE ISCAS*, volume 3, pages 1615–1618, 1988.

[VL92]   K. T. Vu and C.-C. Liu. Shrinking stability regions and voltage collapse in power systems. *IEEE Transactions on Circuits and Systems – I*, 39:271–289, 1992.

[VM98]   C. D. Vournas and G. A. Manos. Modelling of stalling motors during voltage stability studies. *IEEE Transactions on Power Systems*, 1998. Paper PE-975-PWRS-0-06-1997, to appear.

[VPL96]   H. Vu, P. Pruvot, C. Launay, and Y. Harmand. An improved voltage control on large-scale power system. *IEEE Transactions on Power Systems*, 11:1295–1303, 1996.

[VSI75]   V. A. Venikov, V. A. Stroev, V. I. Idelchick, and V. I. Tarasov. Estimation of electrical power system steady-state stability. *IEEE Transactions on Power Apparatus and Systems*, 94:1034–1040, 1975.

[VSP96]   C. D. Vournas, P. W. Sauer, and M. A. Pai. Relationships between voltage and angle stability of power systems. *International Journal of Electrical Power and Energy Systems*, 18(8):493–500, 1996.

[VSZ91]   V. Venkatasubramanian, H. Schättler, and J. Zaborszky. A taxonomy of the dynamics of the large power system with emphasis on its voltage stability. In Fink [Fin91], pages 9–44.

[VSZ93]   V. Venkatasubramanian, H. Schättler, and J. Zaborszky. The varied origins of voltage collapse in the large power systems. In *12th IFAC Congress*, volume 5, pages 451–458, Sydney, Australia, 1993.

[VVC95]   C. D. Vournas and T. Van Cutsem. Voltage oscillations with cascaded load restoration. In *IEEE/KTH Stockholm Power Tech Proceedings (power systems)*, pages 173–178, 1995.

[VWP93]   T. Van Cutsem, L. Wehenkel, M. Pavella, B. Heilbronn, and M. Goubin. Decision tree approaches to voltage security assessment. *IEE Proceedings - Part C*, 140:189–198, 1993.

[WC68]   B. M. Weedy and B. R. Cox. Voltage stability of radial power links. *IEE Proceedings*, 115:528–536, 1968.

[WCP94]   L. Wehenkel, T. Van Cutsem, M. Pavella, B. Heilbronn, and P. Pruvot. Machine learning, neural networks and statistical pattern recognition for voltage security: a comparative study. *Engineering Intelligent Systems*, 2:233–245, 1994.

[Wee79]   B. M. Weedy. *Electric Power Systems*. John Wiley & Sons., 3rd edition, 1979.

[Weh97]   L.A. Wehenkel. *Automatic learning techniques in power systems*. Kluwer Academic Publishers, 1997.

[Wig90]   S. Wiggins. *Introduction to Applied Nonlinear Dynamical Systems and Chaos*. Springer Verlag, 1990.

[WS90]   L. Wang and A. Semlyen. Application of sparse eigenvalue techniques to the small-signal stability analysis of large power systems. *IEEE Transactions on Power Systems*, 5:635–642, 1990.

[WSD92]   B. R. Williams, W. R. Schmus, and D. C. Dawson. Transmission voltage recovery delayed by stalled air conditioner compressors. *IEEE Transactions on Power Systems*, 7:1173–1181, 1992.

[XM94]   W. Xu and Y. Mansour. Voltage stability analysis using generic dynamic load models. *IEEE Transactions on Power Systems*, 9:479–493, 1994.

[XVM97]   W. Xu, E. Vaahedi, Y. Mansour, and J. Tamby. Voltage stability load parameter determination from field tests on B.C. Hydro's system. *IEEE Transactions on Power Systems*, 12:1290–1297, 1997.

[YHS94]   N. Yorino, S. Harada, H. Sasaki, and M. Kitagawa. Use of multiple load flow solutions to approximate closest loadability limit. In Fink [Fin94], pages 627–634.

[YSM92]   N. Yorino, H. Sasaki, Y. Masuda, Y. Tamura, M. Kitagawa, and A. Oshimo. An investigation of voltage instability problems. *IEEE Transactions on Power Systems*, 7:600–611, 1992.

[ZR69]   J. Zaborszky and J. W. Rittenhouse. *Electric Power Transmission*. The Rensselaer Bookstore, 1969.

INDEX